SPACE FLIGHT DYNAMICS FROM THE GROUND UP

Background and Practical Knowledge for Orbital and Attitude Computations and Space Flight Planning

Kenneth J. Ernandes

Published By:

Integrated
Spaceflight
Services

Integrated Spaceflight Publications Office
Boulder, Colorado

Integrated Spaceflight Publications Office
3360 Mitchell Lane, Suite C
Boulder, Colorado 80301

First Printing 2024

ISBN 978-0-9971472-5-4

Integrated Spaceflight Services Website: www.integratedspaceflight.com

Printed and bound in the United States of America

CONTENTS

Acknowledgements

The author would like to express his gratitude to Dr. Jason Reimuller for his help and encouragement in transforming this book from a concept to reality. His vision in creating the International Institute for Astronautical Sciences (IIAS) provides opportunities for both the advancement of human space flight and for a variety of talented and motivated individuals to pursue related goals. I would also express appreciation to Chris Lundeen for stepping up in critical leadership roles in IIAS and for providing his encouragement.

The author also expresses his appreciation to John Lundy, Dr. Trupti Mahendrakar, Drew Takeda, and Diallo Wallace for reviewing this work and providing constructive criticism.

Finally, I express my appreciation to my wife Sharon Kay Gallup for not only her constructive review, but also for her encouragement during the extensive time needed to prepare the manuscript and associated graphics. Her perspective as an education professional was invaluable in structuring this book.

AUT VIAM INVENIAM AUT FACIAM

Preface

As the title implies, this book introduces space flight dynamics from a ground up perspective. While the subject matter covered focuses on topics most applicable to human space flight, the material is readily applied to flight dynamics for all space missions.

The topic organization divides the subject matter, when appropriate, into three discrete strata: the descriptive level, the equation or algorithm level, and the derivation level. The purpose is to allow the reader to assimilate the subjects at the point of interest, which initially may only be at the descriptive level, but then might later evolve to deeper dives into the more technical aspects. Separating the derivations from the equations and algorithms provides the practical ability to easily find how to address a particular problem or application, while still demonstrating (at the next level) the physical and mathematical basis. Thus, the reader may readily "cut to the chase" and reference the methods to perform various computations, while at other times being able to delve into rigorously answering "why" things are the way they are.

The System Internal (SI) units of measure and their derivatives are used throughout since they are the standard within the scientific community. In certain circumstances, other units are introduced collaterally to provide numeric values more familiar to readers who do not routinely use SI units.

Space flight dynamics is applied mathematics. As such, it is presumed that readers pursuing the derivational level have certain prerequisite mathematical knowledge including vector and matrix mathematics, differential and integral calculus, and differential equations. But as is typically the case, knowledge wanes when it falls into disuse. Thus, the appendices provide a basic review and reference, with the hope the reader is often spared the inconvenience of needing to refer to an alternate source.

Obviously, no attempt was made to comprehensively cover all major aspects of space flight dynamics at all three levels of depth. Instead, the effort was focused on providing a solid introduction and covering common subject areas at a level accommodating typical practitioners. Thus, the depth that the more advanced topics are covered tends to be more of an introduction that provides a basic level of expertise that allows a segue to books dedicated to that advanced subject.

Table of Figures

List of Tables

Acronyms and Abbreviations

AD	Attitude Determination
ADBARV	Spherical Orbital Parameter Set
AIAA	American Institute of Aeronautics and Astronautics
AMSAT	Radio Amateur Satellite Corporation
AOS	Acquisition of Signal
BCE	Before Common Era
CCSDS	Consultative Committee for Space Data Systems
CMG	Control Moment Gyroscope
dB	Decibels
DCM	Direction Cosine Matrix
DOY	Day of Year
ECEF	Earth-Centered, Earth-Fixed
ECI	Earth-Centered Inertial
EIRP	Effective Isotropic Radiated Power
EKF	Extended Kalman Filter
EME2000	Earth Mean Equinox and Equator of J2000 Epoch
FIR	Finite Impulse Response
FOG	Fiber Optic Gyroscope
FORTRAN	Formula Translation
GCRF	Global Celestial Reference Frame
GEO	Geostationary Earth Orbit
GMAT	General Mission Analysis Tool
GNSS	Global Navigation Satellite System
GPS	Global Positioning System
IAU	International Astronomical Union
ICRF	International Celestial Reference Frame
IGY	International Geophysical Year
IIAS	International Institute for Astronautical Sciences
IIR	Infinite Impulse Response
IMU	Inertial Measurement Unit
ISS	International Space Station
ITRF	International Terrestrial Reference Frame
JD	Julian Date
JDN	Julian Day Number
JPL	Jet Propulsion Laboratory
kg	Kilogram
km	Kilometer
LEO	Low Earth Orbit
LOI	Lunar Orbital Insertion
LOS	Loss of Signal
LSDC	Least-Squares Differential Correction
LVLH	Local Vertical, Local Horizontal
m	Meter

MEKF	Multiplicative Extended Kalman Filter
MEO	Medium Earth Orbit
MOD	Mean [Equinox and Equator] of Date
MRP	Modified Rodrigues Parameters
MSIS	Mass Spectrometer Incoherent Scatter
N	Newton
NASA	National Aeronautics and Space Administration
NAVSTAR	Navigation Satellite Tracking and Ranging
NM	Nautical Mile
NOAA	National Oceanic and Atmospheric Administration
NORAD	North American Aerospace Defense Command
NRL	Naval Research Laboratory
ODE	Ordinary Differential Equation
PID	Proportional-Integral-Derivative
RCS	Reaction Control System
RFI	Radio Frequency Interference
RIC	Radial-Intrack-Crosstrack
RLG	Ring Laser Gyroscope
RMS	Root Mean Square
RPM	Revolutions per minute
RPO	Rendezvous and Proximity Operations
s	Second
SGP4	Simplified General Perturbations 4
SOFA	Standards of Fundamental Astronomy
SOI	Sphere of Influence
SVD	Singular Value Decomposition
TAI	International Atomic Time (Temps Atomique International)
TEME	True Equator, Mean Equinox
TLE	Two-Line Element
TLI	Trans-Lunar Injection
TOD	True [Equinox and Equator] of Date
TPI	Terminal Phase Initiation
TT	Terrestrial Time
UT	Universal Time
UTC	Coordinated Universal Time

Nomenclature

This section identifies the symbology used in the equations. Symbols with a bar ($\bar{\square}$) indicate dimensioned vector quantities; symbols with a caret ($\hat{\square}$) over the top indicate unit vectors. Symbols in bold (such as *A*) indicate matrices. Symbols with a single dot ($\dot{\square}$) over the top indicate a first time derivative; symbols with a double dot ($\ddot{\square}$) over the top indicate a second time derivative.

α	Semi-major axis reciprocal
ϕ	Flight path angle
γ	Complementary flight path angle
ε	Specific mechanical energy
μ	Celestial body gravitational parameter
θ	True anomaly
A	Area
$\bar{a}$	Acceleration
a	Semi-major axis
b	Semi-minor axis
D	Parabolic eccentric anomaly
E	Elliptical eccentric anomaly
F	Hyperbolic eccentric anomaly
$\bar{F}$	Force
G	Universal gravity constant
$\bar{h}$	Specific angular momentum
M	Mean anomaly
m	Mass
$\bar{L}$	Angular momentum
$\bar{P}$	Momentum
p	Conic parameter (semi-latus rectum)
$\bar{r}$	Position
T	Period
$\bar{v}$	Velocity

1 The Foundations of Space Flight Dynamics

Space flight dynamics is applied mathematics and physics. The subject matter encompasses the tools used to describe, predict, and plan space vehicle translational and rotational motion. The foundations of space flight presented in this chapter provide the most prevalent mathematical and physical concepts used to develop the topics in later chapters.

Orbital motion may appear mysterious to some because it does not relate to the everyday experiences of an observer who is surface bound by gravity. When one's experience is limited to a surface bound environment, it can be difficult to perceive how a non-thrusting spacecraft or even a piece of "space junk" can continue to circle the Earth or other celestial body essentially in perpetual motion. A common question is *"why doesn't the satellite fall out of the sky?"* Understanding this question's answer is this chapter's focus.

Gravity, motion from a central acceleration, the distinction between speed and velocity, and Newton's laws, are the initial topics covered. Their development starts from a simplified circular orbit. Linear and angular momentum are then introduced as key underpinnings for orbital motion. Basic concepts from conic section geometric shapes are the final foundational topic introduced. Topics are covered descriptively at Level I and characterized mathematically at Level II. Level III justifies the level II equations, typically through derivations.

1.1 Gravity

Intuition might suggest that gravity is orbital motion's enemy. Nothing could be further from the truth. Without gravity the satellite could not orbit.

1.1.1 Gravity [Level I – Descriptive]

Gravity is an attractive force – any two objects with mass have a gravitational attraction between them. The larger their masses, the stronger the force. Also, two objects that are close together have a much stronger gravitational attraction than is between equally massive objects that are further apart. If you double the distance between two masses, the gravitational attraction becomes 4 times weaker. If you double the distance again, the gravitational attraction will become yet another 4 times weaker (or 16 times weaker than it was at the original distance). Thus, the gravitational force between any two masses drops off with what mathematicians call the "inverse square" of the distance between them. (The term "inverse square" means the force gets *weaker* by the distance multiplied by itself, also known as the distance, squared.) Note that there is a physical reason for the inverse square weakening, which will be explained in Level II.

Gravitational force weakens as you move away from the Earth (or any celestial body). Table and Figure 1-1 provide the gravitational force relative to that of the Earth's surface for various altitudes.

Table 1-1 Relative Gravitational Force for Various Earth Altitudes		
Altitude	Example Satellite Types	Relative Surface Gravitational Force
Earth's Surface	None	100%
400 km (216 NM)	Human-Occupied Spacecraft	88.5%
1000 km (540 NM)	Navigation and Weather	75.6%
20000 km (10800 NM)	GNSS Navigation	5.8%
36000 km (19440 NM)	Communications and Weather	2.3%

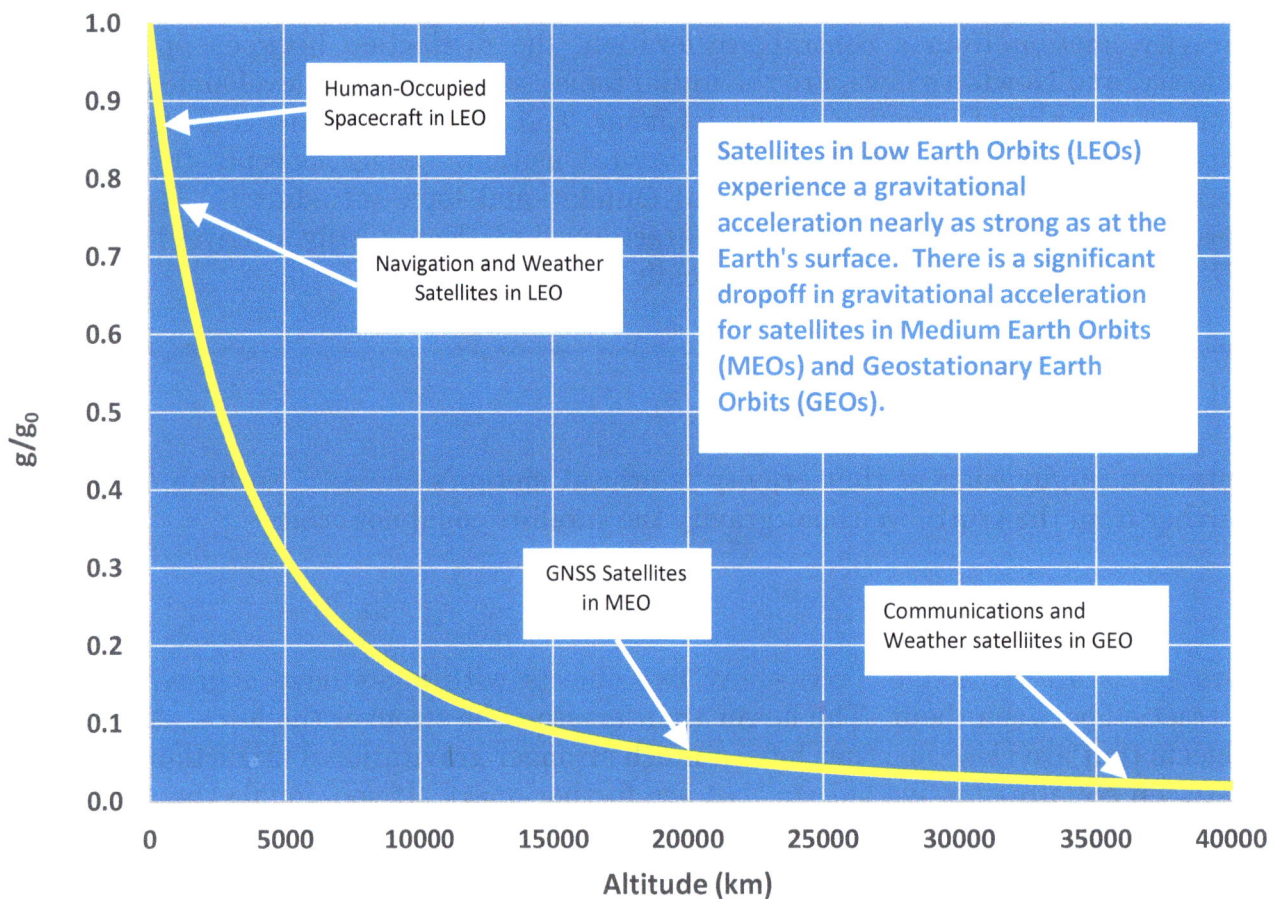

Human-Occupied Spacecraft in LEO

Satellites in Low Earth Orbits (LEOs) experience a gravitational acceleration nearly as strong as at the Earth's surface. There is a significant dropoff in gravitational acceleration for satellites in Medium Earth Orbits (MEOs) and Geostationary Earth Orbits (GEOs).

Navigation and Weather Satellites in LEO

GNSS Satellites in MEO

Communications and Weather satelliites in GEO

Figure 1-1 Earth Gravitational Acceleration versus Altitude

The example altitudes in Table 1-1 and Figure 1-1 correspond to Low Earth Orbit (LEO) satellite missions, Medium Earth Orbits (MEOs) at which Global Navigation Satellite System (GNSS) operate, and Geostationary Earth Orbits (GEOs) in which Communications and Weather satellites orbit.

Something noteworthy is that about 89% of the surface gravity strength remains in the altitude regime which human-occupied spacecraft traditionally operate in LEO. This may initially seem contradictory when viewing spacecraft occupants floating weightless inside their spacecraft. This is because mass and weight are not the same thing.

1.1.1.1 Weight

Weight and mass are often used interchangeably. This convention, though technically incorrect, is reasonable on a planetary surface since an object of a certain mass will have virtually the same weight everywhere on that planet's surface. The ground or any structure where you are standing or sitting resists the gravitational attraction between you and a celestial body such as the Earth. The *force* between you and the ground or supporting structure is your *weight*.

1.1.1.2 Mass

Mass, on the other hand, may be considered an amount of substance. For example, if a person's weight on Earth is 72 kg, they would weigh about one sixth of that (or 12 kg) on the Moon. The attraction to the Moon at its surface is only one sixth the attraction between the person and the Earth at its surface. The person's mass did not change, but the force needed for support is different because the gravitational attraction is different.

Looking one step further, consider the case of a falling object. An object falling in a gravity field is not being supported by the ground or any other structure. Since there is no force resisting the gravitational attraction, the falling object is weightless and continues to fall in the gravity field. The falling object's mass is the same, but it has no weight while it is falling because there is no supporting force.

Key Terms:

Mass: the amount of a substance making up an object. Note also that mass also describes an object's inertia, which is its resistance to having its state of motion changed, as will be developed in the momentum section.
Weight: the force two objects with mass exert with each other when in contact, due to gravitational attraction.

➢ **Transition**: *You may continue this section with Gravity at Level II, or you may skip to Level I of the next topic called Speed and Velocity.*

1.1.2   Gravity [Level II – Equations]

Gravitational force can be mathematically characterized for two objects, knowing their masses and separation distance.   Figure 1-2 illustrates the mutual gravitational attraction of two objects.

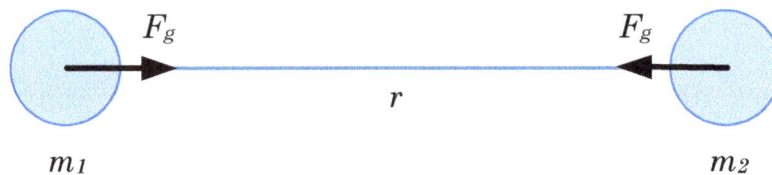

Figure 1-2 Gravitational Force Between Two Masses

The gravitational force between two objects is directly proportional to the product of their masses and inversely proportional to the distance between their centers of mass. Equation 1.1, which is Isaac Newton's gravitational law, expresses this force mathematically:

$$F_g = \frac{G m_1 m_2}{r^2} \qquad\qquad 1.1$$

In equation 1.1, the parameter $G \approx 6.67 \times 10^{-11}\ N \cdot m^3 \cdot kg^{-2}$ is the universal gravitational constant, m_1 and m_2 are the objects' masses (kg), and r is the distance (m) between the objects' centers of mass.

Given an Earth mass ($M_E \approx 5.976 \times 10^{24}\ kg$), equation 1.2 is the gravitational force between the Earth and another object of mass m at a distance r. Note the use of an Earth-specific gravitational parameter ($\mu \approx 398600.4415\ km^3 \cdot s^{-2}$), which is the product of the G and M_E, corrected for km units, using km units for the distance and kg units for the mass.

$$F_g = \frac{\mu}{r^2} m \qquad\qquad 1.2$$

Equation 1.3 is the Earth's gravitational force on a one-kilogram object at distance r.

$$\frac{F_g}{1kg} = \frac{\mu}{r^2} \qquad\qquad 1.3$$

This is the gravitational field expressed as the force per unit mass which, as will be seen later, also describes how two objects accelerate toward each other.

Table 1-2 lists the gravitational force the Earth imparts per unit mass at the same altitudes given in table 1-1. Note that in the computation. The distance r is the sum of the altitude and the mean Earth radius (approximately 6371 km).

Table 1-2 Computed Earth Gravitational Force for Various Altitudes		
Altitude (km)	$r(km)$	$F/1\,kg\,(km^2 \cdot s^{-1})$
0	6371	9.82×10^{-3}
400	6771	8.69×10^{-3}
1000	7371	7.34×10^{-3}
20000	26371	5.73×10^{-4}
36000	42371	2.22×10^{-4}

➢ **Transition**: *You may continue this section with Gravity at Level III, or you may skip to Level I of the next topic called Speed and Velocity.*

1.1.3 Gravity [Level III – Derivation]

Equation 1.1 is Newton's empirical gravitational law and is considered a "first principles" assumption for orbital mechanics. However, the inverse square deduction of gravitational force over distance can be derived using a mental exercise.

This non-rigorous description begins by recognizing that two objects separated by a distance r may be oriented in any direction relative to one another. If one object is considered primary and of a uniform spherical composition, we may consider that its gravitational field reaches out equally in all directions. The secondary object displaced at the distance r will experience the same gravitational force from the primary, regardless of the direction of its displacement.

Because the gravitational strength is the same for a constant radius, the primary object's gravitational field strength may be viewed as a set of thin spheres centered on the object. Since the field extends the same in all directions, gravity must spread out uniformly over the spherical surface. Equation 1.4 is the formula for a sphere's surface area.

$$A = 4\pi r^2 \qquad\qquad 1.4$$

If gravity is like a substance spread equally over the spherical surface, the amount present in any unit area is the total gravity divided by the total area. If the total gravity (based on the object's mass) is constant, the only variable quantity is the r^2 in the denominator. Hence an inverse square field.

The reader may question what became of the 4π from the area formula. It did not go away, but rather is part of the constant μ gravitational parameter seen in equation 1.3.

An analogy may be considered by inflating a spherical balloon. The balloon's surface area increases with the square of its radius. Since the balloon is composed of a finite quantity of material, the spherical shell's thickness will thus decrease as the inverse square of the radius.

1.2   Speed versus Velocity

Understanding orbital motion requires knowing the difference between *speed* and *velocity*. Velocity is often incorrectly considered a fancy technical term for speed. But velocity is more than just speed; velocity is a speed with direction. Since velocity has a direction, it is known in technical terms as a *vector* quantity. Speed, which has no specific direction, is known in technical terms as a *scalar* quantity.

Key Terms:

Scalar: a non-directed quantity.
Vector: a directed quantity.
Velocity: a speed with a direction.

1.3   Why Satellites Orbit

The answer to the question of why satellites orbit is most easily answered by examining circular orbits. This permits a relatively easy explanation that can later be expanded and generalized to all trajectory types.

1.3.1   Circular Orbits [Level I – Descriptive]

A satellite orbits because of the influence of gravity acting on its velocity. This section illustrates the simplest example: a circular orbit. Referring to Figure 1-3 at position ❶, the velocity is directed horizontally, while gravity induces a vertical fall toward the celestial body center. The orbiting satellite is initially moving horizontally at a fast speed. So fast, in fact, that it out-runs gravity. By the time the satellite reaches position ❷ in the figure, it has fallen only the short distance shown by the small arrow at the end of the initial velocity. Gravity also has a turning effect on the velocity. Notice also at position ❷ that the velocity has changed direction.

For the satellite to travel in a closed circular path, the velocity must have both the correct speed and direction. For a circular orbit, the velocity's direction must be perpendicular to the gravitational force. The speed needs to be balanced against gravity, so the combination of the along track movement and the radial gravitational free fall forms a closed circular path.

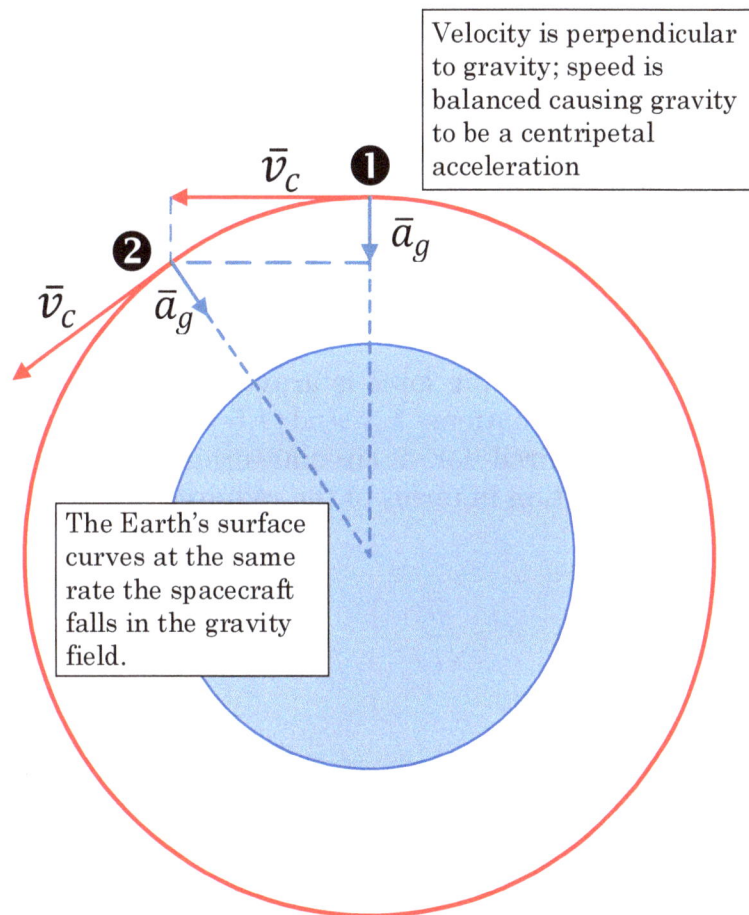

Figure 1-3 Orbital Flight: Velocity and Gravity

If the velocity's speed or direction has differences from what is required for a circular orbit, the trajectory will deviate from a circular path. Small deviations typically produce an elliptical orbit. The amount that the path deviates from circular depends on how much the velocity's speed or direction differs from what is needed for a circular orbit. This will be addressed in more detail later.

An important concept to keep in mind is that *an orbiting satellite is constantly falling*.

➢ **Transition**: *You may continue this section with circular orbits at Level II, or you may skip to Level I of the next topic called Newton's Laws.*

1.3.2   Circular Orbits [Level II – Equations]

This section provides a brief examination of a circular orbit's in-plane characteristics.

For an object to travel in a circular path at a radius r, you need a force directed radially inward toward the circular path's center. This force needs to cause what is called a *centripetal acceleration* on the object. Equation 1.5 is the scalar formula for a centripetal acceleration (a_c) with velocity having a speed (v) directed perpendicular to the circular path's instantaneous radius (r).

$$a_c = \frac{v^2}{r} \qquad\qquad 1.5$$

For a circular orbit, the accelerating force is gravity (as provided by equation 1.3). Setting the right sides of equations 1.3 and 1.5 equal defines gravity as the centripetal acceleration required for a circular orbit. Equation 1.6 solves the centripetal acceleration equation in terms of the required orbital speed for a circular orbit.

$$v = \sqrt{\frac{\mu}{r}} \qquad\qquad 1.6$$

Key Term:

> **Centripetal acceleration**: a central force acceleration balanced with an object's velocity such that the object follows a circular path.

➢ **Transition**: *You may continue this section with centripetal accelerations at Level III, or you may skip to Level I of the next topic called Newton's Laws.*

1.3.3   Motion from a Central Acceleration [Level III – Derivation]

Equation 1.5 that describes centripetal acceleration can be derived from analysis of two-dimensional equations of motion in a Cartesian system (i.e., having three mutually perpendicular coordinate axes) and related to a polar system. The moving object has a position vector $\bar{r}$ with components along the stationary Cartesian $\hat{x}$ and $\hat{y}$ axes illustrated in Figure 1-4. In polar coordinates, the position vector $\bar{r}$ has a magnitude r and is oriented at an angle θ relative to the $\hat{x}$ axis, using a standard right hand coordinate convention.

The figure also has a set of polar axes that rotate, following the object. These are the radial ($\hat{r}$) axis that describes motion toward or away from the x-y origin and the transverse ($\hat{\theta}$) axis that describes motion perpendicular to the radial axis.

Equations 1.7 and 1.8 describe the radial and transverse axes orientations in the x-y coordinate frame, as a function of the angle θ.

$$\hat{r} = \cos\theta\hat{x} + \sin\theta\hat{y} \qquad\qquad 1.7$$

$$\hat{\theta} = -\sin\theta\,\hat{x} + \cos\theta\hat{y} \qquad\qquad 1.8$$

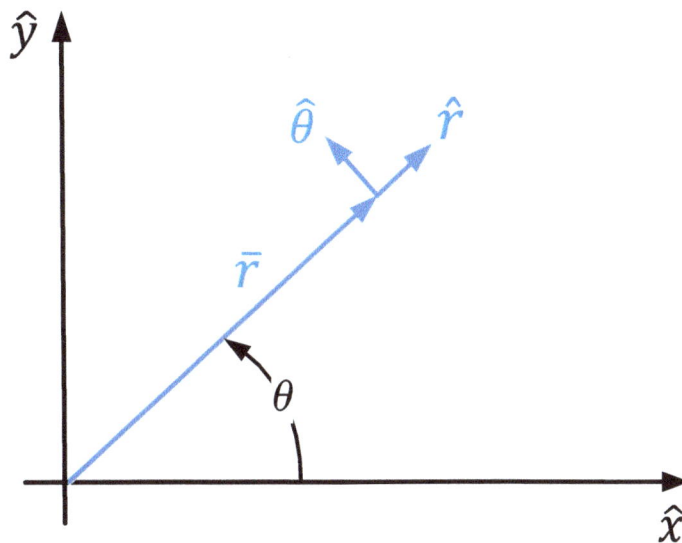

Figure 1-4 Two-Dimensional Motion in Polar Coordinates

Expressing the polar axes relative to the fixed frame provides a convenient way to characterize how their orientations change over time, as given in equations 1.9 and 1.10. The dot ($\dot{}$) accent used in these equations shall henceforth be an equivalent shorthand for a derivative with respect to time.

$$\frac{d\hat{r}}{dt} = -\dot{\theta}\sin\theta\,\hat{x} + \dot{\theta}\cos\theta\hat{y} = \dot{\theta}\hat{\theta} \qquad\qquad 1.9$$

$$\frac{d\hat{\theta}}{dt} = -\dot{\theta}\cos\theta\,\hat{x} - \dot{\theta}\sin\theta\,\hat{y} = -\dot{\theta}\hat{r} \qquad 1.10$$

Equation 1.11 is the position vector ($\bar{r}$) in polar coordinates.

$$\bar{r} = r\hat{r} \qquad 1.11$$

The velocity vector ($\bar{v}$) is the first derivative of position with respect to time.

$$\bar{v} = \frac{d\bar{r}}{dt} = \dot{r}\hat{r} + r\frac{d\hat{r}}{dt} \qquad 1.12$$

The velocity vector is simplified by substituting equation 1.9 into equation 1.12.

$$\bar{v} = \dot{r}\hat{r} + r\dot{\theta}\hat{\theta} \qquad 1.13$$

The acceleration vector ($\bar{a}$) is the first derivative of velocity with respect to time (and thus the second derivative of position with respect to time).

$$\bar{a} = \frac{d\bar{v}}{dt} = \ddot{r}\hat{r} + \dot{r}\frac{d\hat{r}}{dt} + \left(r\ddot{\theta} + \dot{r}\dot{\theta}\right)\hat{\theta} + r\dot{\theta}\frac{d\hat{\theta}}{dt} \qquad 1.14$$

Equation 1.15 is the acceleration vector simplified by substituting the results of equation 1.9 and 1.10 into equation 1.14.

$$\bar{a} = \left(\ddot{r} - r\dot{\theta}^2\right)\hat{r} + \left(r\ddot{\theta} + 2\dot{r}\dot{\theta}\right)\hat{\theta} \qquad 1.15$$

These general polar motion equations are now limited to the case of the centripetal acceleration defining circular motion, by restricting the magnitude $\bar{r}$ to a constant. This restriction zeros the radial magnitude's first ($\dot{r} = 0$) and second ($\ddot{r} = 0$) time derivatives. Furthermore, since the case is limited to a central force (i.e., purely radial) acceleration, the transverse ($\hat{\theta}$) axis acceleration is also zero. The remaining result in equation 1.16 is the centripetal acceleration in polar coordinates.

$$\bar{a}_c = -r\dot{\theta}^2\hat{r} \qquad 1.16$$

The negative sign is consistent with the centripetal acceleration pointing radially inward toward the origin. Thus, equation 1.17 is the centripetal acceleration magnitude.

$$a_c = r\dot{\theta}^2 \qquad 1.17$$

The velocity's transverse component ($r\dot{\theta}$) is all that remains since we have a constant radius ($\dot{r} = 0$). Therefore, by substitution ($v^2 = r^2\dot{\theta}^2$), equation 1.18 is the centripetal acceleration in terms of the magnitudes r and v, which verifies equation 1.5.

$$a_c = \frac{v^2}{r} \qquad\qquad 1.18$$

In the preceding equations, the first time derivative of radius is determined using the dot product definition.

$$r = \sqrt{\bar{r} \cdot \bar{r}} = (\bar{r} \cdot \bar{r})^{1/2} \qquad\qquad 1.19$$

$$\dot{r} = \frac{dr}{dt} = \frac{1}{2}(\bar{r} \cdot \bar{r})^{-1/2}(\bar{r} \cdot \dot{\bar{r}} + \dot{\bar{r}} \cdot \bar{r}) \qquad\qquad 1.20$$

$$\dot{r} = \frac{\bar{r} \cdot \dot{\bar{r}}}{r} = \hat{r} \cdot \bar{v} \qquad\qquad 1.21$$

The above result makes intuitive sense, indicating that the time rate of change of radius is the velocity's radial component. Likewise, the second time derivative of radius is computed below using the quotient and chain rules.

$$\ddot{r} = \frac{d}{dt}\left(\frac{\bar{r} \cdot \dot{\bar{r}}}{r}\right) = \frac{r(\bar{r} \cdot \ddot{\bar{r}} + \dot{\bar{r}} \cdot \dot{\bar{r}}) - \dot{r}(\bar{r} \cdot \dot{\bar{r}})}{r^2} \qquad\qquad 1.22$$

$$\ddot{r} = \hat{r} \cdot \bar{a} + \frac{v^2}{r^2} + \dot{r}^2 \qquad\qquad 1.23$$

Key Terms:

Cartesian Coordinates: a system of three mutually perpendicular coordinate axes (typically but not exclusively having x-, y-, and z-axes).
Magnitude: a vector's quantity component (independent of direction).
Polar Coordinates: a system with a radial offset from the coordinate origin and having one or more offset angles.
Radial: the direction toward or away the coordinate system origin, especially in polar coordinates.
Transverse: the direction perpendicular to radial, especially in polar coordinates.

1.4 Newton's Laws of Motion

Newton's three laws of motion provide a basis for much of classical physics. These laws will be examined with emphasis on how they apply to space flight dynamics.

1.4.1 Newton's Laws [Level I – Descriptive]

Isaac Newton's three physical laws of motion provide an important foundation for orbital dynamics. These laws can be paraphrased as follows:

Newton's First Law: _Inertia_ – an object in motion travels at a constant speed in a straight line unless a net external force changes its speed or direction; an object at rest remains at rest unless it is disturbed by a net external force.

Newton's Second Law: _Momentum, Force, Mass, and Acceleration_ the rate of change of momentum is proportion to a net applied external force. If the object has a constant mass, acceleration equals the net applied external force, divided by the object's mass.

Newton's Third Law: _Action-Reaction_ – for every action there is an equal and opposite reaction.

1.4.1.1 Inertia

Newton's law of _inertia_ states that an object resists any change to its motion (or lack thereof). If it is sitting still, a force must be applied to get it moving, which is a concept that should be easy to grasp. The other part of Newton's First Law, namely that an object in motion will continue traveling indefinitely in a straight line might be less intuitive. In daily experiences, surface-bound observers see objects in motion having the tendency of coming to rest. But the reason objects come to rest is because of environmental forces, such as friction, that slow them down and bring them to a stop. The law of inertia indicates that there are external forces present when objects in motion are observed coming to rest.

1.4.1.2 Momentum, Force, and Mass

Momentum is the product of mass and velocity. Newton's Second Law states that an object's momentum changes in proportion to applied force; otherwise, momentum does not change. The notion that momentum involves both mass and velocity complicate general conclusions characterizing how forces change an object's motion.

This is easier to understand for objects with constant mass. Given the constant mass simplification, Newton's Second Law states that larger masses are more difficult to get moving or bring to a stop than smaller masses. It is likewise more difficult to speed up, slow down, or change the direction of a more massive object than it is for an object with less mass.

The constant mass simplification cannot always be applied to space flight dynamics. Thrusting events are the most noteworthy exception. Thrust is produced by a momentum exchange caused by propellant (mass) expulsion. Thus, a rate of change of mass occurs during thrusting events. This will be addressed in more detail later.

1.4.1.3 Acceleration

Acceleration is a by-product of momentum changes. Accelerations are time rates of change of velocity. Since both are vectors, acceleration may cause a change in speed, direction, or both. For example, a satellite in a circular orbit has a constant speed, but its velocity direction changes because gravity imparts centripetal acceleration. That a satellite with a constant speed is accelerating might not be intuitive because we usually think of acceleration as a speed rather than velocity change.

A centripetal acceleration has a constant velocity magnitude (or speed). But when velocity is viewed as having separate components against the fixed $\hat{x}$ and $\hat{y}$ directions, as illustrated in Figure 1-5, speed changes are also observable. Beginning at position ❶ the velocity is completely in the $\hat{y}$ direction, with a zero value in the $\hat{x}$ direction. Sometime later at position ❷ (and at all intermediate points) the speed remains constant, but the direction changes. However, the object is now moving slower in the $\hat{y}$ direction but is concurrently moving faster in the $-\hat{x}$ direction. The total speed has not changed, but the amounts in each axis have been reapportioned. By looking at the velocity components along the fixed $\hat{x}$ and $\hat{y}$ directions separately, the changes (or accelerations) are more obvious. From this point of view, it becomes clear that an object that has a constant speed, but whose velocity changes direction is accelerating.

Another area that may cause discomfort is referring to a velocity decrease as an acceleration. Acceleration typically implies an increase, and the term deceleration is used for velocity decreases; deceleration being the opposite of acceleration. This distinction can be a useful qualifier in certain circumstances. In mathematical terms, this same distinction is made using the negative (-) sign in front of an acceleration to indicate a deceleration. A deceleration occurs when the acceleration vector opposes the velocity direction. Deceleration, negative acceleration, and acceleration opposing velocity all mean the same thing and can be used interchangeably.

1.4.1.4 Action and Reaction

Newton's Third Law stating that for every action there is an equal and opposite reaction is the basis of rocket propulsion. In this context, a rocket motor produces forward thrust by expelling propellant rearward out of its exhaust. The greater the mass of the expelled propellant and the higher the exhaust speed, the more forceful is the thrust.

This process may be pictured by a thought experiment that can also be assessed physically. If a person on roller skates hurled a massive object such as a bowling ball in one direction as the *action*, they would accelerate in the opposite direction as the *reaction*. The faster the person threw the bowling ball, the more they would accelerate in the opposite direction.

The other observation is the bowling ball would accelerate much more since it is less massive than the person. This relates back to Newton's laws. The person and the ball begin at rest. Action is taken by applying a force on the ball that causes an equal and opposite momentum exchange upon release. The equal momentum exchange results in a greater acceleration to the smaller mass than is imparted on the larger mass. Hence a greater speed for the smaller mass and less speed imparted on the larger mass.

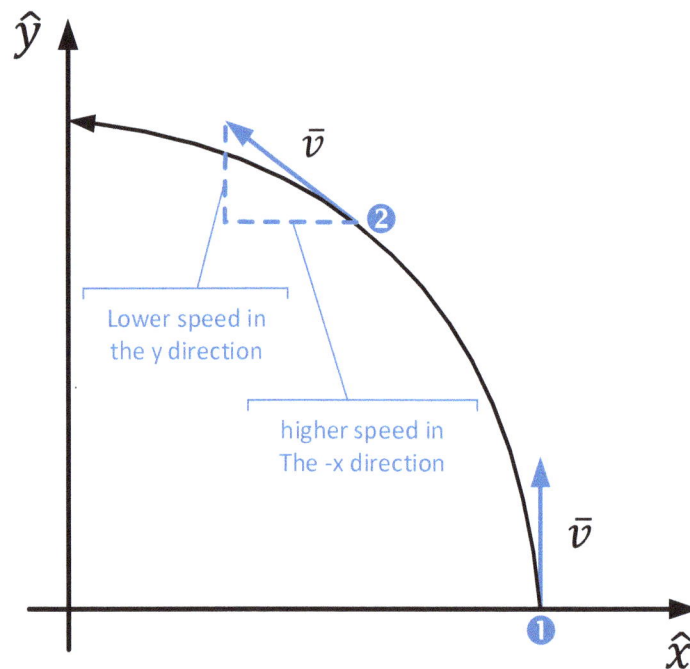

Figure 1-5 Acceleration by Changing Velocity Direction

The "*equality*" in the action is in the momentum exchange between the person and the bowling ball. Thus, momentum is conserved as each receives an equal but opposite amount of momentum in the exchange.

Key Terms:

> **Acceleration**: the rate of change of velocity; acceleration can be a speed or direction change.
> **Inertia**: resistance to change. An object's mass is inertia representing its resistance to linear acceleration.
> **Momentum**: the product of mass and velocity, indicating how much force is required to slow the object down.

➤ **Transition**: *You may continue this section with Newton's laws at Level II, or you may skip to Level I of the next topic called Angular Momentum.*

1.4.2   Newton's Laws [Level II – Equations]

The previous section was a verbal description of Newton's laws. This section briefly lists the mathematical equations associated with each of these laws.

1.4.2.1   Inertia

Inertia was initially Galileo's idea that was adopted by Newton. Inertia is a simplification of Newton's Second Law, asserting that if there is no net force acting on a body (i.e., the sum of all external forces is zero), there is no momentum change and thus no acceleration. With no acceleration, there is no change in speed or direction. Hence, an object at rest remains at rest and an object in motion will continue traveling at a constant speed in a straight line. Equation 1.24 expresses Newton's First Law mathematically.

$$if \ \sum_i \bar{F}_i = 0 \ \ then \ \ \bar{a} = 0 \qquad\qquad 1.24$$

1.4.2.2   Momentum, Force, Mass, and Acceleration

Equation 1.25 expresses Newton's Second Law in its pure form as the rate of change of momentum (P). Equation 1.26 expresses Newton's Second Law for a constant mass, while equation 1.27 expresses acceleration as the ratio of force to [a constant] mass.

$$\bar{F} = \frac{d\bar{P}}{dt} = \frac{d}{dt}(m\bar{v}) = m\frac{d\bar{v}}{dt} + \bar{v}\frac{dm}{dt} \qquad\qquad 1.25$$

$$\bar{F} = m\bar{a} \ \ (when \ mass \ is \ constant) \qquad\qquad 1.26$$

$$\bar{a} = \frac{\bar{F}}{m} \ \ (when \ mass \ is \ constant) \qquad\qquad 1.27$$

1.4.2.3   Action and Reaction

Equation 1.28 directly expresses the law of action and reaction using forces. Equation 1.29 expresses the law of action and reaction as a momentum exchange (or conservation of momentum).

$$\bar{F}_A = -\bar{F}_R \qquad\qquad 1.28$$

$$m_A \bar{v}_A = -m_R \bar{v}_R \qquad\qquad 1.29$$

1.5   Angular Momentum

Angular momentum is rotational momentum.

1.5.1   Angular Momentum [Level I – Descriptive]

The previous section introduced the concept of momentum, stating an object's momentum depends on both its mass and velocity. This section extends this idea, introducing rotational (or angular momentum) and its analog to Newton's Second Law.

Angular momentum is the rotational counterpart to momentum in a straight line (which is also called linear or translational momentum). Newton's First and Second Laws discuss how objects moving in a straight-line resist change to their momentum (speed or direction). Likewise, rotating objects like tops, gyroscopes, and even Frisbees® have rotational inertia that resists change to their rotational momentum. For this reason, a top or gyroscope tends to keep its spin axis pointed in the same direction, seemingly by magic. Understanding the basic concepts of rotational motion will provide insight into orbital motion.

Consider a ball at the end of a string, moving in a circular path as shown in Figure 1-6. The mass has a velocity that, in the absence of interference from the string, would have it travel in a straight line. The string's tension is the centripetal acceleration that causes the ball to travel in a circle. The time it takes the ball to go around the circle is the length of the circular path (i.e., circumference), divided by the ball's speed. For rotational motion, the rotation rate (the ball's speed divided by the circumference) is the quantity of interest. It is expressed by the number of times the ball goes around the circle, typically in revolutions per minute (RPM).

Figure 1-7 develops this idea a little further. If half the string's length were to be reeled in, the ball's circular path would only be half as long as the original circumference. The ball's speed would stay the same since nothing was done to change it. However, since the ball only needs to travel half the distance that it did previously, it goes around the circle twice as many times in a minute as it did with the larger circle. By halving the string length, the rotation rate doubled. Likewise, if the string length is doubled (and the object's speed stays the same), the rotation rate reduces to half its original value. This principle is known as the law of *Conservation of Angular Momentum*.

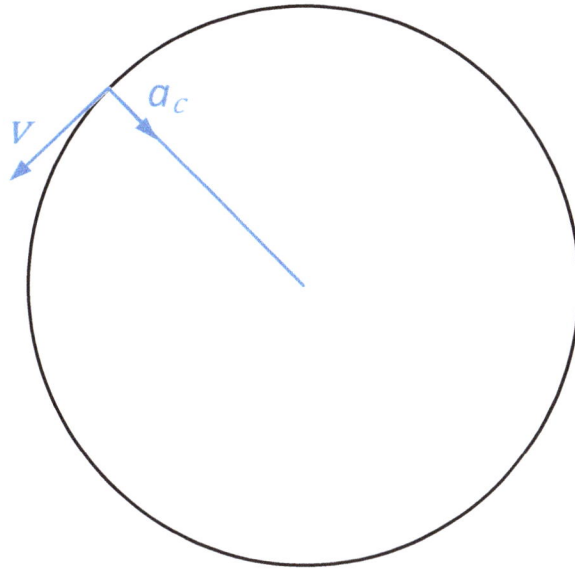

Figure 1-6 Angular Momentum Example

Figure skaters use conservation of angular momentum. Skaters typically begin a spin with their arms extended fully outward. Once the spin maneuver begins, the skater pulls in his or her arms and begins rotating very rapidly. They exit the maneuver by extending the arms fully, causing a slowdown to the original rotation rate. This spin maneuver works the same as the experiment with the ball on the end of the string.

There is one other noteworthy aspect of the ball and string experiment. When half the string length is retracted, the ball needs to turn twice as tightly as it did with the larger circle. Since, for this experiment the speed does not change, the ball is forced to make a tighter turn at the same speed. Doubling the turn rate while keeping the same speed requires double the centripetal acceleration (i.e., the string tension also doubles).

Key Terms:

Acceleration: the rate of change of velocity; acceleration can be a speed or direction change.
Inertia: resistance to change. An object's mass is inertia representing its resistance to linear acceleration.
Momentum: the product of mass and velocity, indicating how much force is required to slow the object down.

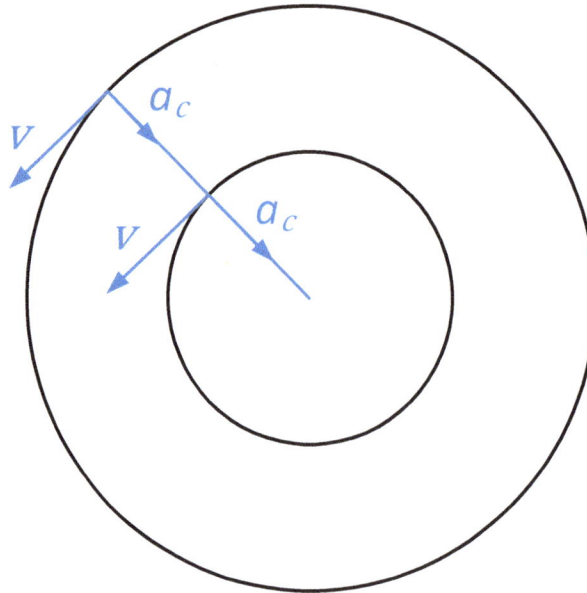

Figure 1-7 Angular Momentum when Shortening a String

➢ **Transition**: *You may continue this section with Angular Momentum at Level II, or you may skip to Level I of the next topic called Conic Sections.*

1.5.2   Angular Momentum [Level II – Equations]

Angular momentum ($\bar{L}$) results from a component of velocity perpendicular to the radial direction (i.e., a non-zero transverse velocity component). Any radial velocity component does not contribute to angular momentum. Equation 1.30 computes angular momentum as the vector cross product of the position vector ($\bar{r}$) and the momentum ($\bar{P}$) expressed as the product of mass and velocity ($m\bar{v}$).

$$\bar{L} = \bar{r} \times m\bar{v} \qquad\qquad 1.30$$

As a principal feature of the cross product, the angular momentum direction is perpendicular to both the position and velocity vectors. Equation 1.31 is the angular momentum's magnitude.

$$L = rvm \sin \gamma \qquad\qquad 1.31$$

The angle γ is between the position and velocity vectors as shown in Figure 1-8. Angle γ is the complement to the flight path angle (ϕ), also shown in Figure 1-8. The flight path angle is the more common of the two angles used in orbital mechanics computations.

Equation 1.32 computes the angular momentum magnitude in terms of the flight path angle.

$$L = rvm \cos \phi \qquad\qquad 1.32$$

In orbital mechanics a variant called *specific angular momentum* is commonly used. Specific angular momentum is angular momentum per unit mass and is denoted by the symbol $\bar{h}$ instead of $\bar{L}$. Equation 1.33 is the formula for computing specific angular momentum. Equation 1.34 computes the specific angular momentum magnitude in terms of the flight path angle.

$$\bar{h} = \bar{r} \times \bar{v} \qquad\qquad 1.33$$

$$h = rv \cos \phi \qquad\qquad 1.34$$

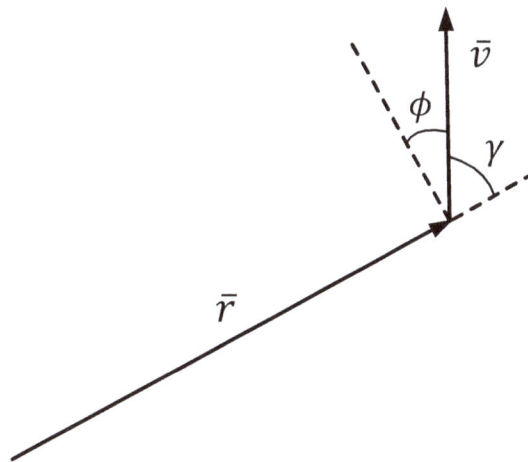

Figure 1-8 Geometry for Computing Angular Momentum

1.6   Conic Sections

All curved free-flight trajectories may be described by a class of geometric shapes called conic sections. This section will characterize conic section shapes, geometries, and basic mathematical properties.

1.6.1   Conic Sections [Level I – Descriptive]

Free-flight (i.e., non-thrusting) trajectories following either a curved path defined by a conics section, or go in a straight line (i.e., are *rectilinear*). Conic sections are a family of geometric shapes, obtained by a plane slicing through a cone as illustrated in Figure 1-9. The four types of conic sections are the circle, the ellipse, the parabola, and the hyperbola.

The angle the plane makes slicing through the cone determines which conic section will be obtained:

- Circle - slicing with the plane parallel to the cone's base (or equivalently perpendicular to the cone's axis).
- Ellipse – slicing with the plane at an oblique angle, shallower than the cone's opposite edge.
- Parabola – slicing with the plane at an angle parallel to the cone's opposite edge.
- Hyperbola – slicing with the plane at an angle steeper than the cone's opposite edge.

Conic sections have two basic categories. The closed shape category includes the circle and ellipse. The open shape category includes the parabola and hyperbola. Closed conics form a complete loop, which is the subset of trajectories called *orbits*, which may repeat indefinitely. The open conics are non-repeating since they have sufficient velocity to escape the gravitational influence of the celestial body they encounter. Open conics are therefore also referred to as escape trajectories.

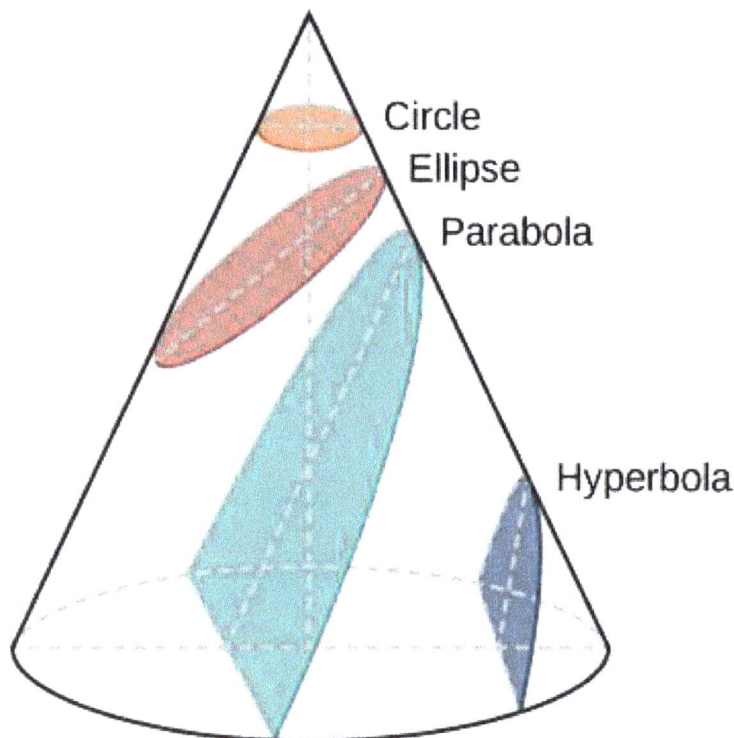

Figure 1-9 Conic Section Geometry

Key Terms:

Conic Sections: are a family of geometric curves that describe the trajectory of a spacecraft in free (non-thrusting) flight.

Orbits: are closed path conic sections (or trajectories), including circular and elliptical flight paths.

Trajectories: are the flight paths of celestial bodies or spacecraft in free (non-thrusting) flight.

Escape Trajectories are open-ended conic sections in which there is sufficient velocity to escape the gravitational influence of the celestial body they encounter. Parabolas and hyperbolas are escape trajectories.

➤ **Transition**: *You may continue this section with Conic Sections at Level II, or you may skip to Level I of the next chapter called Two-Body Trajectory Characteristics.*

1.6.2 Conic Sections [Level II – Equations]

Conic sections have common mathematical and geometric characteristics that are illustrated in this section. Common features of each are eccentricity, foci, a symmetry line, a latus rectum, and a conic parameter.

1.6.2.1 Elliptical Geometry

The ellipse illustrated in Figure 1-10 is presented first since it is the most general closed conic. The *apse line* is aligned with the major symmetry line or semi-major axis (a), which is half the ellipse's longest dimension. The shortest dimension, half of which is called the semi-minor axis (b), is perpendicular to the semi-major axis. The semi-major and semi-minor axes cross at the ellipse's geometric center.

The two foci (f), which lie on the apse line, are separated by the linear eccentricity (c). The latus rectum is a line running perpendicular to the symmetry line, through a focus, meeting up with the conic curve. The conic parameter (p), is half the latus rectum, running from a focus perpendicular to the symmetry line to the ellipse. The conic parameter is also known as the semi-latus rectum.

Equation 1.35 is the ellipse's formula in Cartesian coordinates.

$$\frac{x^2}{a} + \frac{y^2}{b} = 1 \qquad\qquad 1.35$$

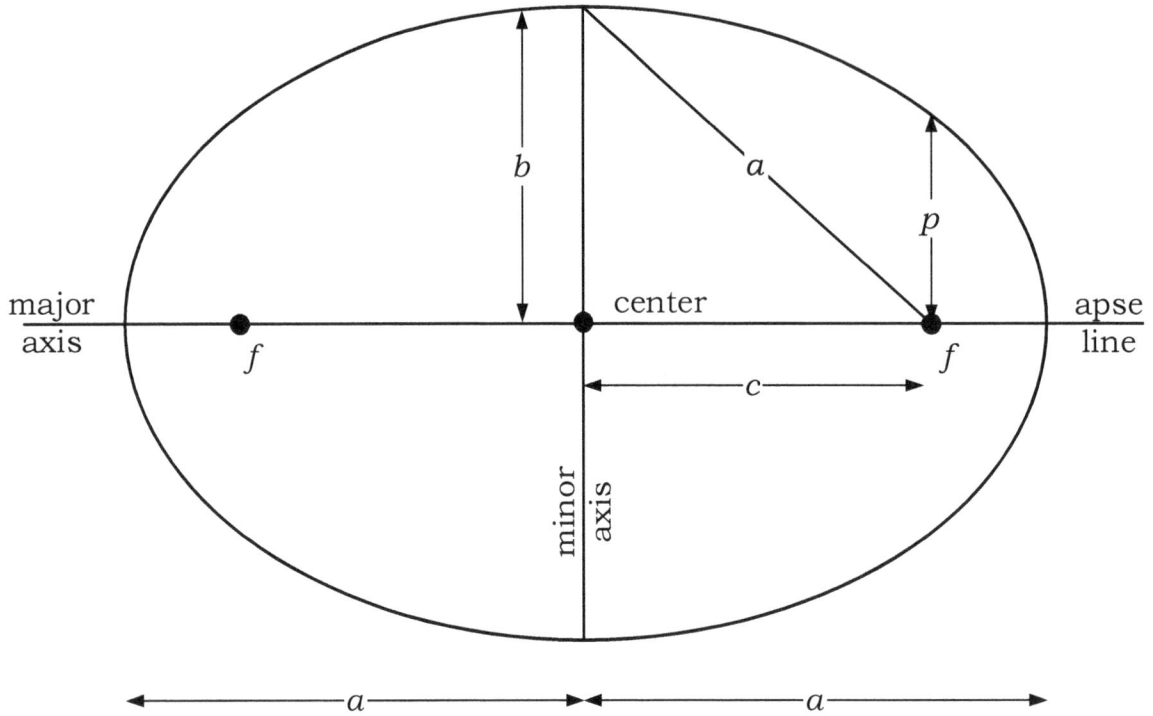

Figure 1-10 Ellipse Geometry

Equation 1.36 provides the relationship between the semi-major axis (a), the semi-minor axis (a), and the linear eccentricity (a).

$$a^2 = b^2 + c^2 \qquad\qquad 1.36$$

Equation 1.37 shows the nondimensional eccentricity (e) is the ratio of the linear eccentricity (c) to the semi-major axis (a). An ellipse's eccentricity is between zero and one.

$$e = \frac{c}{a} \qquad\qquad 1.37$$

Equation 1.38 computes the conic parameter (i.e., semi-latus rectum) from the semi-major axis and eccentricity.

$$p = a(1 - e^2) \qquad\qquad 1.38$$

Equation 1.39 computes the ellipse's semi-minor axis in terms of semi-major axis and eccentricity, using equations 1.36 and 1.37.

$$b = a\sqrt{1 - e^2} \qquad\qquad 1.39$$

1.6.2.2 Circular Geometry

Circles are the simplest conic sections since they are degenerate ellipses. Circles have a zero-eccentricity value ($e = 0$), since the two foci meet at the center. Circles have a constant radius, equal to their semi-major axis ($r = a$). Because of their constant radius nature, circles have no unique symmetry line, and their conic parameter / semi-latus rectum likewise equals their semi-major axis ($p = a$). Equation 1.40 is the circle's formula in Cartesian coordinates.

$$x^2 + y^2 = a^2 \qquad\qquad 1.40$$

1.6.2.3 Hyperbolic Geometry

Hyperbolas represent the most general open or escape conics. The apse line is aligned with the major symmetry line and the semi-major axis (a) is the [negative] distance along the symmetry from the conic to the hyperbola's center. The hyperbola's eccentricity is always greater than unity ($e > 1$).

The center is the location where there is symmetry between two equivalent hyperbolas. Within the context of space flight dynamics, one of the hyperbolas would represent the actual trajectory, while the other is a false pseudo trajectory.

Equation 1.41 is the hyperbola's formula in Cartesian coordinates.

$$\frac{x^2}{a} - \frac{y^2}{b} = 1 \qquad\qquad 1.41$$

Equations 1.42 and 1.43 compute the angle between the major axis and the asymptotes (β) and the angle between the asymptotes (δ), respectively from the eccentricity.

$$\beta = cos^{-1}(1/e) \qquad\qquad 1.42$$

$$\delta = 2\sin^{-1}(1/e) \qquad\qquad 1.43$$

Equation 1.44 is the hyperbola's conic parameter (p) in terms of its semi-major axis (a) and eccentricity (e). Equation 1.45 is the hyperbola's semi-minor axis (b), also in terms of semi-major axis and eccentricity.

$$p = a(e^2 - 1) \qquad\qquad 1.44$$

$$b = \Delta = a\sqrt{e^2 - 1} \qquad\qquad 1.45$$

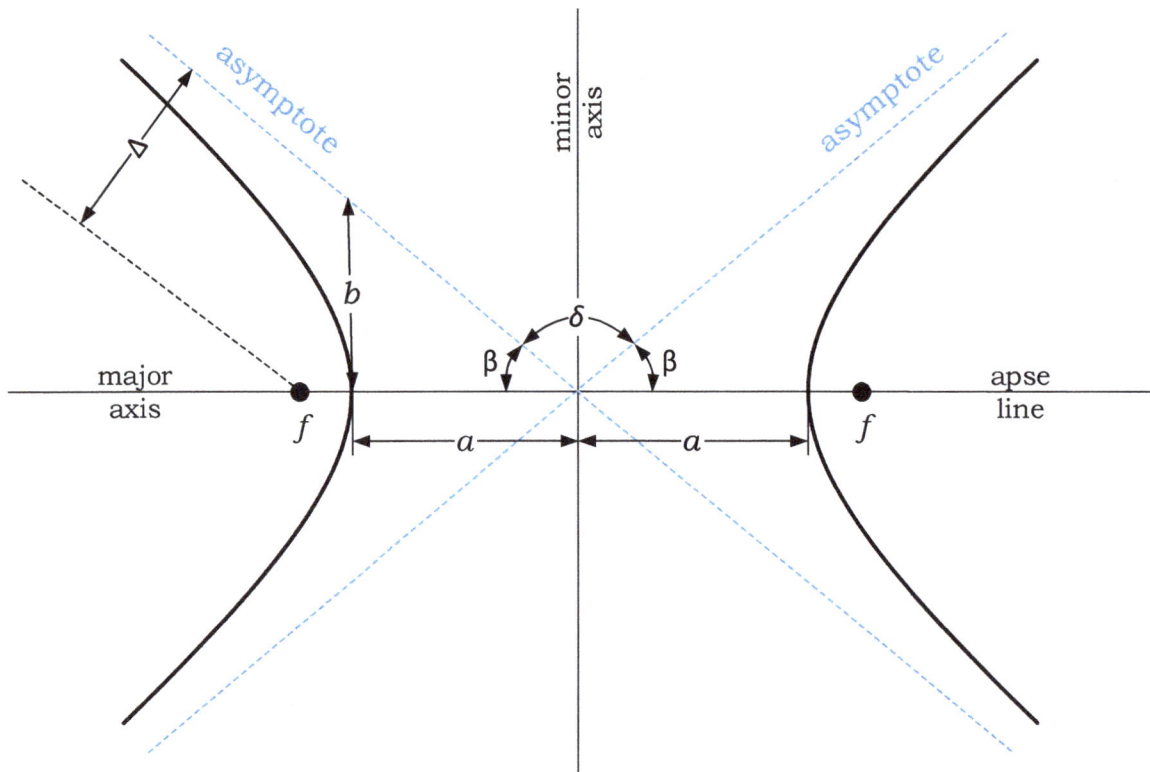

Figure 1-11 Hyperbola Geometry

1.6.2.4 Parabolic Geometry

Like the circle, the parabola is the hyperbola's degenerate form. Also like the circle, the parabola has a defined eccentricity ($e = 1$). A parabola's semi-major axis is infinite ($a = \infty$).

The parabola's conic parameter (p) is twice its linear eccentricity (c). Equation 1.46 is the parabola's formula in Cartesian coordinates.

$$y^2 = 4px \qquad\qquad 1.46$$

Table 1-3 Characteristics of Conic Types		
Conic Type	**Semi-Major Axis**	**Eccentricity**
Circle	Equal to the radius	$e = 0$
Ellipse	Half the longest dimension	$0 < e < 1$
Parabola	Infinite	$e = 1$
Hyperbola	Distance to hyperbola center	$e > 1$

Key Terms:

Apse Line: is the conic's major line of symmetry.
Eccentricity: is a dimensionless parameter describing the elongation of a conic or alternatively its deviation from a circle.
Foci: the direction toward or away the coordinate system origin, especially in polar coordinates.
Conic Parameter: the direction perpendicular to radial, especially in polar coordinates.

References

1. Sears, Francis W and Zemansky, Mark W., *University Physics*, Second Edition, © 1955 Addison-Wesley Publishing Company, Inc., Library of Congress Catalog No. 55-5026.
2. Bate, Roger R. et al., *Fundamentals of Astrodynamics*, © 1971 Dover Publications, Inc., ISBN 0-486-60061-0.
3. Roy, Archie E., *Orbital Motion*, Third Edition, © 1988 by author, ISBN 0-85274-229-0.
4. Seidelmann, P. Kenneth (editor), *Explanatory Supplement to the Astronomical Almanac*, © 1992 University Science Books, ISBN 0-935702-68-7.
5. Symon, Keith R. *Mechanics*, Third Edition, © 1971 Addison-Wesley, ISBN...
6. Kaplan, Wilfred, and Lewis, D.J., *Calculus and Linear Algebra*, Combined Edition, © 1971 John Wiley & Sons, Inc., ISBN 0-471-45687-X.
7. Beyer, William H. (Editor), *CRC Standard Mathematical Tables*, 27th Edition, © 1984 Chemical Rubber Company (CRC) Press, ISBN 0-8493-0627-2.
8. Spiegel, Murray R., *Schaum's Outline Series Theory and Problems of Mathematical Handbook of Formulas and Tables*, © 1968 McGraw-Hill, Inc., ISBN 07-060224-7.

2 Two-Body Trajectory Characteristics

Understanding the two-body trajectory is key for more advanced orbital mechanics. Two-body trajectories are idealized approximations that would rarely be used in the operational control of a spacecraft. Nevertheless, a comprehensive understanding of them provides insight useful for higher fidelity predictions, including what is usually the first approximations for iterative processes.

2.1 Kepler's Laws

Kepler's Laws of Planetary Motion are the foundation for orbital mechanics. Johannes Kepler published his three planetary laws between 1609 and 1619 which describe the motions of the planets in the solar system. Kepler used Tycho Brahe's extensive and accurate planetary observations with a rigorous mathematical approach. The result replaced the circular orbit and epicycle theory by Nicolaus Copernicus with a more elegant theory involving elliptical orbits that better fit Tycho's data. Kepler's theory was later confirmed in 1687 by Isaac Newton, who using his laws of motion and gravitation, derived consistent mathematical relationships.

Kepler's Laws also describe the motion of spacecraft orbiting the Earth or any other celestial body. Although Kepler's laws are in the context of elliptical conic sections, it is easy to show that a circle is a simplified type of ellipse. And with some relatively minor enhancements (involving Newtonian physics), Kepler's Laws are extensible to open conics as described by parabolic and hyperbolic trajectories.

2.1.1 Kepler's Laws [Level I – Descriptive]

Kepler's laws are stated as:

Kepler's First Law: ***Elliptical Orbits*** – All planets move around the Sun in elliptical orbits, having the Sun at one of the foci.

Kepler's Second Law: ***Equal Area in Equal Time*** – An imaginary line joining the Sun and a planet sweeps out equal areas in equal intervals of time.

Kepler's Third Kaw: ***Harmonic Law*** – The square of the planets orbital periods are directly proportional to the cubes of their mean distances from the Sun.

2.1.1.1 Elliptical Orbits

Most planetary orbits are close to circular. Nevertheless, Kepler's development of elliptical orbital theory came from his inability to reconcile Tycho's observations of the planet Mars to the circular orbit specified by the Copernicus theory. After a laborious analysis, Kepler was able to fit the observational data to an elliptical path, after realizing the Sun needed to be at one focus of the ellipse for the theory to work.

Figure 2-1 shows the geometry of an elliptical orbit with the gravitationally attracting body at what is called the *occupied* focus. The ellipse's other focus is called the unoccupied focus.

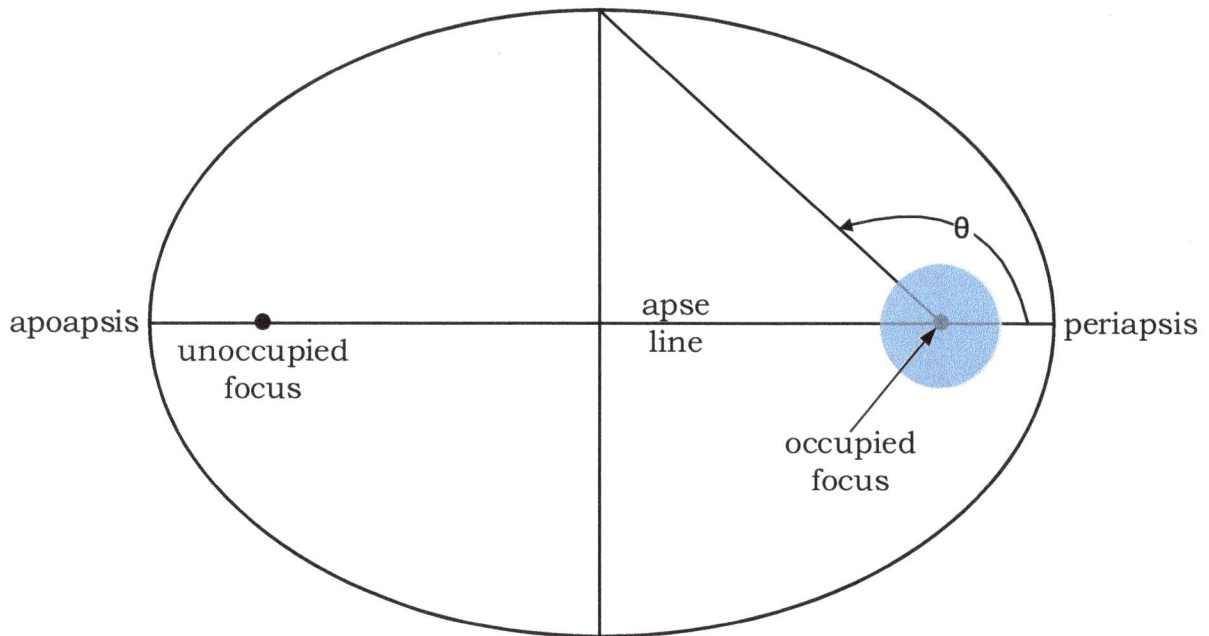

Figure 2-1 Geometry for an Elliptical Orbit

An ellipse is one type of conic section curve applicable to the solar system planets. Newton's analysis indicates the solution for a trajectory in an inverse square gravitational field allows for any conic section as the solution. Thus, while Newton proved Kepler to be correct, his mathematics extended the possibility to allow for circular, parabolic, and hyperbolic trajectories.

2.1.1.2 Equal Area in Equal Time

Orbits sweeping out equal area in equal time is an expression of the law of conservation of angular momentum. The orbiting object moves fastest at its closest radius to the gravitationally attracting body (called periapsis) and moves slowest at its furthest radius (called apoapsis) as shown in Figure 2-1. Likewise, the angular rate varies, being fastest at periapsis and slowest at apoapsis.

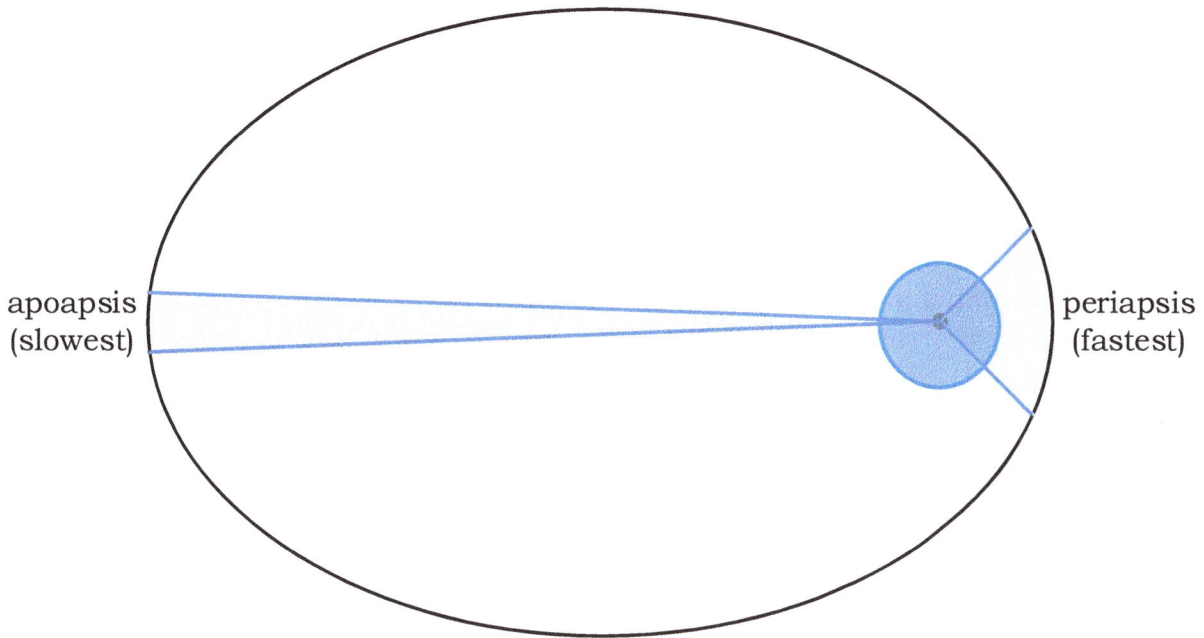

Figure 2-2 Equal Area in Equal Time for Elliptical Orbits

The object orbits in an elliptical path because its velocity is not the correct speed or in the correct direction for its path to be circular. If for example, a spacecraft was inserted at periapsis, the orbital velocity would be horizontal as required for a circular orbit. However, the speed at orbital insertion would have been larger than perfectly balancing gravity (as a centripetal acceleration). The spacecraft speed is too high for gravity to give a circular path. The path is still curved, but initially greater than the gravitational drop, so the spacecraft travels to a higher altitude.

The spacecraft begins gaining altitude with the velocity not being turned quickly enough to remain horizontal. But the vertical velocity component is eroded by gravity the same way a ball thrown straight up slows down as it goes higher. Eventually gravity wins and the vertical velocity component reaches zero. The spacecraft reached apoapsis where once again its velocity is only directed horizontally.

Gaining altitude takes its toll on the spacecraft's speed. Despite the weaker gravitational field at the higher apoapsis altitude, the spacecraft now does not have enough horizontal speed to balance against gravity to have a centripetal acceleration. Thus, the spacecraft begins to lose altitude as it begins falling back toward periapsis.

The increasing gravitational force now causes the spacecraft to gain a downward component of vertical velocity. The downward speed increases the same way any object does in a free fall. Eventually, the spacecraft gains back all the speed lost as it reaches periapsis. This cycle repeats indefinitely as an exchange between kinetic and potential energy.

Kepler's Second Law of equal areas in equal time intervals combines two physical mechanisms already considered: the gravity inverse square law and conservation of angular momentum.

2.1.1.3   Kepler's Harmonic Law

Kepler's Harmonic Law states the square of the orbital period is proportional to the cube of the orbit's semi-major axis. It recognizes both conservation of angular momentum and the fact that larger orbits (i.e., having greater mean distances from their gravitationally attracting body) have a greater orbital circumference. It should be noted that the geometric mean distance from an ellipse to either of its foci is its semi-major axis. The orbital period is the time required to make one complete revolution. Thus, the harmonic law can be re-stated as the square of the orbital period being proportional to the cube of the semimajor axis.

A useful aspect of the harmonic law is that it relates the total system mass (of the two bodies), with the orbital period and orbital semi-major axis. If you know two out of three, you can compute the third parameter.

Key Terms:

Apoapsis: The maximum radius an orbiting body achieves relative to its gravitationally attracting body.
Apse Line: A conic section's major axis of symmetry.
Mean Distance: The mean geometric between a focus and any point on an ellipse is its semi-major axis.
Occupied Focus: The focus of a conic at the location of the gravitationally attracting body.
Periapsis: The maximum radius an orbiting body achieves relative to its gravitationally attracting body.

➤ **Transition**: *You may continue this section with Kepler's Laws at Level II, or you may skip to Level I of the next topic called Conic Sections.*

2.1.2   Kepler's Laws [Level II – Equations]

This section provides the computational equations related to Kepler's laws of orbital motion.

2.1.2.1 Ellipse Polar Radius

Given an elliptical orbit, the orbital radius (r) is computed in equation 2.1 and 2.2 using the polar conic form, relative to the angle from periapsis, called true anomaly (θ) as shown in Figure 2-3. Equation 2.1 is a general form for a conic section, while equation 2.2 uses the ellipse form of the conic parameter (p).

$$r = \frac{p}{1 + e \cos \theta} \qquad\qquad 2.1$$

$$r = \frac{a(1 - e^2)}{1 + e \cos \theta} \qquad\qquad 2.2$$

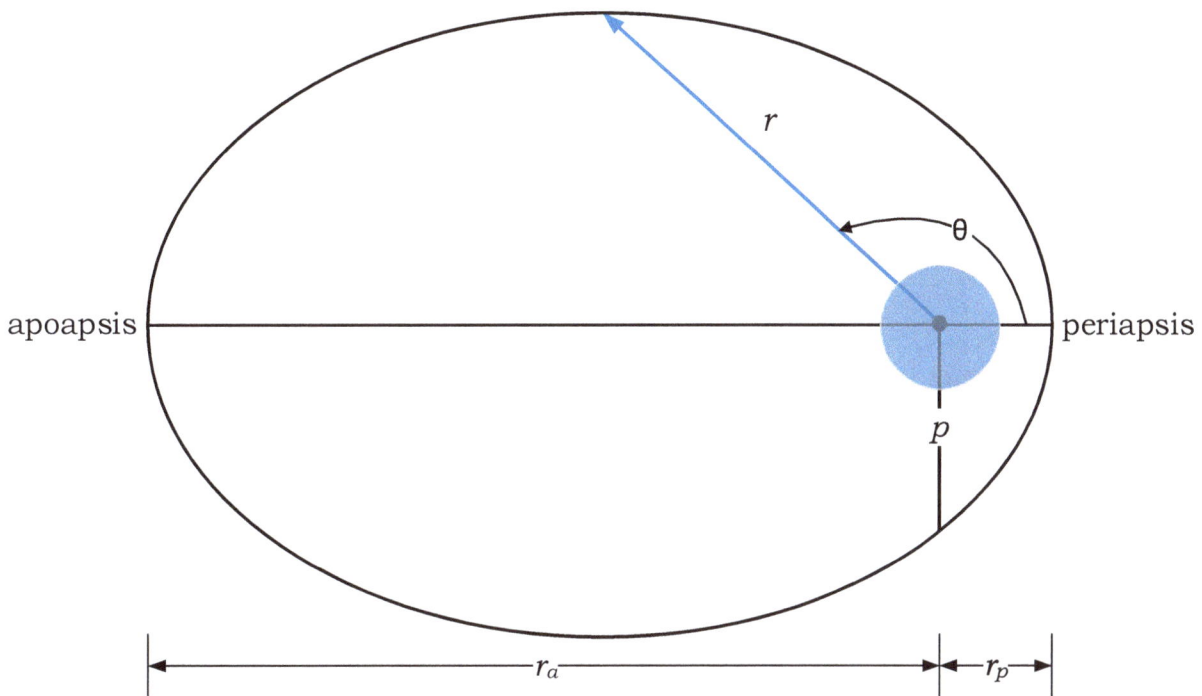

Figure 2-3 Elliptical Orbit Polar Parameters

Since true anomaly is the angle relative to periapsis and measured in the orbital motion direction, it has a zero value ($\theta_p = 0$) at periapsis and has a value of pi radians / 180 degrees value ($\theta_a = \pi\ rad = 180°$) at apoapsis. Equations 2.3 and 2.4 provide the radius of periapsis (r_p) and apoapsis (r_a) respectively, by applying the applicable true anomaly to equation 2.2.

$$r_p = a(1 - e) \qquad\qquad 2.3$$

$$r_a = a(1 + e) \qquad\qquad 2.4$$

2.1.2.2 General Conic Polar Radius

Equation 2.5 is a general expression of radius, applicable to all conic sections, expressed in terms of the trajectory's specific angular momentum (h) and the gravitational parameter (μ) for the central attracting body.

$$r = \frac{h^2}{\mu(1 + e \cos \theta)} \qquad\qquad 2.5$$

Equation 2.6 provides the general expression for a conic parameter in terms of specific angular momentum and the gravitational parameter for the central attracting body. This is inferred by comparison of equation 2.1 with equation 2.5.

$$p = \frac{h^2}{\mu} \qquad\qquad 2.6$$

Thus, equation 2.7 is another general expression for a conic section's polar radius.

$$r = \frac{p}{1 + e \cos \theta} \qquad\qquad 2.7$$

2.1.2.3 Area Rate

The specific angular momentum (h) is *twice the area rate*, as indicated by Kepler's Second Law. Equation 2.8 is the specific angular momentum for all conic section trajectories, based on the conic parameter (p).

$$h = \sqrt{\mu p} \qquad\qquad 2.8$$

Equation 2.9 is the specific angular momentum in terms of the orbital ellipse's semi-major axis (a) and eccentricity (e) as well as the central attracting body's gravitational parameter (μ).

$$h = \sqrt{\mu a(1 - e^2)} \qquad\qquad 2.9$$

Equation 2.10 is the time rate of change of true anomaly with respect to time for a conic trajectory. Equation 2.11 is the true anomaly rate specifically for an elliptical orbit.

$$\dot{\theta} = \frac{h}{r^2} \qquad\qquad 2.10$$

$$\dot{\theta} = \frac{\sqrt{\mu a (1 - e^2)}}{r^2} \qquad\qquad 2.11$$

2.1.2.4   Mean Motion, Period, and Semi-Major Axis

Equation 2.12 describes the mean motion (n), which is the trajectory's mean angular rate in rad/s. For closed trajectories (i.e., orbits) the mean motion's reciprocal can be used to compute the orbital period (T), as provided in equation 2.13.

$$n = \sqrt{\frac{\mu}{a^3}} \qquad\qquad 2.12$$

$$T = 2\pi \sqrt{\frac{a^3}{\mu}} \qquad\qquad 2.13$$

➢ **Transition**: *You may continue this section with Kepler's Laws at Level III, or you may skip to Level I of the next topic called Conic Sections.*

2.1.3   Kepler's Laws [Level III – Derivation]

This section uses Newton's Laws of Motion and Gravitation to derive Kepler's Laws of Orbital Motion.

2.1.3.1   Conic Section Trajectories

The first principles derivation of Kepler's Laws begins with Newton's Law of gravitation for two masses as shown in Figure 2-4. Two objects with masses m_1 and m_2 are separated by distance r. Their positions are described by vectors $\bar{r}_1$ and $\bar{r}_2$ in a non-rotating, non-accelerating (i.e., inertial) coordinate frame.

Two objects attract each other by Newton's Gravitational Law. Vector $\bar{F}_1$ is the gravitational force exerted on m_1 by m_2; vector $\bar{F}_2$ is the gravitational force exerted on m_2 by m_1. Equation 2.14 shows $\bar{F}_1$ being equal and opposite to $\bar{F}_2$. Equation 2.15 is the $\bar{F}_1$ vector from Newton's Law of Gravity.

$$\bar{F}_1 = -\bar{F}_2 \qquad\qquad 2.14$$

$$\bar{F}_1 = \frac{G(m_1 + m_2)}{r^3}\bar{r} \qquad\qquad 2.15$$

Equations 2.16 and 2.17 are vectors $\bar{F}_1$ and $\bar{F}_2$ formulated consistent with Newton's Second Law. Acceleration is expressed as the second time derivative of the position vector.

$$\bar{F}_1 = m_1 \ddot{\bar{r}} \qquad\qquad 2.16$$

$$\bar{F}_2 = -m_1 \ddot{\bar{r}} \qquad\qquad 2.17$$

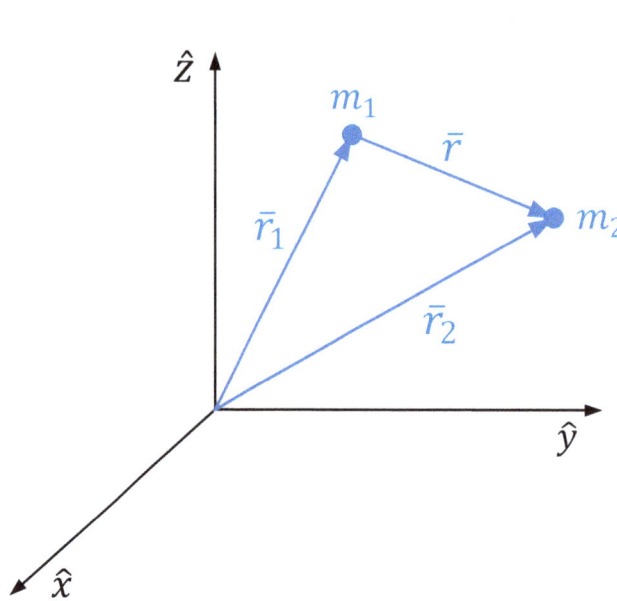

Figure 2-4 Two Masses Positioned in an Inertial Coordinate Frame

Equation 2.18 results from substituting the right side of equation 2.14 into the left side of equation 2.15.

$$\ddot{\bar{r}}_1 = \frac{Gm_2}{r^3}\bar{r} \qquad\qquad 2.18$$

Equation 2.19 is a substitution of the right side of equation 2.14 into the left side of equation 2.15, followed by a substitution of the right side of equation 2.17 into the left side of equation 2.15.

$$-\ddot{\bar{r}}_2 = \frac{Gm_1}{r^3}\bar{r} \qquad\qquad 2.19$$

Equation 2.20 results from adding equations 2.18 and 2.19. Note from Figure 2-4 that vector $\bar{r}$ is the difference between $\bar{r}_2$ and $\bar{r}_1$.

$$-\ddot{\bar{r}} = \frac{G(m_1 + m_2)}{r^3}\bar{r} \qquad\qquad 2.20$$

Equation 2.21 substitutes μ for the universal gravitational constant, multiplied by the sum of the two masses. This case considers the motion of an insignificant mass (m_2) with respect to a dominant mass (m_1). Thus, μ only accounts for the mass of the dominant body, consistent with equation 1.2.

$$\ddot{\bar{r}} + \frac{\mu\bar{r}}{r^3} = 0 \qquad\qquad 2.21$$

It should be noted that the origin in equation 2.21 is at the *barycenter* between the two masses. The barycenter (Greek: βαρύς κέντρον or heavy center) is the center of mass or balance point between the masses. While the difference between that and the center of the gravitationally attracting body is usually negligible for spacecraft, there are circumstances such as the Moon's orbit around the Earth where the difference must be considered.

Equation 2.22 eliminates the second term in equation 2.21 by taking the cross product ($\bar{r}\times$) of both sides and recognizing the cross product of two colinear vectors is zero. Since the right side of equation 2.21 is zero, the vectors on the left side must also be colinear. Thus, the equation is the cross product of position with a radially directed acceleration.

$$\bar{r} \times \ddot{\bar{r}} = 0 \qquad\qquad 2.22$$

Equation 2.23 comes from integrating equation 2.22 with respect to time, with $\bar{h}$ being a constant of integration. This may be verified by taking the time derivative of equation 2.23, recalling the chain rule and that the cross product of a vector with itself is zero.

$$\bar{r} \times \dot{\bar{r}} = \bar{h} \qquad\qquad 2.23$$

Vector $\bar{h}$ is the specific angular momentum. Since $\dot{\bar{r}}$ is the velocity, this result verifies equation 1.28. Furthermore, since $\bar{h}$ is constant the angular momentum is constant, consistent with conservation of angular momentum.

Equation 2.24 is the cross product of equation 2.21 with the angular momentum vector.

$$\ddot{\bar{r}} \times \bar{h} = -\frac{\mu\bar{r}}{r^3} \times \bar{h} \qquad\qquad 2.24$$

Equation 2.25 results from substituting the left side of equation 2.23 into the right side of equation 2.24.

$$\ddot{\bar{r}} \times \bar{h} = -\frac{\mu}{r^3}[\bar{r} \times (\bar{r} \times \dot{\bar{r}})] \qquad 2.25$$

Equation 2.26 expands the bracketed portion of equation 2.25 using the *vector triple cross product identity*. Equation 2.27 is a simplification of equation 2.26. Note that the dot product of $\bar{r}$ with $\dot{\bar{r}}$ equals the scalar r multiplied by its own time rate of change.

$$\ddot{\bar{r}} \times \bar{h} = -\frac{\mu}{r^3}[(\bar{r} \cdot \bar{r})\dot{\bar{r}} - (\bar{r} \cdot \dot{\bar{r}})\bar{r}] \qquad 2.26$$

$$\ddot{\bar{r}} \times \bar{h} = \mu\left[\frac{\dot{\bar{r}}}{r} - \frac{\dot{r}\bar{r}}{r^2}\right] \qquad 2.27$$

Equation 2.28 shows that the expression inside the brackets in equation 2.27 is the time derivative of $\bar{r}$, divided by the scalar r.

$$\frac{d}{dt}\left(\frac{\bar{r}}{r}\right) = \frac{\dot{\bar{r}}}{r} - \frac{\dot{r}\bar{r}}{r^2} \qquad 2.28$$

Thus, equation 2.29 is obtained by integrating equation 2.27, with $\bar{e}$ (i.e., the eccentricity vector) being a constant of integration.

$$\dot{\bar{r}} \times \bar{h} = \frac{\mu}{r}(\bar{r} + r\bar{e}) \qquad 2.29$$

The ability to compute the eccentricity vector will be useful later. Equation 2.30 solves for the eccentricity vector from equation 2.29.

$$\bar{e} = \frac{1}{\mu}\left[\bar{v} \times \bar{h} - \frac{\mu\bar{r}}{r}\right] \qquad 2.30$$

Equation 2.31 results from the dot product of $\bar{r}$ with both sides of equation 2.29.

$$\bar{r} \cdot (\dot{\bar{r}} \times \bar{h}) = \bar{r} \cdot \left[\frac{\mu}{r}(\bar{r} + r\bar{e})\right] \qquad 2.31$$

Equation 2.32 uses the *vector dot and cross product interchange identity* on the left side of equation 2.31. Note that the quantity in parentheses on equation's 2.31 left side is the specific angular momentum.

$$(\bar{r} \times \dot{\bar{r}}) \cdot \bar{h} = \frac{\mu}{r}(r^2 + r\bar{r} \cdot \bar{e}) \qquad 2.32$$

Equation 2.33 is a simplification of equation 2.32, defining θ as the angle between vectors $\bar{r}$ and $\bar{e}$.

$$h^2 = \mu r(1 + e \cos \theta)$$ 2.33

Equation 2.34 results from rearranging terms, solving for radius, which verifies equation 2-5 (the conic section polar parametric equation).

$$r = \frac{h^2}{\mu(1 + e \cos \theta)}$$ 2.34

This mathematically verifies Kepler's First law that a trajectory in an inverse square gravitational field is a conic.

2.1.3.2 Area Rate

The *area rate*, or area swept out by $\bar{r}$ over time can be determined be re-examining motion in polar coordinates. Figure 2-5 shows the radius vector sweeping out an infinitesimal angle θ (i.e., $d\theta$) over an infinitesimal time interval (dt) to create a differential area (dA). Equations 2.35 and 2.36 are the equation for the area rate as the triangular area, divided by dt.

$$\frac{dA}{dt} = \frac{1}{2}r\left(r\frac{d\theta}{dt}\right) = \frac{1}{2}r^2\frac{d\theta}{dt}$$ 2.35

$$\dot{A} = \frac{1}{2}r^2\dot{\theta}$$ 2.36

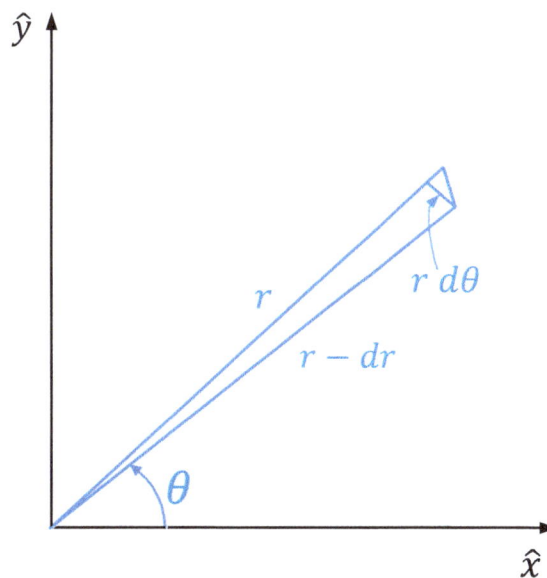

Figure 2-5 Area Swept Out Over Time

Recall the polar equations for $\bar{r}$ and $\bar{v}$ (equations 1.11 and 1.13 respectively). Equation 2.37 computes the angular momentum using the vector cross product of $\bar{r}$ and $\bar{v}$. By cross product definition, the angular momentum is perpendicular to both input vectors, in a perpendicular direction defined in the direction normal to the trajectory plane ($\hat{n}$) or plane of motion.

$$\bar{h} = \bar{r} \times \bar{v} = \begin{vmatrix} \hat{r} & \hat{\theta} & \hat{n} \\ r & 0 & 0 \\ \dot{r} & r\dot{\theta} & 0 \end{vmatrix} = \begin{bmatrix} 0 \\ 0 \\ r^2\dot{\theta} \end{bmatrix} \qquad 2.37$$

Comparing equation 2.36 with equation 2.37 show the angular momentum magnitude is twice the rate of area swept out by vector $\bar{r}$. Since angular momentum is constant, the area rate of change over time is also constant. This verifies Kepler's Second Law mathematically.

2.1.3.3   Semi-Major Axis and Period

The period is the time to orbit once around a closed trajectory (2π radians). The area swept out over a full orbital ellipse is the ellipse area ($A = \pi ab$). Equation 2.38 computes the semi-minor axis and is derivable from geometry.

$$b = a\sqrt{1 - e^2} \qquad 2.38$$

Equation 2.39 computes the angular momentum as the ratio of twice the area rate over the full ellipse, to the period. Equation 2.40 algebraically solves the equation for period, substituting the right side of 2.38 for the semi-minor axis.

$$h = \frac{2\pi ab}{T} \qquad 2.39$$

$$T = \frac{2\pi a^2\sqrt{1 - e^2}}{h} \qquad 2.40$$

Equation 2.41 squares both sides to remove the square root, and substitutes equation 2.9 for the angular momentum magnitude, and simplifies the result.

$$T^2 = \left(\frac{4\pi^2}{\mu}\right) a^3 \qquad 2.41$$

This verifies Kepler's Third Law that the square of the period is proportional to the cube of the semi-major axis. Furthermore, equation 2.41 verifies equation 2.13.

Key Terms:

Period: the time required to make one complete orbital revolution.
Revolution: a complete orbit (360°) about a closed trajectory.

2.2 Energy Law

The energy law governs the trajectory's mechanical energy. It is useful for describing the relationship between speed and radial offset from the gravitationally attracting body.

2.2.1 Energy Law [Level I – Descriptive]

Mechanical energy manifests itself in two forms: potential and kinetic. Potential energy is the energy stored in the trajectory as the result of displacing the spacecraft from the gravitationally attracting body. Potential energy is reclaimed and transformed to kinetic energy as the spacecraft gains speed as it falls in the gravity field toward the gravitationally attracting body.

➤ **Transition**: *You may continue this section with the Energy Law at Level II, or you may skip to the beginning of the next section called In-Plane Phase versus Time.*

2.2.2 Energy Law [Level II – Equations]

The trajectory's total specific energy (ε) (i.e., energy per unit mass) is specified in equation 2.42, which is also known as the *vis viva* (living force) equation.

$$\varepsilon = \frac{v^2}{2} - \frac{\mu}{r} \qquad\qquad 2.42$$

The first term is the specific kinetic energy, with the specific potential energy as the second term. The specific kinetic energy in the first term exchanges with the specific potential energy in the second term, with speed and radius varying accordingly. The combination of kinetic energy and potential energy is the system's total mechanical energy.

Equation 2.43 characterizes the total specific energy (ε) in terms of the gravitational parameter (μ), eccentricity (e), and specific angular momentum (h).

$$\varepsilon = -\frac{\mu^2(1 - e^2)}{2h^2} \qquad\qquad 2.43$$

39

Table 2-1 evaluates equation 2.43 for the various conic types, based on eccentricity. Based on the conventions used, circular and elliptical orbits have negative total energy. Parabolic trajectories are the crossover with zero total energy. Hyperbolic trajectories are the only trajectory type with net positive energy.

Table 2-1 Energy for Conic Types		
Conic Type	**Eccentricity**	**Energy**
Circle	$e = 0$	$\varepsilon = -\dfrac{\mu^2}{2h^2}$
Ellipse	$0 < e < 1$	$\varepsilon = -\dfrac{\mu}{2a}$
Parabola	$e = 1$	$\varepsilon = 0$
Hyperbola	$e > 1$	$\varepsilon = \dfrac{\mu}{2a}$

Key Terms:

> **Kinetic Energy:** is the energy associated with motion.
> **Potential Energy:** is the energy stored in a system, such as in a gravitational field when a mass is displaced from the attracting body.
> **Specific Energy**: energy per unit mass.
> **Specific Mechanical Energy:** the total mechanical energy per unit mass, combining the specific kinetic and potential energies.
> **Vis Viva:** is the "living force" or gravitational energy equation.

➢ **Transition**: *You may continue this section with the Energy Law at Level III, or you may skip to the beginning of the section Chapter called In-Plane Phase versus Time.*

2.2.3   Energy Law [Level III – Derivation]

Since linear momentum ($\bar{p}$) is the product of mass (m) and velocity ($\bar{v}$), velocity may be considered the momentum per unit mass or *specific linear momentum*. Recall also that velocity is the first time derivative of position ($\bar{v} = \dot{\bar{r}}$).

Equation 2.44 is the dot product of a trajectory's acceleration (as specified in equation 2.20) with the specific linear momentum.

$$\ddot{\bar{r}} \cdot \dot{\bar{r}} = -\frac{\mu}{r^3} \bar{r} \cdot \dot{\bar{r}} \qquad\qquad 2.44$$

The dot product in the left term of equation 2.44 is the derivative of half the velocity squared as shown in equations 2.45 through 2.47, recalling the chain rule for derivatives.

$$\ddot{\bar{r}} \cdot \dot{\bar{r}} = \frac{d}{dt}(\dot{\bar{r}} \cdot \dot{\bar{r}}) \qquad\qquad 2.45$$

$$\frac{d}{dt}(\dot{\bar{r}} \cdot \dot{\bar{r}}) = \frac{d}{dt}(v^2) = \dot{\bar{r}} \cdot \ddot{\bar{r}} + \ddot{\bar{r}} \cdot \dot{\bar{r}} = 2\ddot{\bar{r}} \cdot \dot{\bar{r}} \qquad\qquad 2.46$$

$$\ddot{\bar{r}} \cdot \dot{\bar{r}} = \frac{d}{dt}\left(\frac{v^2}{2}\right) \qquad\qquad 2.47$$

Equation 2.48 identifies the dot product in the right term of equation 2.44 as simply the radius vector, multiplied by velocity, per equation 1.21. The remaining equations show the right side being the time derivative of the quantity (μ/r).

$$\bar{r} \cdot \dot{\bar{r}} = r\dot{r} \qquad\qquad 2.48$$

$$\frac{d}{dt}\left(\frac{1}{r}\right) = -\frac{\dot{r}}{r^2} \qquad\qquad 2.49$$

$$-\frac{\mu}{r^3}\bar{r} \cdot \dot{\bar{r}} = -\mu\frac{r\dot{r}}{r^3} = -\mu\frac{\dot{r}}{r^2} = \frac{d}{dt}\left(\frac{\mu}{r}\right) \qquad\qquad 2.50$$

Equations 2.51 shows the derivatives (i.e., the right sides of equations 2.47 and 2.50) substituted into equation 2.44. Equation 2.52 is the integral of equation 2.51 with respect to time.

$$\frac{d}{dt}\left(\frac{v^2}{2} - \frac{\mu}{r}\right) \qquad\qquad 2.51$$

$$\frac{v^2}{2} - \frac{\mu}{r} = \varepsilon \qquad\qquad 2.52$$

The total energy (ε) is the *constant of integration*. Since it is constant, the specific mechanical energy is the same at all points in the trajectory. This result verifies equation 2.42.

The total energy may be characterized by the trajectory parameters. Since angular momentum is conserved, energy may be characterized at periapsis, which exists for all conic types and is convenient since there is only a horizontal velocity component at periapsis. This property allows speed to be related to the angular momentum magnitude as a scalar.

Equation 2.53 is the periapsis radius (r_p), as defined by the general conic radius (i.e., equation 2.34). Equation 2.54 is the periapsis speed (v_p). Equation 2.55 is equation 2.52, using the periapsis radius and speed values.

$$r_p = \frac{h^2}{\mu(1-e)} \hspace{4cm} 2.53$$

$$v_p = \frac{h}{r_p} \hspace{4cm} 2.54$$

$$\varepsilon = \frac{v_p^2}{2} - \frac{\mu}{r_p} \hspace{4cm} 2.55$$

Equation 2.56 is the total energy, after making the applicable substitutions and cancellations. Equation 2.57 simplifies equation 2.56 through factoring.

$$\varepsilon = \frac{\mu^2(1+e)^2}{2h^2} - \frac{\mu^2(1-e)}{h^2} \hspace{3cm} 2.56$$

$$\varepsilon = -\frac{\mu^2(1-e^2)}{2h^2} \hspace{3cm} 2.57$$

2.3 In-Plane Phase versus Time

Most predictions seek to determine the location within the trajectory as a function of time. This section provides the means to determine the in-plane phase for a specified time.

2.3.1 In-Plane Phase versus Time [Level I – Descriptive]

Kepler's Second Law indicates that, except for the circular orbit, the rate of change of the in-plane phase, or true anomaly (θ), varies with time. In general, the higher the eccentricity, the more non-linear is the relationship between time and in-plane phase. Thus, algorithms are devised based on the conic section geometry to compute the phase for a given time.

➤ **Transition**: *You may continue this section with the In-Plane versus Time at Level II, or you may skip to the beginning of the section Computing In-Plane Phase Given Time-of-Flight.*

In-Plane Phase versus Time [Level II – Equations]

The in-plane phase mathematically describes the true anomaly as the orbiting object's angular progression from periapsis. This section provides the equations to convert from the true anomaly angle to the corresponding time since periapsis and is organized by the conic section types (i.e., circles, ellipses, parabolas, and hyperbolas), following the derivation by Curtis (2020).

2.3.1.1   Circular Orbit Phase versus Time

The in-plane phase angle, or true anomaly (θ) for a circular orbit is directly proportional to time in a circular orbit. Equation 2.58 converts the change in true anomaly ($\Delta\theta$) to a time difference (Δt), also using the circular radius (r) and the gravitational parameter (μ) for the attracting body.

$$\Delta t = \Delta\theta \sqrt{\frac{r^3}{\mu}}$$

2.58

Note that equation 2.58 is invertible, allowing $\Delta\theta$ to be computed for a corresponding Δt.

2.3.1.2   Elliptical Orbit Phase versus Time

The true anomaly (θ) for an elliptical orbit may be converted to its corresponding mean anomaly angle (M), which is directly proportional to time. The conversion requires an intermediate auxiliary angle, called elliptical *eccentric anomaly* (E). Figure 2-6 illustrates the geometric relationship in an ellipse between true and eccentric anomalies.

Equation 2.59 is the elliptical orbit relationship between the eccentric anomaly and true anomaly, using orbital eccentricity (e). Equation 2.60, called Kepler's Equation for an Elliptical Orbit, computes the mean anomaly from eccentric anomaly.

$$\tan\frac{E}{2} = \sqrt{\frac{1-e}{1+e}} \tan\frac{\theta}{2}$$

2.59

$$M = E - e\sin E$$

2.60

43

Equation 2.59 is invertible to the extent that it permits eccentric anomaly to be computed from true anomaly and vice versa. Equation 2.60 is not invertible. While it is straight forward to compute mean anomaly from eccentric anomaly, there is no known mathematical inverse process. Instead, the conversion from mean to eccentric anomaly requires an iterative numerical process, which will be covered in the section involving solving Kepler's equation.

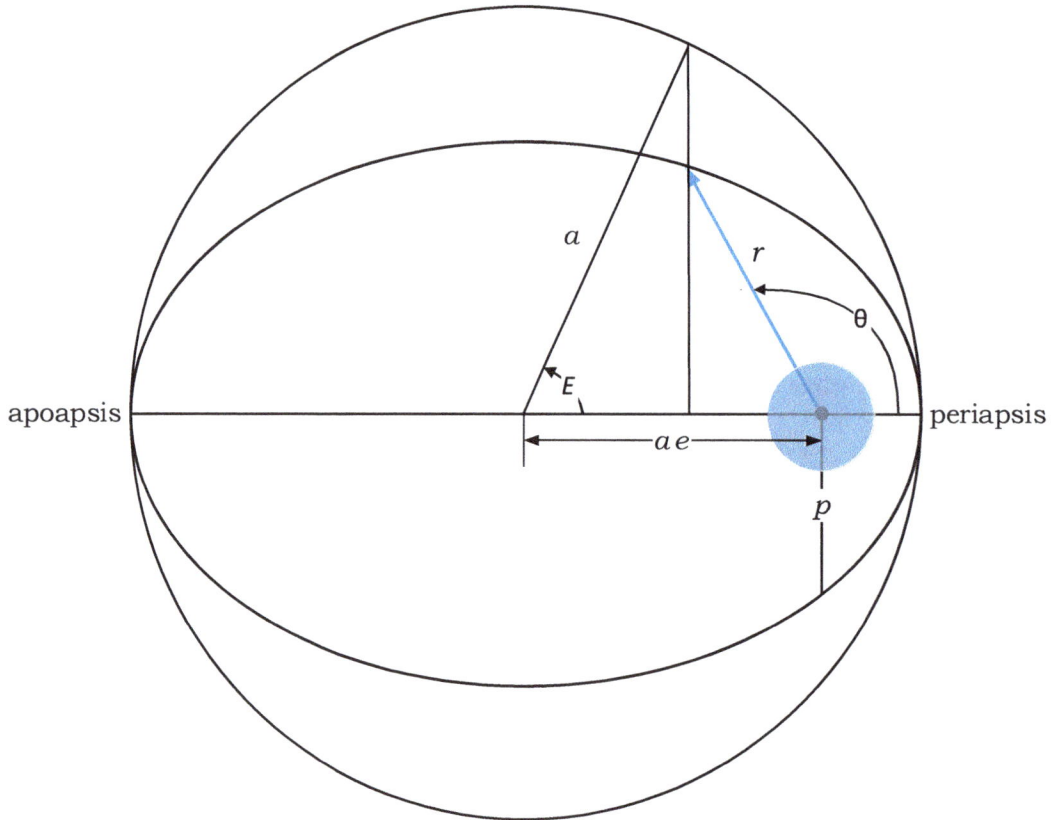

Figure 2-6 True and Eccentric Anomalies in an Elliptical Orbit

It should also be emphasized that equation 2.60 *requires the use of radians angular units* since the $e \sin E$ term is dimensionless. While radians are a proper angular unit of measure, they are effectively dimensionless since they represent the ratio of a circular arc length to the circle's radius.

Equation 2.61 relates time to mean anomaly, angular momentum, and the attracting body's gravitational parameter. Equation 2.62 is Kepler's Time-of-Flight Equation, allowing computation of a time difference (Δt) in direct proportion to a mean anomaly change *(ΔM)*. The use of mean anomaly gives equation 2.61 a direct correspondence with equation 2.58 for the circular orbit.

$$\Delta t = \Delta M \frac{h^3}{\mu^2 (1 - e^2)^{3/2}} \qquad 2.61$$

$$\Delta t = \Delta M \sqrt{\frac{a^3}{\mu}} \qquad 2.62$$

It should be emphasized that the mean anomaly difference in equations 2.61 and 2.62 is the difference between two separately computed mean anomaly angles, each corresponding to a particular true anomaly. Since the mean anomaly at periapsis is zero, a mean anomaly angle substituted for the difference in equations 2.61 and 2.62 produces the elapsed time since periapsis passage.

2.3.1.3   Parabolic Trajectory Phase versus Time

The true anomaly (θ) for a parabolic trajectory may be converted to its corresponding mean anomaly angle (M), which is directly proportional to time. The conversion requires an intermediate auxiliary angle, called parabolic eccentric anomaly (D).

Equation 2.63 converts true anomaly to eccentric anomaly, because a parabolic trajectory's eccentricity equals one. Equation 2.64, which is Kepler's Equation for a Parabolic Trajectory, converts the parabolic eccentric anomaly to its corresponding mean anomaly.

$$D = \tan\frac{\theta}{2} \qquad 2.63$$

$$M = D + \frac{D^3}{3} \qquad 2.64$$

Equation 2.63 may be inverted to determine eccentric anomaly from mean anomaly using an auxiliary variable z as shown in equations 2.65 and 2.66.

$$z = \left(3M + \sqrt{1 + (3M)^2}\right)^{1/3} \qquad 2.65$$

$$\theta = 2\tan^{-1}\left(z - \frac{1}{z}\right) \qquad 2.66$$

Equation 2.67 computes the time of flight from periapsis for a given mean anomaly.

$$t = M\frac{h^3}{\mu^2} \qquad 2.67$$

2.3.1.4 Hyperbolic Trajectory Phase versus Time

Equation 2.68 is the hyperbolic trajectory relationship between the hyperbolic eccentric anomaly (F) and true anomaly (θ), using orbital eccentricity (e). Equation 2.69, called Kepler's Equation for a hyperbolic trajectory, computes the mean anomaly from eccentric anomaly. Equations 2.68 and 2.69 correspond well with their elliptic counterparts (equations 2.59 and 2.60), with a flip in the sign (due to $e>1$) and using hyperbolic trigonometric functions on the eccentric anomaly.

$$\tanh\frac{F}{2} = \sqrt{\frac{e-1}{e+1}}\tan\frac{\theta}{2} \qquad\qquad 2.68$$

$$M = e\,\sinh F - F \qquad\qquad 2.69$$

Equation 2.68 is invertible to the extent that it permits hyperbolic eccentric anomaly to be computed from true anomaly and vice versa. Equation 2.69 is not invertible. While it is straight forward to compute mean anomaly from eccentric anomaly, there is no known mathematical inverse process. Instead, the conversion from mean to hyperbolic eccentric anomaly requires an iterative numerical process, which will be covered in the section involving solving Kepler's equation.

As with its eccentric orbit counterpart, equation 2.69 *requires the use of radians angular units* since the $e\,\sinh F$ term is dimensionless.

Equation 2.70 computes the time of flight from periapsis for a given mean anomaly.

$$t = M\frac{h^3}{\mu^2(e^2-1)^{3/2}} \qquad\qquad 2.70$$

Key Terms:

Eccentric Anomaly: is an orbital phase angle that provides an intermediate transformation between true and mean anomaly.
Mean Anomaly is the time-regularized orbit phase relative to periapsis. This provides the effective or mean angular excursion from periapsis were the angular rate a constant averaged over the trajectory.
True Anomaly: is the geometric orbital phase relative to periapsis in the direction of motion. This represents the true angular excursion from periapsis.

➤ **Transition**: *You may continue this section with the In-Plane versus Time at Level III, or you may skip to the beginning of the section Computing In-Plane Phase Given Time-of-Flight.*

2.3.2   In-Plane Phase versus Time [Level III – Derivation]

Recall that equation 2.10 related the time rate of change of true anomaly rate to specific angular momentum and radius ($\dot{\theta} = h/r^2$). Equation 2.5 relates radius to the in-plane phase or true anomaly ($r = h^2/[\mu(1 + e\cos\theta)]$) and is applicable for all conic trajectories. Equation 2.71 Substitutes for r in the phase time derivative.

$$\frac{\mu}{h^3}dt = \frac{d\theta}{(1 + e\cos\theta)^2} \qquad\qquad 2.71$$

Integrating both sides from periapsis to the specified angle yields:

$$\frac{\mu}{h^3}t = \int_0^\theta \frac{d\theta}{(1 + e\cos\theta)^2} \qquad\qquad 2.72$$

The results for the integral on the right side may be found in a standard mathematical handbook or integral tables, with three distinct solutions based on the relative values of the coefficients.

$$\int \frac{dx}{(a + b\cos x)^2} = \frac{1}{(a^2 - b^2)^{3/2}}\left[2a\tan^{-1}\left(\sqrt{\frac{a-b}{a+b}}\tan\frac{x}{2}\right) - \frac{b\sin x\sqrt{a^2 - b^2}}{a + b\ \cos x}\right] \qquad b < a \qquad 2.73$$

$$\int \frac{dx}{(a + b\cos x)^2} = \frac{1}{a^2}\left(\frac{1}{2}\tan\frac{x}{2} + \frac{1}{6}\tan^3\frac{x}{2}\right) \qquad b = a \qquad 2.74$$

$$\int \frac{dx}{(a + b\cos x)^2} = \frac{1}{(a^2 - b^2)^{3/2}}\left[\frac{b\sin x\sqrt{b^2 - a^2}}{a + b\ \cos x} - a\ ln\left(\frac{\sqrt{b+a} + \sqrt{b-a}\tan\frac{x}{2}}{\sqrt{b+a} - \sqrt{b-a}\tan\frac{x}{2}}\right)\right] \qquad b > a \qquad 2.75$$

The coefficients are identified as: $a = 1$, $b = e$, and $x = \theta$ to conform with the integral in equation 2.72. As such, the first solution ($e < 1$) is applicable to elliptical trajectories, the second solution ($e = 1$) is applicable to parabolic trajectories, and the third solution ($e > 1$) is applicable to hyperbolic trajectories.

2.3.2.1   Elliptical Solution

Equation 2.76 Is the time-of-flight from periapsis for an elliptical orbit.

$$\frac{\mu^2}{h^3}t = \frac{1}{(1-e^2)^{3/2}}\left[2\tan^{-1}\left(\sqrt{\frac{1-e}{1+e}}\tan\frac{\theta}{2}\right) - \frac{e\sin\theta\sqrt{1-e^2}}{1+e\cos\theta}\right] \qquad 2.76$$

Both sides are multiplied by the quantity $(1-e^2)^{3/2}$, leaving the left side as the elliptical mean anomaly (M_e).

$$M_e = \frac{\mu^2}{h^3}(1-e^2)^{3/2}t \qquad 2.77$$

Note that the elliptical mean anomaly is linearly proportional to time. The result becomes:

$$M_e = 2\tan^{-1}\left(\sqrt{\frac{1-e}{1+e}}\tan\frac{\theta}{2}\right) - \frac{e\sin\theta\sqrt{1-e^2}}{1+e\cos\theta} \qquad 2.78$$

The right side of equation 2.78 requires an intermediate transformation. Figure 2-6 introduced an auxiliary circle tangent to the apse points and an auxiliary angle called eccentric anomaly. Eccentric anomaly is geometrically related to true anomaly by:

$$a\cos E = ae + r\cos\theta \qquad 2.79$$

By substituting the orbital radius equation:

$$a\cos E = ae + \frac{a(1-e^2)\cos\theta}{1+e\cos\theta} \qquad 2.80$$

$$\cos E = \frac{e+\cos\theta}{1+e\cos\theta} \qquad 2.81$$

Using algebra and trigonometric identities, the following is also obtained:

$$\sin E = \frac{\sqrt{1-e^2}\sin\theta}{1+e\cos\theta} \qquad 2.82$$

From the ratio of the expressions for $\sin E$ and $\cos E$ an expression relating eccentric anomaly with true anomaly is obtained:

$$\tan\frac{E}{2} = \sqrt{\frac{1-e}{1+e}}\tan\frac{\theta}{2} \qquad 2.83$$

48

This result verifies equation 2.59.

2.3.2.2 Parabolic Solution

The integral solution for the parabolic variant directly determines the parabolic mean anomaly (M_p), with the result known as Barker's equation.

$$M_p = \frac{\mu^2}{h^3} t \qquad\qquad 2.84$$

$$M_p = \frac{1}{2}\tan\frac{\theta}{2} + \frac{1}{6}\tan^3\frac{\theta}{2} \qquad\qquad 2.85$$

Time and true anomaly can be related from solution of Barker's equation as a cubic equation:

$$\frac{1}{6}\left(\tan\frac{\theta}{2}\right)^3 + \frac{1}{2}\left(\tan\frac{\theta}{2}\right) - M_p = 0 \qquad\qquad 2.86$$

The cubic has one real root:

$$\tan\frac{\theta}{2} = z - \frac{1}{z} \qquad\qquad 2.87$$

The parabolic mean anomaly is related to the true anomaly by:

$$z = \left(3M_p + \sqrt{1 + \left(3M_p\right)^2}\right)^{1/3} \qquad\qquad 2.88$$

2.3.2.3 Hyperbolic Solution

The time of flight for the hyperbolic solution is:

$$\frac{\mu^2}{h^3} t = \frac{1}{(e^2 - 1)}\frac{e\sin\theta}{1 + e\cos\theta} - \frac{1}{(e^2 - 1)^{3/2}} \ln\left[\frac{\sqrt{e+1} + \sqrt{e-1}\tan\frac{\theta}{2}}{\sqrt{e+1} - \sqrt{e-1}\tan\frac{\theta}{2}}\right] \qquad\qquad 2.89$$

Both sides are multiplied by the quantity $(e^2 - 1)^{3/2}$, leaving the left side as the hyperbolic mean anomaly (M_h):

$$M_h = \frac{\mu^2}{h^3}(e^2 - 1)^{3/2} t \qquad\qquad 2.90$$

Equation 2.91 is an expression for the hyperbolic mean anomaly in terms of eccentricity and true anomaly.

$$M_h = \frac{e \sin \theta \sqrt{e^2 - 1}}{1 + e \cos \theta} - ln \left[\frac{\sqrt{e + 1} + \sqrt{e - 1} \tan \frac{\theta}{2}}{\sqrt{e + 1} - \sqrt{e - 1} \tan \frac{\theta}{2}} \right] \qquad 2.91$$

Figure 2-7 illustrates the geometric relationship between Cartesian and polar parameters in a hyperbolic trajectory.

Equations 2.92 and 2.93 provide the relationships between the trajectory geometry and the hyperbolic cosine and sine of the eccentric anomaly, based on the hyperbolic Cartesian equation.

$$\cosh F = \frac{x}{a} \qquad 2.92$$

$$\sinh F = \frac{y}{b} \qquad 2.93$$

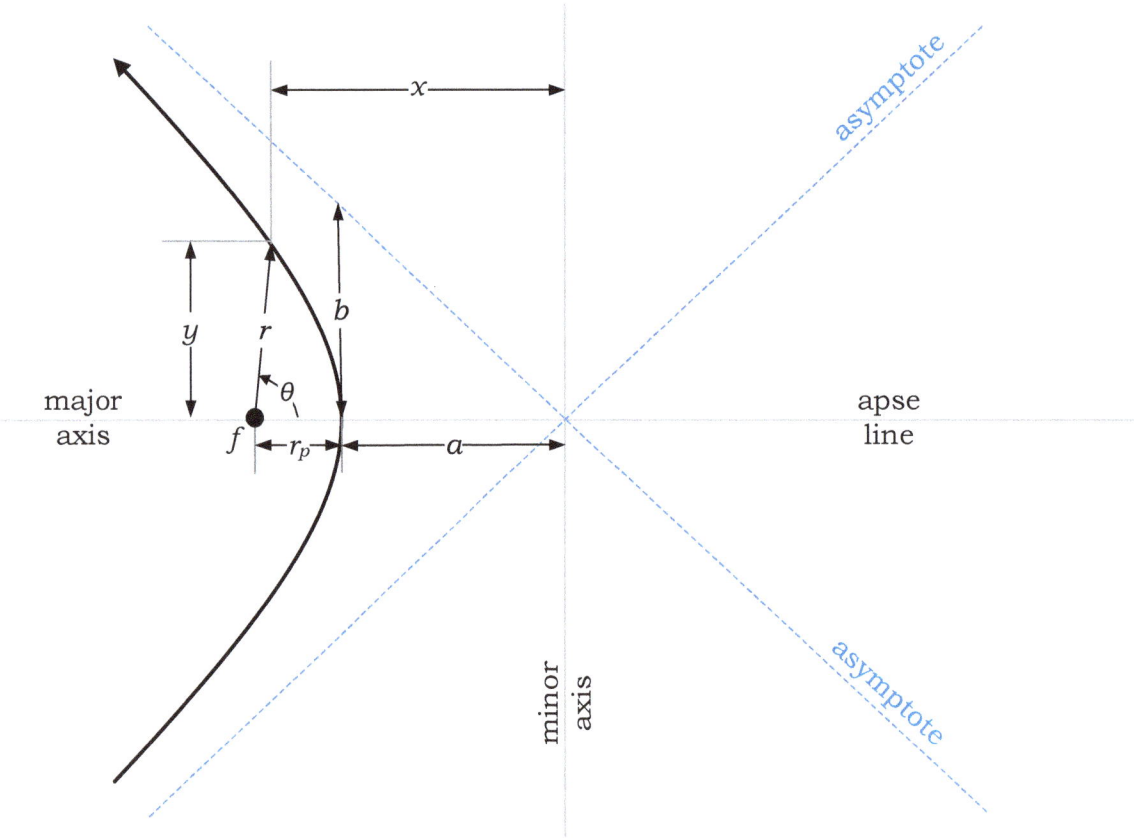

Figure 2-7 Hyperbolic Trajectory Polar Parameters

Equation 2.94 substitutes $r \sin \theta$ for y in equation 2.93 as is apparent in the geometry. The hyperbolic semi-major axis expression (equation 1.45) is substituted for b. The hyperbolic conic parameter (equation 1.44) is substituted into the polar conic expression (equation 2.7) and is in turn, substituted for r.

$$\sinh F = \frac{\sin \theta \sqrt{e^2 - 1}}{1 + e \cos \theta} \qquad 2.94$$

Equation 2.95 provides a relationship between true anomaly and the hyperbolic eccentric anomaly, using the definition of the inverse hyperbolic sine.

$$F = \ln \left(\frac{\sin \theta \sqrt{e^2 - 1} + e + \cos \theta}{1 + e \cos \theta} \right) \qquad 2.95$$

Equation 2.96 substitutes the tangent double angle identities for cosine and sine and further simplifies.

$$F = \ln \left(\frac{e + 1 + (e - 1)\tan^2 \left(\frac{\theta}{2}\right) + 2 \sqrt{e^2 - 1} \tan \left(\frac{\theta}{2}\right)}{e + 1 + (1 - e) \tan^2 \left(\frac{\theta}{2}\right)} \right) \qquad 2.96$$

Factoring out $\sqrt{e - 1}$ from the numerator and denominator of Equation 2.96 produces equation 2.97, which has a more convenient form. The left sides of equations 2.94 and 2.97 are then substituted into equation 2.91, yielding the hyperbolic form of Kepler's Equation in equation 2.97.

$$F = \ln \left(\frac{\sqrt{e + 1} + \sqrt{e - 1} \tan \frac{\theta}{2}}{\sqrt{e + 1} - \sqrt{e - 1} \tan \frac{\theta}{2}} \right) \qquad 2.97$$

$$M = e \sinh F - F \qquad 2.98$$

This result verifies equation 2.69. Equation 2.99 is formulated from equation 2.94, using the hyperbolic trigonometric identity $\cosh^2 x - \sinh^2 x = 1$.

$$\cosh^2 F = 1 + \left(\frac{\sin \theta \sqrt{e^2 - 1}}{1 + e \cos \theta} \right)^2 \qquad 2.99$$

Equations 2.100 through 2.103 put both terms under a common denominator and then factor and simplify the expression.

$$\cosh^2 F = \frac{(1 + e \cos \theta)^2 + \sin^2 \theta (e^2 - 1)}{(1 + e \cos \theta)^2}$$

2.100

$$\cosh^2 F = \frac{e^2 + 2e \cos \theta + \cos^2 \theta}{(1 + e \cos \theta)^2}$$

2.101

$$\cosh^2 F = \frac{(e + \cos \theta)^2}{(1 + e \cos \theta)^2}$$

2.102

$$\cosh F = \frac{e + \cos \theta}{1 + e \cos \theta}$$

2.103

Equation 2.104 combines equations 2.94 and 2.104, using the hyperbolic tangent definition and its double angle identity. Note that the use of the hyperbolic tangent half angle ensures the equation has no quadrant ambiguity.

$$\tanh \frac{F}{2} = \frac{\left[\frac{\sin \theta \sqrt{e^2 - 1}}{1 + e \cos \theta} \right]}{1 + \left[\frac{e + \cos \theta}{1 + e \cos \theta} \right]}$$

2.104

The denominator in equation 2.104 may be simplified as follows.

$$1 + \left[\frac{e + \cos \theta}{1 + e \cos \theta} \right] = \frac{1 + e \cos \theta + e + \cos \theta}{1 + e \cos \theta}$$

2.105

$$1 + \left[\frac{e + \cos \theta}{1 + e \cos \theta} \right] = \frac{(e + 1)(1 + \cos \theta)}{1 + e \cos \theta}$$

2.106

Equation 2.107 updates equation 2.4 using the denominator simplification. Equation 2.108 simplifies the true anomaly to hyperbolic eccentric anomaly conversion, making it resemble its elliptical counterpart in equation 2.83.

$$\tanh \frac{F}{2} = \frac{\sqrt{e^2 - 1}}{(e + 1)} \frac{\sin \theta}{1 + e \cos \theta}$$

2.107

$$\tanh \frac{F}{2} = \sqrt{\frac{e - 1}{e + 1}} \tan \frac{\theta}{2}$$

2.108

This result confirms equation 2.68.

2.4   Computing In-Plane Phase Given Time-Of-Flight

As seen previously, Kepler's Equation provides a relationship between the mean and eccentric anomalies in elliptical orbits and hyperbolic trajectories. The more common solution required is determining the in-plane phasing from a given time since periapsis. In other words, computing the true anomaly at a specified time is more commonly desired than the [easier] inverse computation.

2.4.1   Solving Kepler's Equation [Level I – Descriptive]

There is no known analytic solution to Kepler's Equation for elliptical or hyperbolic trajectories. While sophisticated approximation methods have been proposed, they lack the accuracy needed to serve as anything but a starting value for an iterative numerical solution.

The most common solution involves Newton-Raphson iteration, in which numerically determines a zero value for a well-behaved non-linear function by following the slope of the curve.

Key Term:

Newton-Raphson: is an efficient method for numerically solving for a function's zero values (roots). It is stable for well-behaved functions.

➢ **Transition**: *You may continue this section with Solving Kepler's Equation Level II, or you may skip to the beginning of the section Conic Trajectory Parameters.*

2.4.2   Solving Kepler's Equation [Level II – Equations]

The Newton-Raphson solution to Kepler's equation refines each estimate (x_i) of the function's zero crossing (i.e., x-domain) value using the function's first derivative $[f'(x_i)]$ at the current estimate as the slope. The estimate's update (Δx) is the ratio of the function's value $[f(x_i)]$ to the derivative as shown in Figure 2-8.

Equations 2.109 to 2.111 show how the method makes one iteration in the update process.

$$x_{i+1} = x_i + \Delta x \qquad\qquad 2.109$$

$$\Delta x = -\frac{f(x_i)}{f'(x_i)} \qquad\qquad 2.110$$

$$x_{i+1} = x_i - \frac{f(x_i)}{f'(x_i)} \qquad\qquad 2.111$$

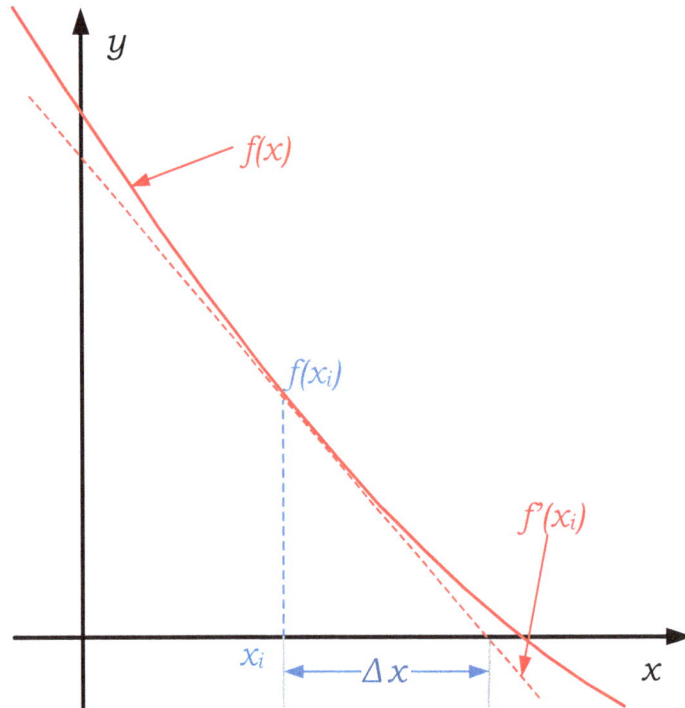

Figure 2-8 Newton-Raphson Update

Newton-Raphson relies on a reasonable initial guess. The process also requires a function that is differentiable and has continuous first derivatives. The process iterates until either $|f(x_i)|$ is less than a pre-defined tolerance or there is a negligible difference (i.e., diminishing returns) returns in the estimate between successive iterations.

2.4.2.1   Solving Kepler's Equation for an Elliptical Orbit

This Kepler's Equation solution solves for the elliptical eccentric anomaly (E) as the independent variable. Equation 2.112 is Kepler's Equation for elliptical orbits set as $f(E) = 0$. Equation 2.113 is the first derivative $f'(E)$.

$$f(E_i) = E_i - e \sin E_i - M \qquad\qquad 2.112$$

$$f'(E_i) = \frac{dF(E_i)}{dE} = 1 - e \cos E_i \qquad\qquad 2.113$$

The process is performed as follows (remembering that all angular values must be in radians):

1. The mean anomaly is a reasonable initial estimate of eccentric anomaly (i.e., $E_0 = M$).
2. Zero the iteration counter: $iter = 0$.
3. Iterate on steps 4 through 7 until the convergence criteria are met. It is prudent to include a counter as part of this criteria to ensure loop exit.
4. For the current iteration (i), compute $f(E_i)$ and $f'(E_i)$ from equations 2.112 and 2.113.
5. Increment the iteration counter: $iter = iter + 1$.
6. Update the estimate as $E_{i+1} = E_i - f(E_i)/f'(E_i)$.
7. Perform the convergence check. The process is converged if any of the below criteria are met.
   a. The absolute value of the function is below a predetermined tolerance ($|f(E_i)| < \epsilon$). The threshold (ϵ) is typically near the floating-point precision.
   b. The absolute value of the update is below the same or similar predetermined tolerance as step 7.b. (i.e., $|f(E_i)/f'(E_i)| < \varepsilon$).
   c. The iteration counter exceeds a predetermined maximum iteration counter (for example: $iter > 20$).
8. If not converged, repeat steps 4 through 7. Once converged, use the current estimate (E_i) as the eccentric anomaly.

2.4.2.2 Solving Kepler's Equation for a Hyperbolic Trajectory

This Kepler's Equation solution solves for the hyperbolic eccentric anomaly (F) as the independent variable. Equation 2.114 is Kepler's Equation for hyperbolic orbits set as $f(F) = 0$. Equation 2.115 is the first derivative $f'(F)$.

$$f(F_i) = e \sinh F_i - F_i - M \qquad\qquad 2.114$$

$$f'(F_i) = \frac{df(F_i)}{dF} = e \cosh F - 1 \qquad\qquad 2.115$$

The process is performed as follows (remembering that all angular values must be in radians):

1. The mean anomaly is a reasonable initial estimate of eccentric anomaly (i.e., $F_0 = M$).
2. Zero the iteration counter: $iter = 0$.
3. Iterate on steps 4 through 7 until the convergence criteria are met. It is prudent to include a counter as part of this criteria to ensure loop exit.
4. For the current iteration (i), compute $f(F_i)$ and $f'(F_i)$ from equations 2.114 and 2.115.
5. Increment the iteration counter: $iter = iter + 1$.

6. Update the estimate as $F_{i+1} = F_i - f(F_i)/f'(F_i)$.
7. Perform the convergence check. The process is converged if any of the below criteria are met.
   a. The absolute value of the function is below a predetermined tolerance ($|f(F_i)| < \epsilon$). The threshold (ϵ) is typically near the floating-point precision.
   b. The absolute value of the update is below the same or similar predetermined tolerance as step 7.b. (i.e., $|f(F_i)/f'(F_i)| < \varepsilon$).
   c. The iteration counter exceeds a predetermined maximum iteration counter (for example: $iter > 50$).
8. If not converged, repeat steps 4 through 7. Once converged, use the current estimate (F_i) as the eccentric anomaly.

➤ **Transition**: *You may continue this section with Solving Kepler's Equation Level III, or you may skip to the beginning of the section Conic Trajectory Parameters.*

2.5 Conic Trajectory Parameters

This section summarizes the equations applicable to each type of conic trajectory. It serves as both a reference for needed parameter calculations as well as a location for various useful mathematical relationships that have not been covered previously.

2.5.1 Conic Trajectory Parameters [Level I – Descriptive]

The conic equations typically compute parameters with counterparts that span the various trajectory types. In some cases, the equations are identical across the trajectory types. Other equations are similar for different trajectory types. In the case of the semi-major axis, considering the hyperbolic semi-major axis as having a negative value enhances the similarity (or exactness) of equations across the conic types. This convention will facilitate the trajectory prediction using the universal variables formulation.

➤ **Transition**: *You may continue this section with conic Sections at Level II, or you may skip to the beginning of the next Chapter called Trajectory Characteristics.*

2.5.2 Conic Trajectory Parameters [Level II – Equations]

This section summarizes key equations applicable to the various conic trajectories. It includes equations applicable to all conic types as well as those unique to the circle, ellipse, parabola, and hyperbola.

2.5.2.1 All Conic Trajectories

Equations 2.116 and 2.117 are the radius and speed equations applicable to all conic trajectories. These are functions of the central body's gravitational parameter (μ), the angular momentum magnitude (h), eccentricity (e), and true anomaly (θ).

$$r = \frac{h^2}{\mu(1 + e\cos\theta)} \qquad\qquad 2.116$$

$$v = \sqrt{\mu\left(\frac{2}{r} - \frac{\mu(1 - e^2)}{h^2}\right)} \qquad\qquad 2.117$$

Equation 2.118 is an expression for periapsis radius (r_p).

$$r_p = \frac{h^2}{\mu(1 + e)} \qquad\qquad 2.118$$

Equations 2.119 and 2.120 are the intrack (v_θ) and radial velocity (v_r) components as a function of the same parameters used to compute radius and speed.

$$v_\theta = \frac{\mu}{h}(1 + e\cos\theta) \qquad\qquad 2.119$$

$$v_r = \frac{\mu}{h}e\sin\theta \qquad\qquad 2.120$$

Equation 2.121 computes the flight path angle (ϕ) from the eccentricity (e) and true anomaly (θ).

$$\phi = \tan^{-1}\left[\frac{e\sin\theta}{1 + e\cos\theta}\right] \qquad\qquad 2.121$$

Equation 2.122 is the time rate of change of true anomaly ($\dot{\theta}$).

$$\dot{\theta} = \frac{h}{r^2} \qquad\qquad 2.122$$

At this point, the convention of considering the hyperbola's semi-major axis (a) as having a negative will be employed. This convention is reasonable when evaluating figures 1-11 and 2-7 and realizing a is on the opposite side of the trajectory from the occupied focus. When this convention is used, several equations take on an imaginary value ($i = \sqrt{-1}$), but this does not impact their usefulness.

Equation 2.123 is the conic equation for semi-major axis, using the negative hyperbolic value convection. Equation 2.124 is the corresponding relationship for the absolute value of the semi-minor axis. Equation 2.125 introduces the parameter α as the *semimajor axis reciprocal*. The α parameter is particularly useful for general formulations since the semi-major axis is infinite for the parabolic trajectory.

$$a = \frac{h^2}{\mu(1 - e^2)} \qquad\qquad 2.123$$

$$|b| = \sqrt{1 - e^2} \qquad\qquad 2.124$$

$$\alpha = \frac{\mu(1 - e^2)}{h^2} = \frac{2}{r} - \frac{v^2}{\mu} \qquad\qquad 2.125$$

Equation 2.126 is the conic parameter (p) – a.k.a. semi-latus rectum – in terms of angular momentum and the central body gravitational parameter.

$$p = \frac{h^2}{\mu} \qquad\qquad 2.126$$

Equation 2.127 is the eccentricity vector, valid for all conic trajectories.

$$\bar{e} = \frac{1}{\mu}\left[\left(v^2 - \frac{\mu}{r}\right)\bar{r} - (\bar{r} \cdot \bar{v})\bar{v}\right] = \frac{\bar{r} \times (\bar{r} \times \bar{v})}{\mu} - \frac{\bar{r}}{r} \qquad\qquad 2.127$$

2.5.2.2 Circular Orbits

The circular orbit has a constant radius, making r, the conic parameter (p), and a equal. Equation 2.128 is the speed in a circular orbit.

$$v = \sqrt{\frac{\mu}{r}} \qquad\qquad 2.128$$

Similarly, most circular orbital parameters are trivial since the trajectory has zero eccentricity. Consequently, the velocity is purely in the intrack ($v_\theta = v\hat{\theta}$) direction, with a zero velocity in the radial direction ($v_r = 0$). This flight path angle is always zero as well. Equation 2.129 is the time rate of change of true anomaly, which is constant for a circular orbit.

$$\dot{\theta} = \sqrt{\frac{\mu}{r^3}} \qquad\qquad 2.129$$

For a circular orbit, the α parameter is the reciprocal of the orbital radius.

2.5.2.3   Elliptical Orbits

Equations 2.130 and 2.131 are alternative radius and speed equations applicable to elliptical orbits.

$$r = \frac{a(1 - e^2)}{1 + e \cos \theta} \tag{2.130}$$

$$v = \sqrt{\mu \left(\frac{2}{r} - \frac{1}{a} \right)} \tag{2.131}$$

Equations 2.132 and 2.133 express the periapsis and apoapsis radii (r_p and r_a respectively) for the elliptical orbit.

$$r_p = a(1 - e) \tag{2.132}$$

$$r_a = a(1 + e) \tag{2.133}$$

Equations 2.134 and 2.135 express the elliptical orbit's intrack and radial velocity components in terms of the central body gravitational parameter, semi-major axis, eccentricity, and true anomaly.

$$v_\theta = \frac{\mu(1 + e \cos \theta)}{a(1 - e^2)} \tag{2.134}$$

$$v_r = \frac{\mu e \sin \theta}{a(1 - e^2)} \tag{2.135}$$

Equation 2.136 is an alternative expression for the elliptical orbit's conic parameter (p). Equation 2.137 is an alternative expression for angular momentum magnitude. Equation 2.138 is an alternative expression for the time rate of change of true anomaly in the elliptical orbit.

$$p = a(1 - e^2) \tag{2.136}$$

$$h = \sqrt{\mu a(1 - e^2)} \tag{2.137}$$

$$\dot{\theta} = \frac{\sqrt{\mu a(1 - e^2)}}{r^2} \tag{2.138}$$

Equation 2.139 is an alternative expression for the elliptical orbit radius in terms of eccentric anomaly (E).

$$r = a(1 - e \cos E) \qquad\qquad 2.139$$

Equation 2.140 is a means of computing eccentricity, given the apoapsis and periapsis radii (r_a and r_p).

$$e = \frac{r_a - r_p}{r_a + r_p} \qquad\qquad 2.140$$

2.5.2.4 Parabolic Trajectories

Equations 2.141 and 2.142 are the radius and speed equations applicable to parabolic trajectories, based on an infinite semi-major axis ($a = \infty$) and eccentricity equal to one ($e = 1$). Note that the parabolic speed for any radius, which is the minimum escape speed, is $\sqrt{2}$ times the speed of a circular orbit of equal radius.

$$r = \frac{h^2}{\mu} \frac{1}{1 + \cos \theta} \qquad\qquad 2.141$$

$$v = \sqrt{\frac{2\mu}{r}} \qquad\qquad 2.142$$

Equation 2.143 is an alternative expression for the parabolic trajectory's periapsis radius (r_p). Note that the periapsis radius is half the parabolic conic parameter (p).

$$r_p = \frac{h^2}{2\mu} \qquad\qquad 2.143$$

Recall that the apoapsis for a parabolic trajectory is at infinity as is its semi-major axis. Consequentially and consistent with equation 2.125, the parabolic trajectory's α parameter is zero.

Equations 2.144 and 2.145 are the parabolic trajectory intrack (v_θ) and radial velocity (v_r) components.

$$v_\theta = \frac{\mu}{h}(1 + \cos \theta) \qquad\qquad 2.144$$

$$v_r = \frac{\mu}{h} \sin \theta \qquad\qquad 2.145$$

Equation 2.146 is the parabolic trajectory's flight path angle, which is always half the true anomaly.

$$\phi = \frac{\theta}{2}$$ 2.146

2.5.2.5 Hyperbolic Trajectories

Given the implicit negative sign convention for a hyperbolic trajectory's semimajor axis, the following equations become identical to their counterparts in the elliptical orbit:

- Radius (r)
- Speed (v)
- Intrack velocity component (v_θ)
- Radial velocity component (v_r)
- Conic parameter (p)
- Angular momentum magnitude (h)
- Time rate of change of true anomaly $(\dot{\theta})$

One difference is the radius (r) in terms of the [hyperbolic] eccentric anomaly (F), provided in equation 2.147.

$$r = a(1 - e \cosh F)$$ 2.147

Figure 2-9 shows some additional parameters associated with hyperbolic trajectories that are useful for planning encounters and flybys with celestial bodies.

The distance between an asymptote and a line parallel to the asymptote that goes through the occupied focus is denoted by the Δ symbol. This is also known as the *aiming radius* and is particularly useful for planning a celestial body's flyby distance at the inbound asymptote. Due to symmetry, the aiming radius has the same relationship with the outbound asymptote. Equation 2.148 computes the aiming radius.

$$\Delta = |a|\sqrt{e^2 - 1}$$ 2.148

The angle between the apse line and either asymptote is denoted by the β symbol. This is useful for orienting the hyperbolic trajectory's periapsis relative an asymptote. Equation 2.149 computes the β angle from the trajectory's eccentricity.

$$\beta = \cos^{-1}\left(\frac{1}{e}\right)$$ 2.149

The angle between the asymptotes is denoted by the δ symbol. This is also known as the *turn angle* since it indicates the change in trajectory direction relative to the celestial body being encountered. Equation 2.150 computes the turn angle from the trajectory's eccentricity.

$$\delta = 2 \sin^{-1}\left(\frac{1}{e}\right) \qquad\qquad 2.150$$

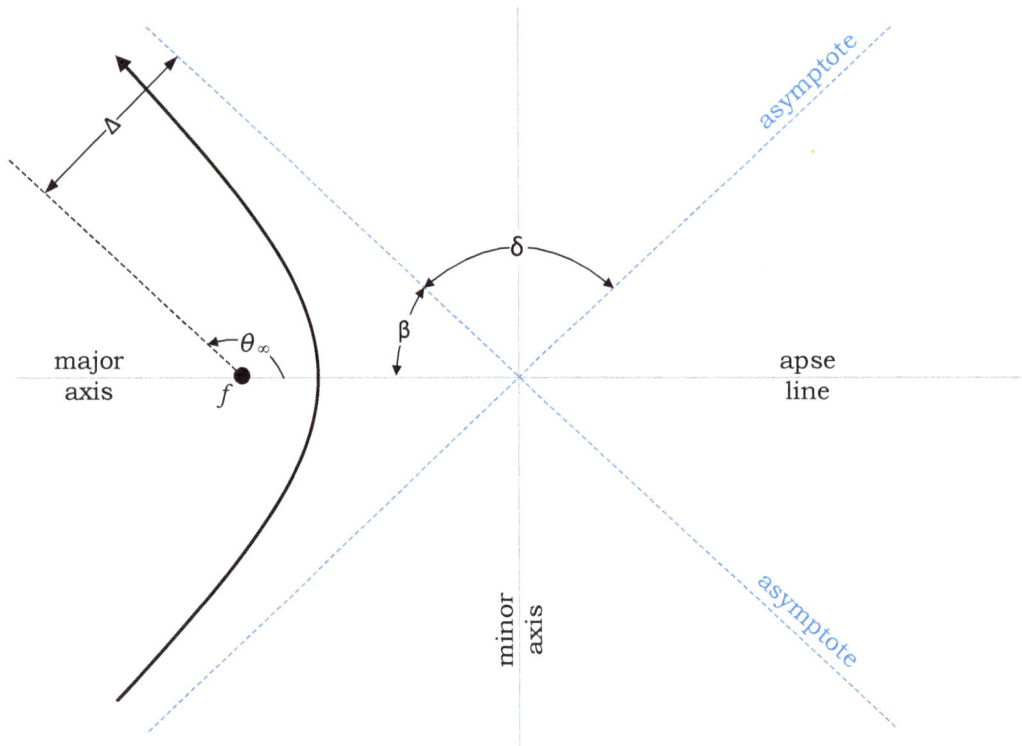

Figure 2-9 Hyperbolic Flyby Parameters

The hyperbolic trajectory's maximum true anomaly is denoted as θ_∞. This is also known as the *true anomaly of the asymptote*. Equation 2.151 computes the true anomaly of the asymptote from the trajectory's eccentricity.

$$\theta_\infty = \cos^{-1}\left(-\frac{1}{e}\right) \qquad\qquad 2.151$$

Since there are two asymptotes, equation 2.151 has two distinct solutions. Each of these solutions may be computed as 2π minus the θ_∞ (in radians) for the other asymptote.

The speed at any point along the hyperbolic trajectory has a relationship to the corresponding minimum escape (or parabolic) speed. This speed is denoted as v_∞

and is also known as the *hyperbolic excess speed*. Equations 2.152 through 2.154 compute the hyperbolic excess speed from various input parameters.

$$v_\infty^2 = v^2 - v_{esc}^2 \qquad\qquad 2.152$$

$$v_\infty = \frac{\mu}{h} e \sin \theta_\infty \qquad\qquad 2.153$$

$$v_\infty = \frac{\mu}{h} \sqrt{e^2 - 1} \qquad\qquad 2.154$$

Key Terms:

> **Aiming Radius**: is the distance between a hyperbolic trajectory's focus and its asymptotes.
> **Hyperbolic Excess Speed**: is the speed a hyperbolic trajectory has over and above escape speed.
> **True Anomalies of the Asymptote**: are a hyperbolic trajectory's minimum and maximum true anomaly values.
> **Turn Angle**: is the angle between a hyperbolic trajectory's asymptotes. This is the angular turn from the inbound to the outbound asymptote.

➢ **Transition**: *You may continue this section with conic Sections at Level III, or you may skip to the beginning of the next Chapter called Trajectory Characteristics.*

2.5.3   Conic Trajectory Parameters [Level III – Derivation]

This section gives derivations for equations that are not previously provided. Some trivial or readily verifiable derivations are omitted.

2.5.3.1   Circular Orbits

The true anomaly rate ($\dot{\theta}$) in equation 2.128 was the only new mathematical expression for the circular orbit. The derivation for this begins with equation 2.122. The angular momentum (h), as expressed in equation 2.126 is substituted, recognizing that the circle's conic parameter (p) equals the radius (r). Equation 2.155 shows the preliminary substitutions. Equation 2.156 shows the equation in simplified form.

$$\dot{\theta} = \frac{\sqrt{\mu r}}{r} \qquad\qquad 2.155$$

$$\dot{\theta} = \sqrt{\frac{\mu}{r}} \qquad\qquad 2.156$$

This result verifies equation 2.127.

2.5.3.2 Elliptical Orbits

Equation 2.123 may be substituted into the second term of equation 2.117:

$$v = \sqrt{\mu\left(\frac{2}{r} - \frac{1}{a}\right)} \qquad\qquad 2.159$$

This result verifies equation 2.131. Equations 2.132 and 2.133 may be verified by setting the true anomaly to the corresponding periapsis ($\theta = 0$) and apoapsis ($\theta = \pi\ rad$) values.

Equation 2.160 results from algebraically manipulating equation 2.113 to produce the μ/h ratio:

$$\frac{\mu}{h} = \frac{h}{a(1 - e^2)} \qquad\qquad 2.160$$

The right side of this expression may be substituted into equations 2.119 and 2.1120 to verify equations 2.134 and 2.135.

Equation 1.161 is equation 2.81 solved algebraically in terms of $\cos\theta$.

$$\cos\theta = \frac{e - \cos E}{e\cos E - 1} \qquad\qquad 2.161$$

The right side is substituted for $\cos\theta$ in the right side of equation 2.79.

$$a\cos E = ae + r\left(\frac{e - \cos E}{e\cos E - 1}\right) \qquad\qquad 2.162$$

Equations 2.163 through 2.167 are algebraic manipulations to produce a radius equation in terms of eccentric anomaly.

$$a(\cos E - e)(e\cos E - 1) = r(e - \cos E) \qquad\qquad 2.163$$

$$r = a\left[\frac{e\cos^2 E - \cos E - e^2\cos E + e}{e - \cos E}\right] \qquad\qquad 2.164$$

$$r = a \left[\frac{e \cos^2 E - e^2 \cos E}{e - \cos E} + \frac{e - \cos E}{e - \cos E} \right] \qquad 2.165$$

$$r = a \left[\frac{e \cos E (\cos E - e)}{e - \cos E} + 1 \right] \qquad 2.166$$

$$r = a(1 - e \cos E) \qquad 2.167$$

This result verifies equation 2.139.

2.5.3.3 Parabolic Trajectories

The parabolic flight path angle results from an eccentricity equal to one ($e = 1$) in equation 2.121.

$$\phi = \tan^{-1} \left[\frac{\sin \theta}{1 + \cos \theta} \right] \qquad 2.168$$

The value inside the brackets is the trigonometric identity for the $\tan \theta/2$, which is negated by the inverse tangent outside the brackets. Thus, equation 2.169 reduces the parabolic flight path angle to half the true anomaly.

$$\phi = \frac{\theta}{2} \qquad 2.169$$

This result verifies equation 2.146.

2.5.3.4 Hyperbolic Trajectories

The true anomaly of the asymptote (θ_∞) is computed by recognizing its value as $\pi - \beta$ radians by geometric inspection of figure 2-9. Thus, the cosine of the angular difference trigonometric identity provides a means of computing its value.

$$\cos \theta_\infty = \cos(\pi - \beta) = \frac{1}{e} \cos \pi + \frac{1}{e} \sin \pi \qquad 2.170$$

$$\cos \theta_\infty = -\frac{1}{e} \qquad 2.171$$

This result verifies equation 2.151.

The hyperbolic excess speed (v_∞) can be determined from two factors. First, that the parabolic speed has a zero value at infinity and thus the hyperbolic trajectory's speed at infinity is the hyperbolic excess speed (as its symbol implies). Second, the velocity at infinity is aligned with the asymptote, which is purely radial at an infinite distance. Thus, the equation for the radial speed component (2.120) defines the hyperbolic excess speed, when evaluated at the true anomaly of the asymptote (θ_∞). This result verifies equation 2.153. Equation 2.154 may be verified by trigonometric substitution of the $e \sin \theta_\infty$ term, using the $\cos \theta_\infty$ result from equation 2.171.

2.6   In-Plane Position and Velocity Relationships

Section 1.5 showed that specific angular momentum is the vector cross product of position and velocity. Since angular momentum is conserved (i.e., constant), the position and velocity must lie in a fixed plane that is perpendicular to the angular momentum. The planar confinement of trajectories provides convenience in certain computational relationships.

2.6.1   The Perifocal Coordinate Frame

The perifocal coordinate frame is fundamental since it defines both the plane and the periapsis directions, which are attributes common to all trajectories.

2.6.1.1   Perifocal Coordinates [Level I – Descriptive]

The perifocal coordinate frame, illustrated in Figure 2-9, places the trajectory in its fundamental plane. It is a Cartesian coordinate frame with three mutually perpendicular (i.e., orthonormal) axes: $\hat{p}$, $\hat{q}$, and $\hat{w}$.

The $\hat{p}$- and $\hat{q}$-axes are in the trajectory plane and the $\hat{w}$-axis is normal (i.e., perpendicular) to the trajectory plane. The $\hat{p}$-axis is aligned with periapsis and the $\hat{q}$-axis is aligned with the semi-latus rectum (at the $\theta = 90°$ true anomaly). The $\hat{w}$-axis, being perpendicular to the trajectory plane, is aligned with the angular momentum.

Key Term:

Perifocal Coordinates: are a coordinate frame with the fundamental plane aligned with the trajectory plane and the principal direction aligned with the periapsis.

➢ **Transition**: *You may continue this section with Perifocal Coordinates at Level II, or you may skip to the beginning of the next section called Lagrange Coefficients.*

Equations 2.172 And 2.173 Represent the position and velocity vectors in the perifocal system in terms of their respective magnitudes (r, v) and true anomaly (θ).

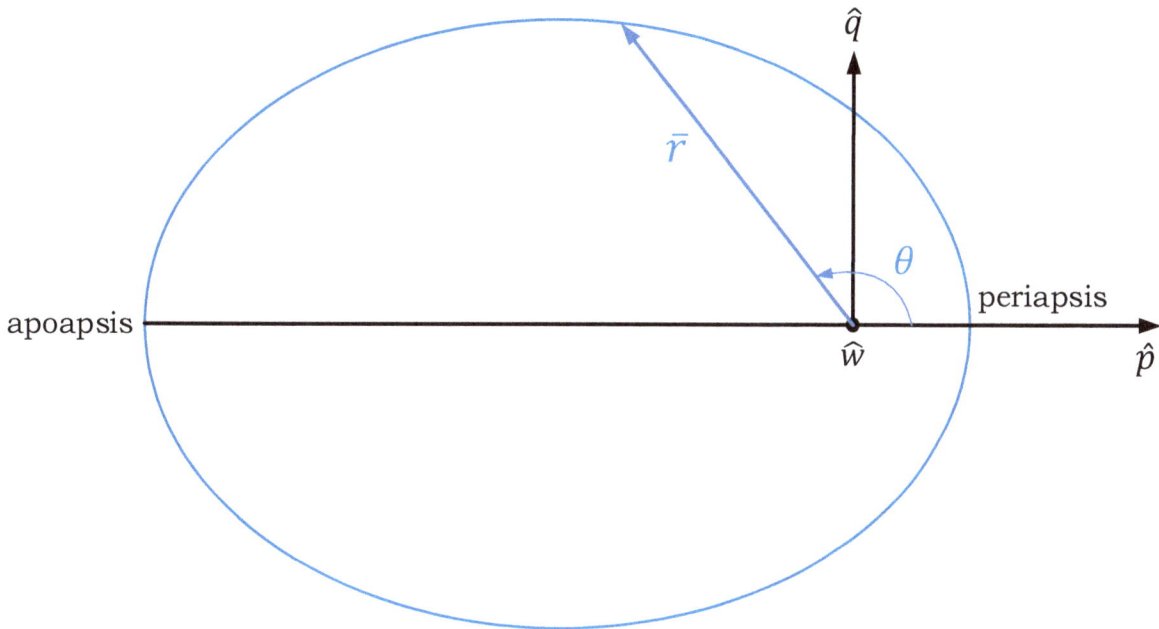

Figure 2-10 Perifocal Coordinate Frame

$$\bar{r} = \frac{h^2}{\mu(1 + e \cos \theta)} (\cos \theta \, \hat{p} + \sin \theta \, \hat{q}) \qquad\qquad 2.172$$

$$\bar{v} = \frac{\mu}{h} (-\sin \theta \, \hat{p} + [e + \cos \theta]\hat{q}) \qquad\qquad 2.173$$

➤ **Transition**: *You may continue this section with Perifocal Coordinates at Level III, or you may skip to the beginning of the next section called Lagrange Coefficients.*

2.6.1.3   Perifocal Coordinates [Level III – Derivation]

Position and velocity may be defined in the perifocal system using variables x and y and their time derivatives.

$$\bar{r} = x\hat{p} + y\hat{q} \qquad\qquad 2.174$$

$$\bar{v} = \dot{x}\hat{p} + \dot{y}\hat{q} \qquad\qquad 2.175$$

Equation 2.176 defines the $\hat{p}$ and $\hat{q}$ position components using variables x and y plus geometry. Consequently, the corresponding velocity components are $\dot{x}$ and $\dot{y}$ as determined in equation 2.177.

$$x = r\cos\theta \quad y = r\sin\theta \qquad\qquad 2.176$$

$$\dot{x} = \dot{r}\cos\theta - r\dot{\theta}\sin\theta \quad \dot{y} = \dot{r}\sin\theta + r\dot{\theta}\cos\theta \qquad\qquad 2.177$$

As was seen previously, $\dot{r} = \hat{r} \cdot \bar{v}$. The radial velocity component is defined in equation 2.120. Similarly, the intrack velocity component is defined by equation 2.119. Substituting these expressions into equation 2.178 defines the velocity components in terms of angular momentum (h), the central body gravitational parameter (μ), eccentricity (e), and true anomaly (θ).

$$\dot{x} = -\frac{\mu}{h}\sin\theta \quad \dot{y} = \frac{\mu}{h}(e + \cos\theta) \qquad\qquad 2.178$$

These results confirm the perifocal velocity equation.

2.6.2   Lagrange Coefficients

Giuseppe Lodovico Lagrangia was an Italian mathematician, physicist, and astronomer who was later French naturalized and thus is more commonly known as Joseph-Louis Lagrange. He discovered useful coefficients analytically describing a trajectory's positions and velocities by linear combinations.

2.6.2.1   Lagrange Coefficients [Level I – Descriptive]

The Lagrange coefficients operate within the perifocal coordinate system's trajectory plane. They consist of four parameters, two of which compute a *coupled* position vector in the trajectory as a linear combination of the current position and velocity vectors. The remaining two coefficients, which are the time rates of change of the first two, compute the *coupled* velocity vector, also as a linear combination of the current position and velocity vectors.

It is important to stress that a set of four Lagrange coefficients couples two sets of position and velocity vectors with each other. If the coupled vector sets are known, the coefficients are readily computed. However, the useful case occurs when one pair of vectors is unknown, but computable from the Lagrange coefficients by leveraging information about how the vector sets are coupled.

Techniques exist for computing the Lagrange coefficients for two types of coupling conditions: an angular (or geometric) separation of position or a temporal separation between the vector sets. For either of these coupling conditions, the coefficients can be determined and thus the other corresponding vector set may be computed.

Key Term:

> **Lagrange Coefficients**: are a set of four values that linearly relate a trajectory's position and velocity vectors to another position and velocity set at another place in the trajectory. The coupling between the vector sets may be either an angular or temporal offset.

➤ **Transition**: *You may continue this section with Lagrange Coefficients at Level II, or you may skip to the beginning of the next section called Lagrange Coefficients for a True Anomaly Offset.*

2.6.2.2   Lagrange Coefficients [Level II – Equations]

The Lagrange coefficients $(f, g, \dot{f}, \text{and } \dot{g})$ relate an initial position $(\bar{r}_i)$ and velocity vector $(\bar{v}_i)$ set to another position $(\bar{r})$ and velocity $(\bar{v})$ vector set on a trajectory as linear combinations of each other.

$$\bar{r} = f\bar{r}_i + g\bar{v}_i \qquad\qquad 2.179$$

$$\bar{v} = \dot{f}\bar{r}_i + \dot{g}\bar{v}_i \qquad\qquad 2.180$$

While the usefulness is not immediately obvious, applications include trajectory prediction and fitting trajectories to available information. These applications need further development within their own context.

➤ **Transition**: *You may continue this section with Lagrange Coefficients at Level III, or you may skip to the beginning of the next section called Lagrange Coefficients for a True Anomaly Offset.*

2.6.2.3   Lagrange Coefficients [Level III – Derivation]

Equations 2.181 and 2.182 describe position and velocity as components against the in-plane perifocal frame basis vectors ($\hat{p}$ and $\hat{q}$).

$$\bar{r} = x\hat{p} + y\hat{q} \tag{2.181}$$

$$\bar{v} = \dot{x}\hat{p} + \dot{y}\hat{q} \tag{2.182}$$

An arbitrary initial position and velocity are described in the same manner.

$$\bar{r}_i = x_i\hat{p} + y_i\hat{q} \tag{2.183}$$

$$\bar{v}_i = \dot{x}_i\hat{p} + \dot{y}_i\hat{q} \tag{2.184}$$

Angular momentum is computed from the initial conditions.

$$\bar{h} = \bar{r}_i \times \bar{v}_i = \begin{vmatrix} \hat{p} & \hat{q} & \hat{w} \\ x_i & y_i & 0 \\ \dot{x}_i & \dot{y}_i & 0 \end{vmatrix} = (x_i\dot{y}_i - y_i\dot{x}_i)\hat{w} \tag{2.185}$$

The $\hat{q}$ axis is then determined algebraically from the initial position.

$$\hat{q} = \frac{1}{y_i}\bar{r}_i - \frac{x_i}{y_i}\hat{p} \tag{2.186}$$

This result is then substituted into the velocity equation and simplified.

$$\bar{v}_i = \dot{x}_i\hat{p} + \dot{y}_i\left(\frac{1}{y_i}\bar{r}_i - \frac{x_i}{y_i}\hat{p}\right) = \frac{y_i\dot{x}_i - x_i\dot{y}_i}{y_i}\hat{p} + \frac{\dot{y}_i}{y_i}\bar{r}_i \tag{2.187}$$

Notice that the numerator of the $\hat{p}$ term is $-h$.

$$\bar{v}_i = -\frac{h}{y_i}\hat{p} + \frac{\dot{y}_i}{y_i}\bar{r}_i \tag{2.188}$$

Solving for $\hat{p}$ – in terms of the initial position and velocity

$$\hat{p} = \frac{\dot{y}_i}{h}\bar{r}_i - \frac{y_i}{h}\bar{v}_i \tag{2.189}$$

This result for $\hat{p}$ is now substituted into the previous result for $\hat{q}$. After simplifying:

$$\hat{q} = \frac{h - x_i\dot{y}_i}{hy_i}\bar{r}_0 + \frac{x_i}{h}\bar{v}_i \tag{2.190}$$

Substitute the previous expression for $h = x_i\dot{y}_i - y_i\dot{x}_i$ in the numerator of the left term.

$$\hat{q} = -\frac{\dot{x}_i}{h}\bar{r}_i + \frac{x_i}{h}\bar{v}_i \qquad\qquad 2.191$$

Perifocal basis vectors $\hat{p}$ and $\hat{q}$ are now defined in terms of the initial orbital state. Now substitute the new values for $\hat{p}$ and $\hat{q}$ into the original state vector equation for a final position and velocity vector set.

$$\bar{r}_f = x_f \left(\frac{\dot{y}_i}{h}\bar{r}_i - \frac{y_i}{h}\bar{v}_i \right) + y_f \left(-\frac{\dot{x}_i}{h}\bar{r}_i + \frac{x_i}{h}\bar{v}_i \right) \qquad\qquad 2.192$$

$$\bar{v}_f = \dot{x}_f \left(\frac{\dot{y}_i}{h}\bar{r}_i - \frac{y_i}{h}\bar{v}_i \right) + \dot{y}_f \left(-\frac{\dot{x}_i}{h}\bar{r}_i + \frac{x_i}{h}\bar{v}_i \right) \qquad\qquad 2.193$$

Simplifying the above gives an intermediate equation set:

$$\bar{r}_f = \frac{x_f\dot{y}_i - y_f\dot{x}_i}{h}\bar{r}_i + \frac{-x_fy_i + y_fx_i}{h}\bar{v}_i \qquad\qquad 2.194$$

$$\bar{v}_f = \frac{\dot{x}_f\dot{y}_i - \dot{y}_f\dot{x}_i}{h}\bar{r}_i + \frac{-\dot{x}_fy_i + \dot{y}_fx_i}{h}\bar{v}_i \qquad\qquad 2.195$$

The fractional values are the Lagrange coefficients $(f, g, \dot{f}, and\ \dot{g})$.

$$\bar{r}_f = f\bar{r}_i + g\bar{v}_i \qquad\qquad 2.196$$

$$\bar{v}_f = \dot{f}\bar{r}_i + \dot{g}\bar{v}_i \qquad\qquad 2.197$$

There are thus a set of four coefficients that couple the position and velocity of one state in the trajectory with another trajectory state. These coefficients are computed by:

$$f = \frac{x_f\dot{y}_i - y_f\dot{x}_i}{h} \qquad g = \frac{-x_fy_i + y_fx_i}{h} \qquad\qquad 2.198$$

$$\dot{f} = \frac{\dot{x}_f\dot{y}_i - \dot{y}x_i}{h} \qquad \dot{g} = \frac{-\dot{x}_fy_i + \dot{y}_fx_i}{h} \qquad\qquad 2.199$$

Conservation of angular momentum imposes a useful constraint on the Lagrange coefficients. Equation 2.200 Is the definition of angular momentum in terms of the Lagrange coefficients. Equation 2.201 uses the cross product distributive property and equation 2.202 eliminates colinear cross products and factors out the scalars.

$$\bar{h} = \bar{r}_f \times \bar{v}_f = (f\bar{r}_i + g\bar{v}_i) \times (\dot{f}\bar{r}_i + \dot{g}\bar{v}_i) \qquad\qquad 2.200$$

$$\bar{h} = (f\bar{r}_i \times \dot{f}\bar{r}_i) + (f\bar{r}_i \times \dot{g}\bar{v}_i) + (g\bar{v}_i \times \dot{f}\bar{r}_i) + (g\bar{v}_i \times \dot{g}\bar{v}_i) \qquad 2.201$$

$$\bar{h} = f\dot{g}(\bar{r}_i \times \bar{v}_i) + \dot{f}g(\bar{v}_i \times \bar{r}_i) \qquad 2.202$$

Equation 2.203 combines terms using the cross product anti-commutativity principle; equation 2.204 substitutes angular momentum for the cross product.

$$\bar{h} = (f\dot{g} - \dot{f}g)(\bar{r}_i \times \bar{v}_i) \qquad 2.203$$

$$\bar{h} = (f\dot{g} - \dot{f}g)\bar{h}_i \qquad 2.204$$

Equation 2.205 accounts for since angular momentum being conserved ($\bar{h} = \bar{h}_0$), so it is factored out of the above expression.

$$f\dot{g} - \dot{f}g = 1 \qquad 2.205$$

This result shows that the Lagrange coefficients are coupled. Thus, any of the coefficients may be determined from the other three.

2.7 Lagrange Coefficients for a True Anomaly Position Offset

The Lagrange coefficients for a true anomaly offset are used to compute the position and velocity vectors having an in-plane angular offset relative to a known initial position and its corresponding velocity. This section provides the method for determining the corresponding Lagrange coefficients for this coupling arrangement and then computing the trajectory's position and velocity at the angular offset.

2.7.1 Lagrange for True Anomaly [Level I – Description]

The Lagrange coefficients for a true anomaly offset are used to compute the position and velocity vectors having a true anomaly difference relative to a known initial position and its corresponding velocity. This method is useful when the vectors are desired with an angular geometric offset from a known set of position and velocity.

➢ **Transition**: *You may continue this section with Lagrange Coefficients for True Anomaly at Level II, or you may skip to the beginning of the next section called Orbit Prediction Using Universal Variables.*

2.7.2 Lagrange Coefficients for True Anomaly [Level II – Equations]

There are cases where it is desirable to determine a trajectory's position and velocity state separated by a geometric offset (i.e., $\Delta\theta$) from an initial position ($\bar{r}_i$) and velocity ($\bar{v}_i$) state. This process is developed by defining the Lagrange coefficients in terms of the true anomaly offset ($\Delta\theta$).

When $\bar{r}_i$ and $\bar{v}_i$, and are known, a set of final position ($\bar{r}_f$) and velocity ($\bar{v}_f$) vectors coupled by a true anomaly offset ($\Delta\theta$) can be computed, provided the central body gravitational parameter (μ) is also known.

Equations 2.206 through 2.209 compute the Lagrange coefficients ($f, g, \dot{f}, \dot{g}$).

$$f = 1 - \frac{\mu r_f}{h^2}(1 - \cos\Delta\theta) \qquad\qquad 2.206$$

$$g = \frac{r_i r_f}{h}\sin\Delta\theta \qquad\qquad 2.207$$

$$\dot{f} = \frac{\mu(1 - \cos\Delta\theta)}{h\sin\Delta\theta}\left[\frac{h^2}{\mu}(1 - \cos\Delta\theta) - \frac{1}{r_i} - \frac{1}{r_f}\right] \qquad\qquad 2.208$$

$$\dot{g} = 1 - \frac{\mu r_i}{h^2}(1 - \cos\Delta\theta) \qquad\qquad 2.209$$

Three of four of these equations rely on knowledge of the final radius magnitude (r_f). Since the change in true anomaly ($\Delta\theta$) is known, r_f can be computed from equation 2.211 once angular momentum (h) is computed from the known conditions.

$$\bar{h} = \bar{r}_i \times \bar{v}_i \qquad\qquad 2.210$$

$$r_f = \frac{h^2}{\mu\left(1 + \left[\frac{h^2}{\mu r_i} - 1\right]\cos\Delta\theta - \frac{h(\hat{r}_i \cdot \bar{v}_i)}{\mu}\sin\Delta\theta\right)} \qquad\qquad 2.211$$

Once the Lagrange coefficients can be computed, the final position and velocity vectors at a $\Delta\theta$ offset can be computed.

$$\bar{r}_f = f\bar{r}_i + g\bar{v}_i \qquad\qquad 2.212$$

$$\bar{v}_f = \dot{f}\bar{r}_i + \dot{g}\bar{v}_i \qquad\qquad 2.213$$

➢ **Transition**: *You may continue this section with Lagrange Coefficients for True Anomaly at Level III, or you may skip to the beginning of the next section called Orbit Prediction Using Universal Variables.*

2.7.3   Lagrange for True Anomaly [Level III – Derivation]

Relating the Lagrange coefficients to true anomaly begins with the perifocal equations 2.176 and 2.178, which are in turn substituted into equations 2.194 and 2.195. Equation 2.176 Makes the substitution for the f coefficient.

$$f = \frac{x_f \dot{y}_i - y_f \dot{x}_i}{h} = \frac{1}{h}\left[r_f \cos\theta_f \left(\frac{\mu}{h}[e + \cos\theta_i]\right) - r_f \sin\theta_f \left(\frac{\mu}{h}\sin\theta_i\right)\right] \qquad 2.204$$

$$f = \frac{\mu r_f}{h}\left[e \cos\theta_f + \cos\theta_f \cos\theta_i + \sin\theta_f \sin\theta_i\right] \qquad 2.205$$

Equation 2.206 simplifies this expression using the cosine difference trigonometric identity.

$$f = \frac{\mu r_f}{h}\left[e \cos\theta_f + \cos(\theta_f - \theta_i)\right] \qquad 2.206$$

The definition $\Delta\theta = \theta_f - \theta_i$ provides a further simplification of the equation for f.

$$f = \frac{\mu r_f}{h^2}\left[e \cos\theta_f + \cos\Delta\theta\right] \qquad 2.207$$

The conic radius equation can be solved algebraically for $e\cos\theta_f$ as follows:

$$e \cos\theta_f = \frac{h^2}{\mu r_f} - 1 \qquad 2.208$$

Equation 2.209 Substitutes this result, providing an equation for f in terms of the true anomaly difference ($\Delta\theta$).

$$f = 1 - \frac{\mu r_f}{h^2}(1 - \cos\Delta\theta) \qquad 2.209$$

The same process is used for the g coefficient, which is simplified using algebra and the sine difference trigonometric identity.

$$g = \frac{-x_f y_i + y_f x_i}{h} = \frac{1}{h}\left[(-r_f \cos\theta_f)(r_i \cos\theta_i) + (r_f \sin\theta_f)(r_i \cos\theta_i)\right] \qquad 2.210$$

$$g = \frac{r_i r_f}{h} \left[\sin \theta_f \cos \theta_i - \cos \theta_f \cos \theta_i \right] \qquad 2.211$$

$$g = \frac{r_i r_f}{h} \sin \Delta\theta \qquad 2.212$$

The process like deriving the f coefficient is repeated for the $\dot{g}$ coefficient. Equation 2.178 provides the expressions for $\dot{x}_f$ and $\dot{y}$.

$$\dot{g} = \frac{-\dot{x}_f y_i + \dot{y}_f x_i}{h} = \frac{1}{h} \left[\left(\frac{\mu}{h} \sin \theta_f \right) (r_i \cos \theta_i) + \left(\frac{\mu}{h} (e + \cos \theta_f) \right) (r_i \cos \theta_i) \right] \qquad 2.213$$

$$\dot{g} = \frac{\mu r_i}{h^2} \left[e \cos \theta_i + \sin \theta_f \cos \theta_i - \cos \theta_f \cos \theta_i \right] \qquad 2.214$$

$$\dot{g} = \frac{\mu r_i}{h^2} \left[e \cos \theta_i + \sin \Delta\theta \right] \qquad 2.215$$

$$\dot{g} = 1 - \frac{\mu r_i}{h^2} \sin \Delta\theta \qquad 2.216$$

The $\dot{f}$ equation relies on the conservation of angular momentum relationship between the Lagrange coefficients in equation 2.205.

$$\dot{f} = \frac{f\dot{g} - 1}{g} \qquad 2.217$$

$$\dot{f} = \frac{h}{r_i r_f \sin \Delta\theta} \left[\left(1 - \frac{\mu r_f}{h^2} (1 - \cos \Delta\theta) \right) \left(\frac{r_i r_f}{h} \sin \Delta\theta \right) - 1 \right] \qquad 2.218$$

$$\dot{f} = \frac{\mu(1 - \cos \Delta\theta)}{h \sin \Delta\theta} \left[\frac{\mu}{h^2} (1 - \cos \Delta\theta) - \frac{1}{r_i} - \frac{1}{r_f} \right] \qquad 2.219$$

Since the equations for the Lagrange coefficients rely on knowledge of the final position magnitude (r_f), this must first be inferred from the initial trajectory state and the true anomaly offset. The key equation is the conic radius equation, to which the $\Delta\theta$ true anomaly offset is applied in equation 2.220.

$$r_f = \frac{h^2}{\mu[1 + e \cos(\theta_i + \Delta\theta)]} \qquad 2.220$$

However, since the initial true anomaly (θ) is not known, this must also be inferred from the initial conditions. The $\cos(\theta + \Delta\theta)$ term can be expanded using the cosine angular sum trigonometric identity:

$$r_f = \frac{h^2}{\mu[1 + e\cos\theta_i\cos\Delta\theta - e\sin\theta_i\sin\Delta\theta]} \qquad 2.221$$

The quantity $e\cos\theta$ can be inferred from the initial radius from the conic radius equation. The quantity $e\sin\theta$ can likewise be inferred from the initial conditions using the radial velocity equation.

$$e\cos\theta_i = \frac{h^2}{\mu r_i} - 1 \qquad 2.222$$

$$e\sin\theta_i = \frac{h(\hat{r}_i \cdot \bar{v}_i)}{\mu} \qquad 2.223$$

Once substituted into equation 2.224, the equation 2.211 can be verified.

$$r_f = \frac{h^2}{\mu\left(1 + \left[\frac{h^2}{\mu r_i} - 1\right]\cos\Delta\theta - \frac{h(\hat{r}_i \cdot \bar{v}_i)}{\mu}\sin\Delta\theta\right)} \qquad 2.224$$

2.8   Trajectory Prediction Using Universal Variables

The most common trajectory prediction problem is, given an initial position and velocity, finding a trajectory's position and velocity vectors at a future (or past) time. Such predictions were made in the preceding time-of-flight and in-plane phase determinations, with distinctly different computations depending on the trajectory's conic section type.

This section takes a uniform approach, encapsulating the computations into a single process that is agnostic to the trajectory's conic type. This is especially advantageous when a trajectory is on the borderline of being parabolic in that the computations are numerically stabilized.

2.8.1   Universal Variables [Level I – Descriptive]

The universal variables prediction provides a method to compute a trajectory's position and velocity vectors at a desired time, based on the current position and velocity vectors and the central body gravitational parameter. Because of the highly non-linear nature of conic sections, such a prediction requires far more sophisticated mathematics than would be done by extrapolation using linear or quadratic deduced reckoning. Instead, the method leverages the commonality of mathematical characteristics of conic sections to provide a universal method for predicting a trajectory state as a function of time.

The inputs for the universal variables trajectory prediction are the current trajectory state as an initial position ($\bar{r}_i$), velocity ($\bar{v}_i$), time (t_i), the central body gravitational parameter (μ), and the prediction time (t_f). The results are the final position ($\bar{r}_f$) and velocity ($\bar{v}_f$) at the prediction time.

> ➤ **Transition**: *You may continue this section with Universal Variables at Level II, or you may skip to the beginning of the next Chapter called Trajectory Orientation.*

2.8.2   Universal Variables [Level II – Equations]

The universal variables prediction may be modularized into two steps: the universal time-of-flight and the update using the Lagrange coefficients. The intense portions are in the universal time-of-flight computations.

2.8.2.1   Universal Time-of-Flight

The first step in determining the universal time-of-flight is computing the reciprocal of the conic's semi-major axis from the velocity equation. Use of the semimajor axis reciprocal (α) avoids the parabolic trajectory's infinite semi-major axis value.

$$\alpha = \frac{2}{r_i} - \frac{v_i^2}{\mu} \qquad\qquad 2.225$$

The α value determines the trajectory's conic type (i.e., positive is elliptical, zero is parabolic, and negative is hyperbolic). The difference between the desired and current time is computed below.

$$\Delta t = t_f - t_i \qquad\qquad 2.226$$

For circular and elliptical orbits ($\alpha > 0$), a large Δt may exceed one revolution and possibly go multiple revolutions. For these cases, a geometrically equivalent Δt corresponding to less than one revolution is determined.

Equation 2.227 determines the integer number of complete revolutions (n_{rev}) spanned by Δt.

$$n_{rev} = int\left(\frac{|\Delta t|\sqrt{\mu}}{2\pi a^{3/2}}\right) \qquad\qquad 2.227$$

Thus, only for the circular or elliptical orbits ($\alpha > 0$), the geometrically equivalent Δt is adjusted as follows:

$$\Delta t = \Delta t - sign(\Delta t)\left(\frac{2\pi a^{3/2}}{\sqrt{\mu}}n_{rev}\right) \qquad\qquad 2.228$$

Equation 2.229 provides the initial estimate for the *universal anomaly* (χ_0) which has $\sqrt{r}$ units (i.e., if the position has km units, χ_0 has $\sqrt{km}$ units.)

$$\chi_0 = \Delta t|\alpha|\sqrt{\mu} \qquad\qquad \alpha \geq 0 \qquad 2.229$$

$$\chi_0 = sign(\Delta t)\sqrt{-\frac{1}{\alpha}}ln\left[\frac{-2\mu\alpha\Delta t}{\bar{r}_0 \cdot \bar{v}_0 + sign(\Delta t)\sqrt{-\frac{\mu}{\alpha}}(1 - \alpha r_0)}\right] \qquad \alpha < 0 \qquad 2.230$$

The universal anomaly is solved iteratively, using the Newton-Raphson method. The solution steps are listed below, beginning at iteration $k = 0$:

1. Compute the dimensionless Stumpff function argument as $z_k = \alpha\chi_k^2$.
2. Evaluate the Stumpff functions $C(z_k)$ and $S(z_k)$ as provided at the end of this section.
3. Estimate the zero value for the universal anomaly form of Kepler's equation:

$$f(\chi_k) = \frac{\bar{r}_i \cdot \bar{v}_i}{\sqrt{\mu}}\chi_k^2 C(z_k) + (1 - \alpha r_i)\chi_k^3 S(z_k) + r_i\chi_k - \Delta t\sqrt{\mu} \qquad\qquad 2.231$$

4. Compute the derivative of the universal Kepler equation as provided below:

$$f'(\chi_k) = \frac{\bar{r}_i \cdot \bar{v}_i}{\sqrt{\mu}}\chi_k[1 - \alpha\chi_k^2 S(z_k)] + (1 - \alpha r_i)\chi_k^2 C(z_k) + r_i \qquad\qquad 2.232$$

5. Compute the adjustment to the universal anomaly:

$$\Delta\chi_k = -\frac{f(\chi_k)}{f'(\chi_k)} \qquad\qquad 2.233$$

6. Perform the convergence check. The Process is converged if any of the conditions below are satisfied:

a. The absolute value of the universal Kepler function is below a predetermined (ϵ) threshold, i.e., $|f(\chi_k)| < \epsilon$. The ϵ threshold is typically somewhat larger than floating point machine precision.

b. The universal anomaly update would provide diminishing returns, i.e., $|\Delta\chi_k| < \epsilon$.

c. The iteration counter exceeds the predetermined maximum iteration ($k > k_{max}$) counter (where the maximum iterations is a specified value such as $k_{max} = 20$).

7. If not converged, update the universal anomaly by $\chi_{k+1} = \chi_k + \Delta\chi_k$, increment the iteration counter as $k = k + 1$, and repeat Steps 1-7 above.

8. Once converged, use the current estimate of universal anomaly to compute the Lagrange coefficients.

2.8.2.2   Lagrange Coefficients for Universal Anomaly Trajectory Propagation

Equations 2.234 through 2.237 compute the Lagrange coefficients in terms of the Stumpff functions using the universal anomaly.

$$f = 1 - \frac{\chi^2}{r_i}C(\alpha\chi^2) \qquad\qquad 2.234$$

$$g = \Delta t - \frac{\chi^3}{\sqrt{\mu}}S(\alpha\chi^2) \qquad\qquad 2.235$$

$$\dot{f} = \frac{\sqrt{\mu}}{r_i r_f}[\alpha\chi^3 S(\alpha\chi^2) - \chi] \qquad\qquad 2.236$$

$$\dot{g} = 1 - \frac{\chi^3}{r_f}C(\alpha\chi^2) \qquad\qquad 2.237$$

Note that the velocity-associated Lagrange coefficients ($\dot{f}$ and $\dot{g}$) depend on knowledge of the magnitude of the predicted position (r_f). Thus, the computed position ($\bar{r}_f$) and its magnitude (r_f) need to be computed first from equation 2.211 before $\dot{f}$ and $\dot{g}$ can be computed for the velocity update.

The Lagrange coefficients update uses the same equations as in previous processes.

$$\bar{r}_f = f\bar{r}_i + g\bar{v}_i \qquad\qquad 2.238$$

$$\bar{v}_f = \dot{f}\bar{r}_i + \dot{g}\bar{v}_i \qquad\qquad 2.239$$

The α parameter that was computed at the beginning of the universal time-of-flight process is the reciprocal of the semi-major axis. Its value provides knowledge of the trajectory's conic type as shown table 2-2.

Table 2-2 Conic type from α Parameter Value	
Condition	Conic Type
$\alpha > 0$	Circle or Ellipse
$\alpha = 0$	Parabola
$\alpha < 0$	Hyperbola

2.8.2.3 Stumpff Functions

Functions $C(z)$ and $S(z)$ are Stumpff functions, named for German astronomer Karl Stumpff. The dimensionless parameter is computed as: $z = \alpha\chi^2$. The functions are computed by different methods depending on the value of z.

Equation 2.240 computes the value for $C(z)$; equation 2.241 computes the value for $S(z)$.

$$C(z) = \frac{1 - \cos\sqrt{z}}{z} \qquad\qquad z > 0$$

$$C(z) = \frac{1}{2} \qquad\qquad z = 0 \qquad\qquad 2.240$$

$$C(z) = \frac{\cosh\sqrt{-z} - 1}{-z} \qquad\qquad z < 0$$

$$S(z) = \frac{\sqrt{z} - \sin\sqrt{z}}{z^{3/2}} \qquad\qquad z > 0$$

$$S(z) = \frac{1}{6} \qquad\qquad z = 0 \qquad\qquad 2.241$$

$$S(z) = \frac{\sinh\sqrt{-z} - \sqrt{-z}}{z^{3/2}} \qquad\qquad z < 0$$

The Stumpff functions may alternatively be computed by the infinite series in equations 2.242 and 2.243.

$$C(z) = \sum_{k=0}^{\infty} (-1)^k \frac{z^k}{(2k+3)!} = \frac{1}{2} - \frac{z}{24} + \frac{z^2}{720} - \frac{z^3}{40420} + \cdots \qquad 2.241$$

$$S(z) = \sum_{k=0}^{\infty} (-1)^k \frac{z^k}{(2k+2)!} = \frac{1}{6} - \frac{z}{120} + \frac{z^2}{5040} - \frac{z^3}{362880} + \cdots \qquad 2.242$$

Key Terms:

Stumpff Functions: are transcendental functions used in trajectory analysis in conjunction with the universal variables orbit prediction method.

Universal Anomaly: is a transcendental counterpart to mean anomaly, used to propagate position and velocity states in a conic trajectory.

References

1. Curtis, Howard D., *Orbital Mechanics for Engineering Students*, Fourth Edition, © 2020 Elsevier Ltd., ISBN 978-0-08-102133-0.
2. Roy, Archie E., *Orbital Motion*, Third Edition, © 1988 by author, ISBN 0-85274-229-0.
3. Battin, R.H., *An Introduction to the Mathematics and Methods of Astrodynamics*, AIAA Education Series, © 1987 by author, American Institute of Aeronautics and Astronautics (AIAA), ISBN 0-930403-25-8.
4. Herrick, Samuel, *Astrodynamics*, © 1971 by author, Van Nostrand Reinhold Company, Library of Congress Card Catalog Number 78-125199.
5. Baker, Robert M, Jr. and Makemson, Maude W., *Introduction to Astrodynamics*, Second Edition, © 1967 Academic Press, Inc., Library of Congress Card Catalog Number 67-14534.
6. Wie, Bong, *Space Vehicle Dynamics and Control*, AIAA Education Series, © 1988 American Institute of Aeronautics and Astronautics (AIAA), ISBN 1-56347-261-9.
7. Gurfil, Pini and Seidelmann, P. Kenneth, *Celestial Mechanics and Astrodynamics: Theory and Practice*, © 2016 Springer-Verlag, ISBN 978-3-662-50368-3.
8. Chobotov, Vladimir A (Editor), *Orbital Mechanics*, AIAA Education Series, © 1991 American Institute of Aeronautics and Astronautics (AIAA), ISBN 1-56734-007-1.
9. Bate, Roger R. et al., *Fundamentals of Astrodynamics*, © 1971 Dover Publications, Inc., ISBN 0-486-60061-0.
10. Danby, J.M.A., *Fundamentals of Celestial Mechanics*, © 1992 by author, Willmann-Bell, Inc., ISBN 0-943396-20-4.
11. McCalla, Thomas R., *Introduction to Numerical Methods and FORTRAN Programming*, © 1967 john Wiley & Sons, ISBN 0-471-58125-9.

3   Trajectory Orientation

Thus far, trajectories the presentations have exploited used the planar nature of undisturbed trajectories and thus represented them in two dimensions. This chapter will extend the representation to orient trajectories in three-dimensional space using coordinate systems.

It should be noted that there is no loss of generality in the three-dimensional representation. All equations and processes, including predictions made with Lagrange coefficients extend seamlessly in three dimensions.

3.1   Coordinate Systems

A Coordinate system has parameters that uniquely describe the locations or other physical quantities in space. It is natural, for example, to express vectors in coordinate systems. Three elements common to coordinate systems are an *origin*, *fundamental plane*, and *principal direction*.

The origin is the coordinate system's center. The origin defines the coordinate system's "zero point" from which all measurements are expressed. Origins are typically located at a defined reference point, such as celestial bodies or spacecraft's center-of-mass.

The fundamental plane is a reference plane from which parameters can be conveniently expressed in the coordinate system. The fundamental plane often contains an element of symmetry such as a celestial body's equator.

The principal direction is the zero orientation from the origin through the fundamental plane. While the principal direction may often be chosen arbitrarily (or preferentially), it needs to be sufficiently well-defined to lack ambiguity. An example fundamental direction in geographic coordinates is the *Prime Meridian*, which is the zero-longitude meridian passing through the Royal Observatory in Greenwich, England.

Key Terms:

Fundamental Plane: is a plane passing through a coordinate system's origin that is oriented with a definable reference.
Origin: is a coordinate system's center where the three Cartesian axes meet.
Prime Meridian: is the zero-longitude north-south line passing through the Royal Observatory in Greenwich, England.
Principal Direction: is a definable reference direction in a coordinate system's fundamental plane.

3.1.1 Coordinate System Types [Level I – Descriptive]

The two most common types of coordinate systems used in space flight dynamics are *Cartesian* coordinates and *spherical* coordinates. Cartesian coordinates are references to three [typically right-handed] coordinates axes radiating from the origin. Spherical coordinates consist of a radius from the origin oriented by two angles: one in the fundamental plane and one perpendicular to the fundamental plane in the principal direction's plane. It is common to express vectors of the same coordinate system in both the Cartesian and spherical categories. The methods to convert between the Cartesian and spherical expression will be covered in the sections that follow.

3.1.2 Cartesian Coordinates

Cartesian coordinates, named after mathematician René Descartes, have a set of perpendicular three-dimensional axes. Thes axes are typically, but not exclusively labeled x, y, and z and follow a right-handed convention. The right-handed convention provides an unambiguous axis set arrangement. The arrangement begins with straightening the fingers on the right hand to indicate the first (x-axis) direction. The second (y-axis) is perpendicular to the x-axis, in the directions the finger joints naturally would bend. The thumb indicates the z-axis direction for this set. Figure 3-1 illustrates a right-handed Cartesian coordinate system.

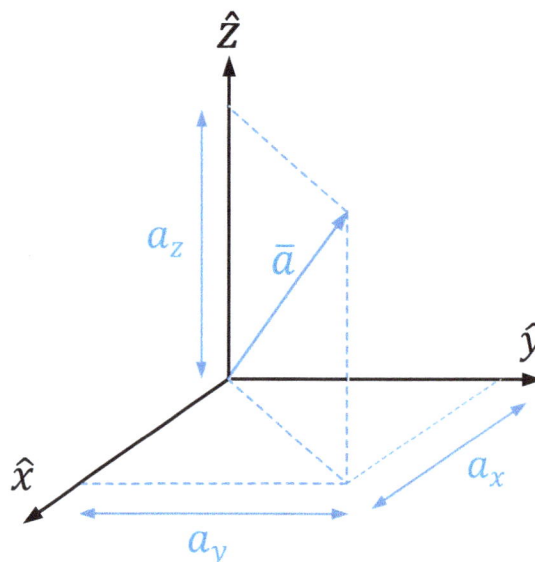

Figure 3-1 Cartesian Coordinate System

The origin is located at the coordinate axes meeting point. The plane formed by the first two axes (x, y) is the fundamental plane and the first axis (x) defines the principal direction. A vector expressed in Cartesian coordinates has components representing its projection along each of the coordinate axes.

Key Term:

3.1.3   Spherical Coordinates

Spherical coordinates provide an alternative way to express a vector. A spherical coordinate vector representation has the vector magnitude (or norm) and two orientation angles. The first angle (α), called right ascension in astronomy, is the measured from the principal direction to the vector's projection in the fundamental plane, as shown in figure 3-2. Traditional coordinate definition measures the second angle between the vector and the out-of-plane coordinate axis (z in this case). However, in astronomical conventions, the second angle (δ), called declination in astronomy, is measured between the vector and the fundamental plane.

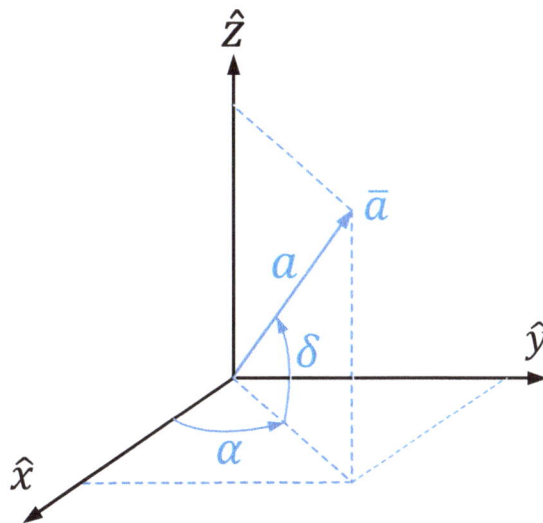

Figure 3-2 Spherical Coordinate System

Key Terms:

3.1.4 Geodetic Coordinates

The Earth is an oblate spheroid meaning its equatorial radius is greater than its polar radius (by approximately 21 km). The elliptical profile causes a slight shift to local vertical, which is closely aligned to the gravity gradient, that in turn causes a mathematical re-definition of the coordinates.

3.1.4.1 Geodetic Coordinates [Level I – Descriptive]

Coordinates fixed to the Earth's surface have correspondence to spherical coordinates. Longitude (λ) corresponds to right ascension while latitude (L) corresponds to declination. Longitude is measured east or west of the *Prime Meridian* which passes through the Royal Observatory in Greenwich, England. Latitude is measured north or south of the equator.

Geocentric coordinates have a direct correspondence to the spherical counterparts, with the presumption of the Earth being spherical in shape. Thus, there is a geocentric latitude (L_c) that maps directly to declination (δ). But conforming to the oblate spheroid's elliptical profile requires the creation of a geodetic latitude (L_d) that aligns with the local vertical as shown in figure 3-3. (There is no deviation for longitude.) When not explicitly specified, latitude should be presumed to be geodetic to conform to cartographic standards.

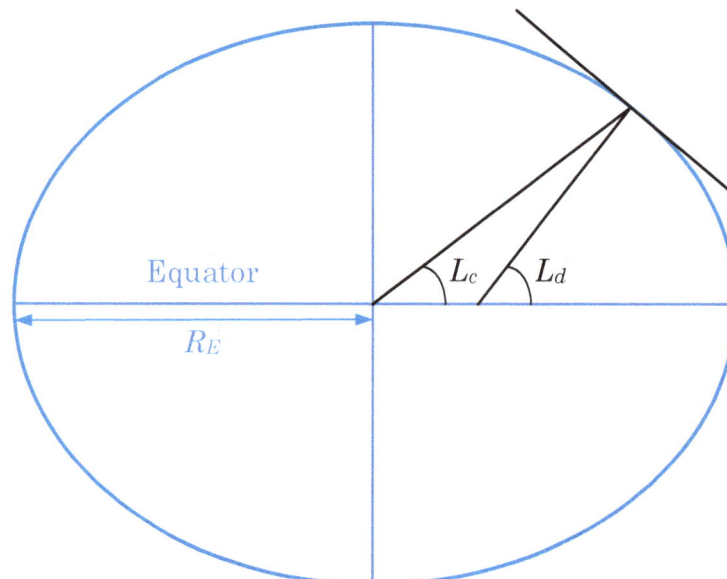

Figure 3-3 Geocentric versus Geodetic Latitude

The geocentric latitude is thus measured from the ellipsoid center to a surface location. In contrast, the geodetic latitude is aligned with local vertical, which is perpendicular to the local horizon.

Key Terms:

> **Geodetic**: coordinates have latitude and height conform to the ellipsoidal shape.
> **Latitude**: represents the angular excursion north or south of the equator and are illustrated by lines parallel to the equator.
> **Longitude**: represents the angle east or west of the zero-degree longitude line (meridian) passing through the Royal Observatory in Greenwich, England.

➤ **Transition**: *You may continue this section with Geodetic Coordinates at Level II, or you may skip to the beginning of the next section called Celestocentric Coordinates.*

3.1.4.2 Geodetic Coordinates [Level II – Equations]

The geodetic coordinates consist of latitude (L), longitude (λ) and height (h) above the ellipsoid. Earth fixed-Cartesian coordinates may be computed from the geodetic coordinates, provided specific Earth geoidal parameters are available. For the World Geodetic Survey of 1984 (WGS-84), the Earth's equatorial radius is: $R_E = 6378137$ meters. The geoidal flattening factor is: $1/f = 298.257223563$. Equation 3.1 computes the Earth ellipsoidal eccentricity (ecc_e).

$$ecc_e = \sqrt{f(2-f)} \qquad\qquad 3.1$$

The *radius of curvature of prime vertical* (R_n) provides the basis for computing the Cartesian coordinates.

$$R_n = \frac{R_E}{\sqrt{1 - ecc_e^2 \sin^2 L}} \qquad\qquad 3.2$$

Equations 3.3 through 3.5 compute the corresponding Earth-fixed Cartesian components.

$$X = (R_n + h)\cos\lambda\cos L \qquad\qquad 3.3$$

$$Y = (R_n + h)\sin\lambda\cos L \qquad\qquad 3.4$$

$$Z = [R_n(1 - ecc_e^2) + h]\sin L \qquad\qquad 3.5$$

Note that it is less straightforward to compute the geodetic coordinates from the Cartesian components for a non-zero height (h) relative to the ellipsoid. Numerical methods are generally used for this task.

Key Terms:

> **Meridians**: are lines of longitude running north-south from pole to pole.
> **Parallels**: are lines of latitude parallel to the equator.
> **Prime Meridian**: is the zero-degree longitude line (meridian) passing through the Royal Observatory in Greenwich, England.

3.1.5   Celestocentric Coordinates

Celestocentric coordinate systems have their origin at a celestial body's center of mass. The coordinate systems are typically named for the celestial body whose center of mass is at the origin as listed in table 3-1. The fundamental plane for heliocentric coordinates is the *ecliptic*, which is the Earth's orbital plane. The fundamental plane for geocentric coordinates may be either the Earth's equator or the ecliptic. The Fundamental plane for other celestial bodies is typically their equator.

The principal direction generally depends on whether the coordinate systems is inertial (non-rotating, non-accelerating) or fixed to the celestial body surface (i.e., rotating with the body). Rotating coordinate systems have a body-fixed reference, such as along the Prime Meridian in the case for the Earth. Most inertial systems point the principal direction parallel to the *vernal equinox*, which is the intersection of the ecliptic with the Earth's equatorial plane.

Table 3-1 Celestial Coordinate System Origin References	
Central Celestial Body	**Coordinate Reference**
Sun	Heliocentric
Earth (Terra)	Geocentric or Earth-Centered
Moon (Luna)	Selenocentric
Mars	Areocentric or Mars-Centered

Since the Earth's equatorial plane is subject to *precession* with a 26,000-year gyration and *nutation* that imposes an additional 9.2 seconds of arc amplitude every 18.6 years, the equator must be qualified (at a particular date) for precise definition. A mean equator and equinox accounts only for precession, while a true equator and equinox also accounts for nutation.

3.1.6 Specific Celestocentric Coordinate Systems

Table 3-2 lists some traditional geocentric coordinate frames. The 1950 coordinate systems are referenced to what is known as the Besselian epoch for that year. The 2000 coordinate systems are referenced to the epoch on January 1, 2000, at the 12:00 Terrestrial Time (TT) or J2000. The undated Mean and True of date and Earth-Fixed systems use the current date as epoch.

Table 3-2 Traditional Geocentric Coordinate Frames				
Name	**Abbreviation**	**Fundamental Plane**	**Principal Direction**	**Inertial or Rotating**
Mean of 1950 Equatorial	M50 Eq	Mean Equator of B1950	Mean Equinox of B1950	Inertial
Mean of 1950 Ecliptic	M50 Ec	Mean Ecliptic of B1950	Mean Equinox of B1950	Inertial
Mean of 2000 Equatorial	EME2000 Eq	Mean Equator of J2000	Mean Equinox of J2000	Inertial
Mean of 2000 Ecliptic	EME2000 Ec	Mean Ecliptic of J2000	Mean Equinox of J2000	Inertial
Mean of Date Equatorial	MOD Eq	Mean Equator of Date	Mean Equinox of Date	Inertial
Mean of Date Ecliptic	MOD Ec	Mean Ecliptic of Current date	Mean Equinox of Date	Inertial
True of Date Equatorial	TOD Eq	True Equator of Date	True Equinox of Date	Inertial
True of Date Ecliptic	TOD Ec	True Ecliptic of Date	True Equinox of Date	Inertial
Earth-Centered, Earth-Fixed	ECEF	Equator of Date	Prime Meridian	Rotating

The term Earth-Centered Inertial (ECI) is a generic term that encompasses any of the inertial frames in the table. Earth-Centered, Earth-Fixed (ECEF) and similar names refer to rotating counterparts that may or may not include corrections such as for nutation, the Earth's irregular rotation rate, and polar motion. More information regarding transformations between these legacy coordinate systems may be found in International Astronomical Union (IAU) resolutions or from Vallado.

The IAU's 2006 resolutions removed coordinate system dependency on the Earth's precessing and nutating equator in defining celestial and terrestrial coordinate frames. The result was the International Celestial Reference Frame (ICRF) becoming the preferred inertial coordinate frame for the solar system, with the Geocentric Celestial Reference Frame (GCRF) and the International Terrestrial

Reference Frame (ICRF) becoming the preferred Earth-fixed coordinate frame. The ICRF and GCRF have parallel coordinate axes. The difference is the ICRF has the solar system barycenter as its origin while the GCRF has the Earth's center of mass as its origin.

The ICRF has its origin at the solar system barycenter and is virtually parallel with the EME2000 coordinate frame. (There is a small corrective coordinate transformation between ICRF and EME2000.) The ITRF represents the Earth's instantaneous orientation, accounting for the Earth's precession, nutation, irregular rotation rate, and polar axis drift.

When the spherical coordinates define an inertial system, the angles *are* the *right ascension* (α) and *declination* (δ) as used by astronomers. Right ascension is the angle measured eastward from the vernal equinox to the vector's projection in the xy plane. The *vernal equinox* is the direction of intersection of the Earth's equatorial plane and the ecliptic, with the ecliptic being the Earth's orbital plane about the Sun. Declination is the angle the vector is north (positive) or south (negative) of the equator.

The IAU provides a set of well-tested software functions for converting between ICRF, ITRF, and the legacy coordinate systems, plus time conversions and star catalog corrections in their Standards of Fundamental Astronomy (SOFA) libraries. These libraries are available in the FORTRAN and C computing languages and have been translated by third parties to MATLAB as well. SOFA is authoritative and thus should be the source for coordinate conversion. The SOFA source code and associated documentation (including cookbooks) is available at the IAU SOFA web site (https://iausofa.org/).

Key Terms:

Declination: is the angle the vector is north or south of the equator.
Ecliptic: is the Earth's orbital plane about the Sun.
Nutation: is the 18.6-year 9.2 arc second amplitude wobbling (or nodding) of the Earth's axis due primary to lunar tidal perturbations.
Precession: is the 26,000-year conical gyration of the Earth's rotational axis primarily due to lunar, solar, and planetary tidal perturbations.
Right Ascension: is the angle measured eastward from the vernal equinox to the vector's projection in the xy plane.

➢ **Transition**: *You may continue this section with Cartesian Spherical Conversion at Level II, or you may skip to the beginning of the next section called Topocentric Coordinates.*

3.1.7 Cartesian Spherical Conversion [Level II – Equations]

Equation 3.6 expresses the Cartesian vector components directly in terms of the spherical coordinate parameters.

$$\bar{a} = \begin{bmatrix} a_x \\ a_y \\ a_z \end{bmatrix} = \begin{bmatrix} a \cos\alpha \cos\delta \\ a \sin\alpha \cos\delta \\ a \sin\delta \end{bmatrix} \qquad 3.6$$

Equation 3.7 provides the relationship between the Cartesian vector components and the spherical coordinate angles. The third spherical component is the vector's magnitude, which is the vector's norm.

$$\tan\alpha = \frac{a_y}{a_z} \qquad \sin\delta = \frac{a_z}{a} \qquad a = |\bar{a}| \qquad 3.7$$

Since the α angle may be in any of the four quadrants, it is best resolved using a two-argument inverse tangent function. The δ angle is restricted to the $-\pi/2 \leq \delta \leq \pi/2$ range and thus may be recovered using a standard inverse sine function.

➢ **Transition**: *You may continue this section with Cartesian Spherical Conversion at Level III, or you may skip to the beginning of the next section called Topocentric Coordinates.*

3.1.8 Cartesian Spherical Conversion [Level III – Derivation]

The projection of $\bar{a}$ into the fundamental plane ($\bar{a}_{xy}$) has a magnitude $a \cos\delta$. Since it is totally in the x-y plane, its z-component has a zero value. Its projections along the coordinate axes thus become:

$$\bar{a}_{xy} = a \cos\delta \begin{bmatrix} \cos\alpha \\ \sin\alpha \\ 0 \end{bmatrix} = \begin{bmatrix} a \cos\delta \cos\alpha \\ a \cos\delta \sin\alpha \\ 0 \end{bmatrix} \qquad 3.8$$

The final component is the z-axis projection:

$$a_z = a \sin\delta \qquad 3.9$$

These results confirm the relationship in equation 3.6. Equation 3.7 is verified by the trigonometric relationships of the ratios.

3.2 Topocentric Coordinates

The topocentric coordinate system is used extensively to reference and observe trajectories from a celestial body surface.

3.2.1 Topocentric Coordinates [Level I – Descriptive]

A topocentric coordinate system, as its name suggests, has its origin is at a surface location, typically referenced to a site with a telescope or antenna used for tracking or communication. The fundamental plane is the local horizon, and the principal direction is true north, directed toward the celestial body's rotational axis. Figure 3-4 illustrates an example topocentric coordinate system.

The topocentric coordinate system also has applications to describing a trajectory's flight progression. For these cases, the origin will be at location above the surface, but defined relative to the celestial body.

The Cartesian axes are directed to the local East-North-Up (*ENU*) directions, in that order to maintain the right-handed coordinate system convention.

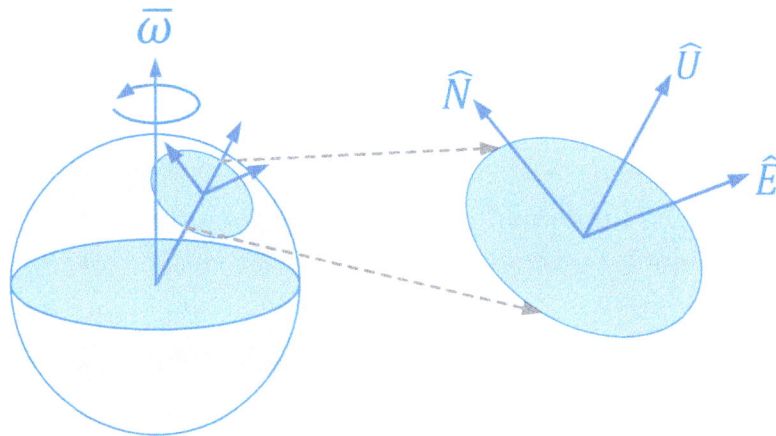

Figure 3-4 Topocentric Coordinate System

Key Term:

> **Topocentric Coordinates:** are a coordinate system on the surface of a celestial body.

➤ **Transition**: *You may continue this section with Topocentric Coordinates at Level II, or you may skip to the beginning of the next section called Transformation from Perifocal to Celestocentric.*

3.2.2   Topocentric Coordinates [Level II – Equations]

The topocentric coordinates are referenced to a celestial body-centered coordinate system. When referenced to a trajectory an inertial system is used (such as GCRF for Earth), while a body-fixed system (such as GCRF) is used when referencing a surface location.

Equation 3.10 through 3.12 define the *ENU* axes from the position vector ($\bar{r}$). The $\hat{k}$ vector defines the orientation of the Earth's rotational axis.

$$\hat{U} = \frac{\bar{r}}{r} = \frac{1}{r}\begin{bmatrix} r_x \\ r_y \\ r_z \end{bmatrix} \qquad\qquad 3.10$$

$$\hat{E} = \frac{1}{\sqrt{r_x^2 + r_y^2}}\begin{bmatrix} -r_y \\ r_x \\ 0 \end{bmatrix} \qquad\qquad 3.11$$

$$\hat{N} = \frac{1}{r\sqrt{r_x^2 + r_y^2}}\begin{bmatrix} -r_x r_z \\ -r_y r_z \\ r_x^2 + r_y^2 \end{bmatrix} \qquad\qquad 3.12$$

Equation 3.13 and 3.14 are the transformation matrices from ECI to ENU in terms of the axis vectors and the right ascension and declination angles, respectively.

$$C_{ECI}^{ENU} = \begin{bmatrix} E_x & E_y & E_z \\ N_x & N_y & N_z \\ U_x & U_y & U_z \end{bmatrix} \qquad\qquad 3.13$$

$$C_{ECI}^{ENU} = \begin{bmatrix} -\sin\alpha & \cos\alpha & 0 \\ -\cos\alpha\sin\delta & -\sin\alpha\sin\delta & \cos\delta \\ \cos\alpha\cos\delta & \sin\alpha\cos\delta & \sin\delta \end{bmatrix} \qquad\qquad 3.14$$

Equations 3.11 and 3.12 are singular along the rotational axis due to division by zero (i.e., the r_x and r_y components are both zero along the z-axis). Thus, equation 3.14 is employed in these cases to define axis orientations at the poles.

The right ascension has no unique value at the poles. Thus, a default value of zero ($\alpha = 0$) is used. The north pole has a $+90°$ declination and the south pole has a $-90°$ declination. Substituting these values into equation 3.8 aligns the $\hat{E}$ with the $+y$-

axis for both poles. This same substitution aligns the $\widehat{N}$ with the $+x$-axis at the south pole and the $-y$-axis at the north pole.

Equation 3.15 redefines the transformation matrix in terms of latitude (L) and longitude (λ) in terms of a reference fixed to the Earth's surface. This is the Earth-fixed transformation used for ground sites.

$$C_{ECEF}^{ENU} = \begin{bmatrix} -\sin\lambda & \cos\lambda & 0 \\ -\cos\lambda\sin L & -\sin\lambda\sin L & \cos L \\ \cos\lambda\cos L & \sin\lambda\cos L & \sin L \end{bmatrix} \qquad 3.15$$

➢ **Transition**: *You may continue this section with Topocentric Coordinates at Level III, or you may skip to the beginning of the next section called Transformation from Perifocal to Celestocentric.*

3.2.3   Topocentric Coordinates [Level III – Derivation]

In the inertial system, the local vertical is aligned with the position vector. Thus, the up ($\widehat{U}$) axis is the normalized position vector. Equation 3.16 determines the $\widehat{U}$ orientation.

$$\widehat{U} = \frac{\bar{r}}{r} \qquad 3.16$$

Note that the Earth-fixed reference has a slightly different definition of vertical, which is perpendicular to the Earth's oblate spheroid shape. This was previously covered in the geodetic coordinates section.

The local north ($\widehat{N}$) and east ($\widehat{E}$) axes lie in the local horizontal plane, which is perpendicular to the up ($\widehat{U}$) direction. The east axis is also perpendicular to the celestial body's rotational ($\hat{k}$) axis. Equations 3.17 and 3.18 determine the $\widehat{E}$ and $\widehat{N}$ orientations.

$$\widehat{E} = \frac{\hat{k}\times\widehat{U}}{|\hat{k}\times\widehat{U}|} \qquad 3.17$$

$$\widehat{N} = \widehat{U}\times\widehat{E} \qquad 3.18$$

Equation 3.16 defines aligns the $\widehat{U}$ direction by normalizing the vector in spherical coordinates as defined in equation 3.6. The $\widehat{E}$ and $\widehat{N}$ are computed likewise, substituting the spherical $\widehat{U}$ components into the cross products and simplifying with trigonometric identities.

3.3   Inertial versus Rotating Coordinates

Trajectories are most naturally represented in inertial coordinate systems. The law of conservation of angular momentum establishes that the trajectory plane has a constant orientation in inertial space.

While it might appear desirable to represent trajectories in celestial body fixed coordinates, celestial bodies rotate in inertial space. Thus, the coordinate systems fixed to these bodies are rotating. Rotating coordinates provide significant complications particularly when representing velocities and accelerations. Thus, rotating coordinates should be avoided when representing trajectories, with only a few special exceptions.

3.3.1   Inertial to Rotating Transformation [Level I – Descriptive]

Position vectors may be readily transformed to and from rotating coordinates, but the transformation matrix is valid only for the instant that the position is represented. Thus, position transformations require a new transformation unique to their time and rotation angle relative to an inertial reference.

A velocity vector represented in a rotating coordinate system differs from its inertial representation in magnitude and usually in direction. This difference is attributable to the intrack velocity of the rotating frame masking a portion of the inertial velocity. This is also the case for acceleration which also experiences additional differences for a non-constant coordinate frame rotation rate.

➢ **Transition**: *You may continue this section with Inertial to Rotating Transformation at Level II, or you may skip to the beginning of the next section called Transformation from Perifocal to Celestocentric.*

3.3.2   Inertial to Rotating Transformation [Level II – Equations]

Consider two coordinate systems with origins at a celestial body's center of mass. Both coordinate systems have the z-axis colinear with the body's rotation rate ($\bar{\omega}$) expressed by equation 3.19.

$$\bar{\omega} = \begin{bmatrix} 0 \\ 0 \\ \omega \end{bmatrix} \qquad\qquad 3.19$$

At a particular reference epoch (t_0) the rotating coordinate system has a reference rotation angle (α_0). Equation 3.20 is the rotation angle as a function of time. Equation 3.21 position transformation from inertial to rotating (C_I^R), which is a z-axis or $R_3(\alpha)$ rotation. Equation 3.22 computes the position in the rotating coordinate system.

$$\alpha(t) = \alpha_0 + \omega(t - t_0) \tag{3.20}$$

$$C_I^R = \begin{bmatrix} \cos\alpha & \sin\alpha & 0 \\ -\sin\alpha & \cos\alpha & 0 \\ 0 & 0 & 1 \end{bmatrix} \tag{3.21}$$

$$\bar{r}_R = C_I^R \bar{r}_I \tag{3.22}$$

Equation 3.23 is the velocity transformed from the inertial to rotating system.

$$\bar{v}_R = C_I^R \bar{v}_I + \bar{\omega} \times \bar{r}_R \tag{3.23}$$

➢ **Transition**: *You may continue this section with Inertial to Rotating Transformation at Level III, or you may skip to the beginning of the next section called Five Term Acceleration Equation.*

3.3.3 Inertial to Rotating Transformation [Level III – Derivation]

The inertial-to-rotating transformation is an artifact of the chain rule for derivatives on equation 3.22.

$$\bar{v}_R = C_I^R \dot{\bar{r}}_I + \dot{C}_I^R \bar{r}_I \tag{3.24}$$

$$\bar{v}_R = C_I^R \bar{v}_I + \omega \begin{bmatrix} -\sin\alpha & \cos\alpha & 0 \\ -\cos\alpha & -\sin\alpha & 0 \\ 0 & 0 & 0 \end{bmatrix} \begin{bmatrix} r_{I,i} \\ r_{I,j} \\ r_{I,k} \end{bmatrix} \tag{3.25}$$

$$\bar{v}_R = C_I^R \bar{v}_I + \omega \begin{bmatrix} r_{R,j} \\ -r_{R,i} \\ 0 \end{bmatrix} \tag{3.26}$$

$$\bar{v}_R = C_I^R \bar{v}_I + \bar{\omega} \times \bar{r}_R \tag{3.27}$$

This result confirms equation 3.23.

3.3.4 Five Term Acceleration Equation

The preceding position and velocity transformations were between an inertial and rotating coordinate system with a common origin. The most generalized kinematic relationship between inertial and non-inertial (or moving) coordinate systems permits the moving system to translate as well as rotate and the rotation rate may vary. The ability to translate means the two coordinate systems no longer share a common origin.

Equation 3.28 provides the position relationship between the inertial and moving coordinate systems. Equation 3.29 provides the velocity relationship and equation 3.30 provides the acceleration relationship.

$$\bar{r}_I = \bar{r}_O + \bar{r}_M \qquad\qquad 3.28$$

$$\bar{v}_I = \bar{v}_O + \bar{v}_M + \bar{\omega} \times \bar{r}_M \qquad\qquad 3.29$$

$$\bar{a}_I = \bar{a}_O + \bar{a}_M + \dot{\bar{\omega}} \times \bar{r}_M + \bar{\omega} \times (\bar{\omega} \times \bar{r}_M) + 2\bar{\omega} \times \bar{v}_M \qquad\qquad 3.30$$

In the above equations, the "O" subscript represents the parameter's offset between the inertial and moving system. The subscript "M" represents the parameter's value relative to the moving coordinate system. Thus $\bar{r}_O$ is the offset in the origins between the two systems, $\bar{v}_O$ is the velocity between the origins, and $\bar{a}_O$ is the translational acceleration between the origins. The $\dot{\bar{\omega}}$ parameter is the moving coordinate system's rotational acceleration.

Note that equation 3.30 is called the *five-term acceleration equation*, because there are five distinct terms relating acceleration between inertial and moving coordinate systems. The complexity of this process should be a sufficient incentive to use inertial coordinate systems for trajectories, except in the circumstances where moving coordinate systems provide a distinct advantage.

3.4 Transformation from Perifocal to Celestocentric

The perifocal coordinate system is transformed to a celestocentric coordinate system using three single axis rotations. The example illustrates transformation to an ECI equatorial coordinate system as illustrated in figure 3-5.

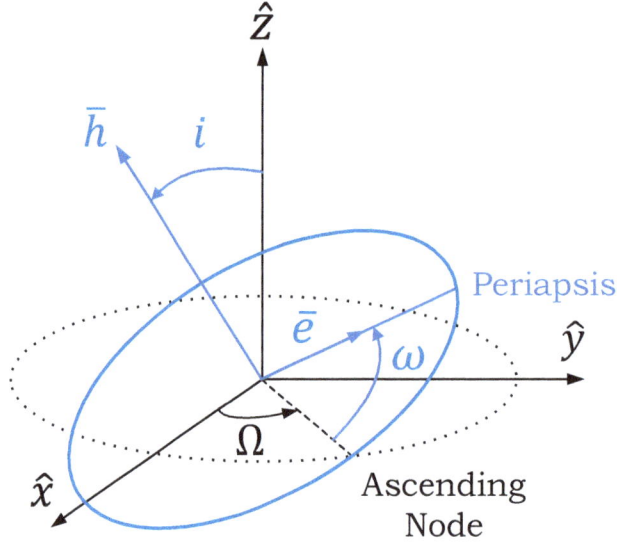

Figure 3-5 Rotations from Perifocal to ECI

Consider the perifocal system initially aligned in the Earth's equatorial plane ($\hat{\omega} = \hat{k}$) with periapsis ($\hat{p}$) pointing toward the *vernal equinox*. The first rotation is about the ECI z-axis by an angle Ω, called right ascension of the ascending node (often abbreviated by the RAAN acronym). This rotation causes periapsis to initially coincide with what will be the spacecraft's south-to-north equatorial crossing (i.e., *ascending node*). Since the rotation is about the z-axis by angle Ω, this is an $\boldsymbol{R_3}(\Omega)$ rotation.

$$\boldsymbol{R_3}(\Omega) = \begin{bmatrix} \cos\Omega & \sin\Omega & 0 \\ -\sin\Omega & \cos\Omega & 0 \\ 0 & 0 & 1 \end{bmatrix} \qquad 3.31$$

The second rotation tilts the trajectory plane by the *inclination* angle (i) to establish the orientation between it and the equatorial plane. This rotation is accomplished about the current x-axis, which is pointing in the ascending node direction. Since the rotation is about the current x-axis by angle i, this is an $\boldsymbol{R_1}(i)$ rotation.

$$\boldsymbol{R_1}(i) = \begin{bmatrix} 1 & 0 & 0 \\ 0 & \cos i & \sin i \\ 0 & -\sin i & \cos i \end{bmatrix} \qquad 3.32$$

The third rotation advances the periapsis location in the trajectory plane to its proper location by an angle called *argument of periapsis* (ω). Since it is a rotation in the current trajectory plane by angle ω, this is an $\boldsymbol{R_3}(\omega)$ rotation.

$$\boldsymbol{R_3}(\omega) = \begin{bmatrix} \cos\omega & \sin\omega & 0 \\ -\sin\omega & \cos\omega & 0 \\ 0 & 0 & 1 \end{bmatrix} \qquad 3.33$$

The net transformation from perifocal (*pqw*) to the *ECI* coordinate system is the three rotations multiplied in the proper order:

$$C_{ECI}^{pqw} = R_3(\omega) \cdot R_1(i) \cdot R_3(\Omega)$$

3.34

$$C_{ECI}^{pqw} = \begin{bmatrix} -\sin\omega \sin\Omega \cos i + \cos\omega \cos\Omega & \sin\omega \cos\Omega \cos i + \cos\omega \sin\Omega & \sin\omega \sin i \\ -\cos\omega \sin\Omega \cos i - \sin\omega \cos\Omega & \cos\omega \cos\Omega \cos i - \sin\omega \sin\Omega & \cos\omega \sin i \\ \sin\Omega \sin i & -\cos\Omega \sin i & \cos i \end{bmatrix}$$

3.35

The position and velocity vectors in *pqw* coordinates may be derived from equations 2.172 and 2.173.

$$\bar{r}_{pqw} = \frac{h^2}{\mu(1 + e\cos\theta)} \begin{bmatrix} \cos\theta \\ \sin\theta \\ 0 \end{bmatrix}$$

3.36

$$\bar{v}_{pqw} = \frac{\mu}{h} \begin{bmatrix} -\sin\theta \\ e + \cos\theta \\ 0 \end{bmatrix}$$

3.37

Equation 3.38 computes the position and velocity in ECI, using the coordinate transformation.

$$\bar{r}_{ECI} = \left(C_{ECI}^{pqw}\right)^T \bar{r}_{pqw} \qquad \bar{v}_{ECI} = \left(C_{ECI}^{pqw}\right)^T \bar{v}_{pqw}$$

3.38

Key Terms:

> **Argument of Periapsis**: is the angle measured from the ascending node to periapsis, in the direction of motion.
> **Ascending Node**: is the trajectory's south-to-north crossing of the central celestial body's equator.
> **Inclination**: is the angle between the trajectory's angular momentum direction and the central celestial body's rotational axis. As such, inclination is the tilt of the trajectory plane relative to the central body's equator.
> **Vernal Equinox**: is the intersection of the ecliptic with the Earth's equator in which the Sun's subpoint transitions from the southern to northern hemisphere. This transition corresponds to the beginning of spring in the Earth's northern hemisphere.

3.5 Keplerian Parameters from Position and Velocity

The Cartesian ECI position and velocity are collectively known as a trajectory *state vector*. They may represent any trajectory type and are an advantageous form for most computations. However, they lack the ability to easily discern trajectory characteristics and are constantly changing over time.

Keplerian parameters have a form from which trajectory characteristics are readily discernable. They are stable since, in the absence of perturbative disturbances, the true anomaly (orbital phase angle) is the only parameter that changes over time.

This section provides the process to convert a trajectory state vector to a set of Keplerian elements as depicted in figure 3-6.

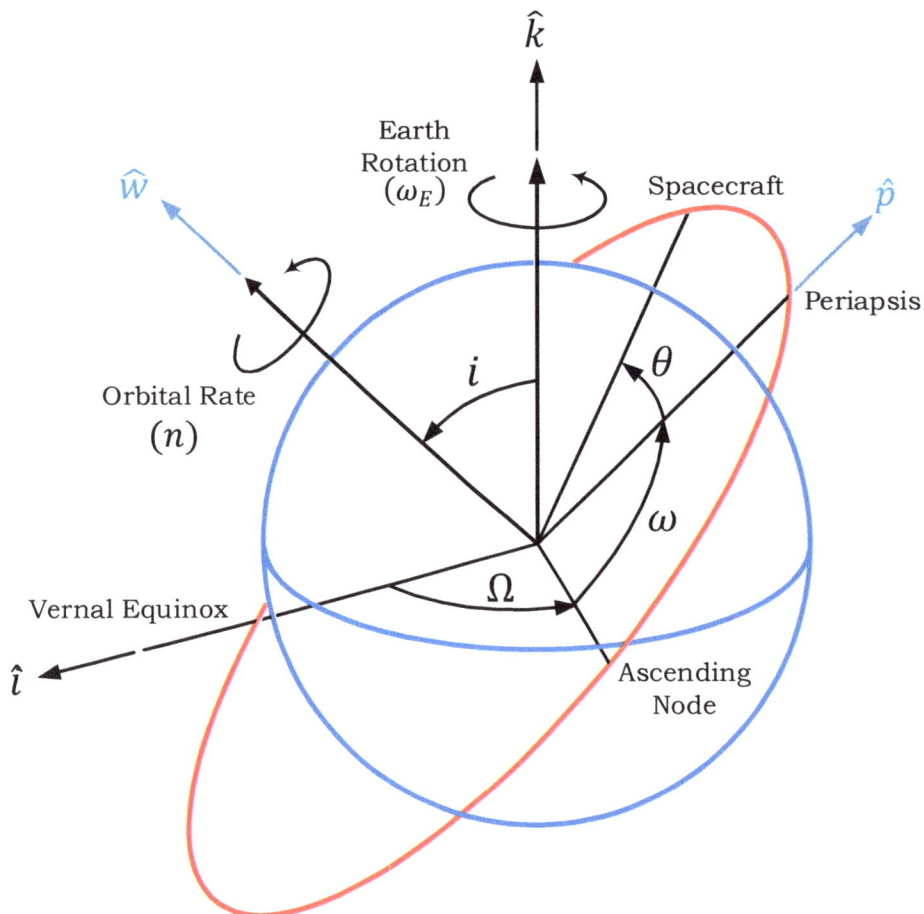

Figure 3-6 Trajectory Oriented in ECI Coordinates

3.5.1 Keplerian Parameters from State Vectors [Level I – Descriptive]

Keplerian orbital parameters may be recovered from position ($\bar{r}$) and velocity ($\bar{v}$) state vectors using the geometry and trajectory equations previously developed. This begins with computing the specific angular momentum ($\bar{h}$), semi-major axis (a), and eccentricity vector ($\bar{e}$) directly from the position and velocity. From there several vector operations determine the orientation angles, with attention to the correct quadrant determination using known physical properties of motion.

➤ **Transition**: *You may continue this section with Keplerian Parameters from State Vectors at Level II, or you may skip to the beginning of the next section called Mathematical Trajectory Descriptions.*

3.5.2 Keplerian Parameters from State Vectors [Level II – Equations]

The angular momentum ($\bar{h}$), eccentricity vector ($\bar{e}$), and semi-major axis (a) are computed from $\bar{r}$ and $\bar{v}$ using equations 1.33, 2.127, and [the reciprocal of] equation 2.225 respectively. The eccentricity is the norm (or magnitude) of the eccentricity vector. The remaining Keplerian parameters are the orientation and phase angles: inclination (i), right ascension of the ascending node (Ω), argument of periapsis (ω), and true anomaly (θ).

The inclination angle (i) is computed directly from the angular momentum vector.

$$i = \cos^{-1}\left(\frac{h_k}{h}\right) \qquad\qquad 3.39$$

The ascending node vector is mutually perpendicular to the angular momentum ($\bar{h}$) and the ECI coordinate system z-axis (or $\hat{k}$). Thus, it is computed by the cross product of these vectors. The Ω angle is computed from the ascending node vector components.

$$\bar{n} = \hat{k} \times \bar{h} \qquad\qquad 3.40$$

$$\tan \Omega = \frac{n_y}{n_x} = \frac{h_x}{-h_y} \qquad\qquad 3.41$$

Since the right ascension of the ascending node has as full 2π radians range, the quadrant should be resolved using a two-argument inverse tangent function.

The argument of periapsis (ω) is the angle between the ascending node and periapsis, in the direction of motion. The ω angle is computed leveraging the properties of the vector dot product.

$$\omega = \cos^{-1}\left(\frac{\bar{n} \cdot \bar{e}}{ne}\right) \qquad if \ e_z \geq 0 \qquad\qquad 3.42$$

$$\omega = 2\pi - \cos^{-1}\left(\frac{\bar{n} \cdot \bar{e}}{ne}\right) \qquad if \ e_z < 0 \qquad\qquad 3.43$$

The quadrant is resolved by recognizing that angle ω is less than π radians when periapsis is in the northern hemisphere. This is identifiable by the eccentricity vector's z-component being positive.

The true anomaly (θ) is the orbital phase, measured as the angle from periapsis to the spacecraft position, in the direction of motion. The angle θ is likewise computed using the properties of the dot product.

$$\theta = \cos^{-1}\left(\frac{\bar{e} \cdot \bar{r}}{er}\right) \qquad if \ v_r \geq 0 \qquad\qquad 3.44$$

$$\theta = 2\pi - \cos^{-1}\left(\frac{\bar{e} \cdot \bar{r}}{er}\right) \qquad if \ v_r < 0 \qquad\qquad 3.45$$

The quadrant is resolved by recognizing that the trajectory's radius is only increasing when radial velocity (v_r) is positive, which occurs in the $0 < \theta < \pi$ angular range (i.e., approaching apoapsis from periapsis).

3.6 Mathematical Trajectory Descriptions

Six individual quantities are needed to provide a mathematical description of a trajectory. Thus far, emphasis has been on the Cartesian position and velocity state vector and classical Keplerian element representation. Other representations exist, each of which may be computed from or converted to the Cartesian state vector representation. These alternative representations have advantages in various contexts.

3.6.1 Keplerian Elements

The classical Keplerian elements include the size and shape, described by semi-major axis (a) and eccentricity (e). The orientation in space is described by the angles depicted in figure 3-4. These include the inclination (i) and right ascension of the ascending node (Ω) that define the trajectory plane orientation. The in-plane angles include the argument of periapsis (ω) that orients periapsis relative to the ascending node and the true anomaly (θ) that is the orbital phase angle relative to periapsis.

3.6.1.1   Keplerian Elements [Level I – Descriptive]

The Keplerian elements are descriptive, with their values characterizing the various trajectory attributes.

Keplerian orbital representations are specified classically described by the following parameters at a specified time or *epoch*:
- semi-major axis (a)
- eccentricity (e)
- inclination (i)
- right ascension of the ascending node (Ω)
- argument of periapsis (ω)
- true anomaly (θ)

➢ **Transition**: *You may continue this section with Keplerian Elements at Level II, or you may skip to the beginning of the next section called Spherical Elements.*

3.6.1.2   Keplerian Elements [Level II – Equations]

Section 3.3 demonstrated the computation of Keplerian elements from the Cartesian state vector. The corresponding state vector may likewise be computed from the Keplerian elements.

In doing this we define a new coordinate system which is an in-plane rotation of the perifocal coordinate system through the true anomaly angle or an $\boldsymbol{R_3(\theta)}$ *rotation*, such that the first axis ($\widehat{U}$) is aligned with position vector ($\bar{r}$). Consequently, the second axis ($\widehat{V}$) is in the locally horizontal intrack direction. The third axis ($\widehat{W}$) is in the angular momentum direction and is the same as that for the *pqw* coordinate system. This *UVW* coordinate system goes by several other names, including Radial-Intrack-Crosstrack (RIC).

The additional $\boldsymbol{R_3(\theta)}$ rotation could be applied to equation 3.16, adding an additional true anomaly (θ) dependency. Instead, the argument of latitude angle (u) is defined as the sum of argument of periapsis and true anomaly as shown in equation 3.46. Equation 3.47 is the transformation matrix from ECI to UVW, substituting the u angle for the ω angle.

$$u = \omega + \theta \qquad\qquad 3.46$$

$$C_{ECI}^{UVW} = \begin{bmatrix} -\sin u \sin \Omega \cos i + \cos u \cos \Omega & \sin u \cos \Omega \cos i + \cos u \sin \Omega & \sin u \sin i \\ -\cos u \sin \Omega \cos i - \sin u \cos \Omega & \cos u \cos \Omega \cos i - \sin u \sin \Omega & \cos u \sin i \\ \sin \Omega \sin i & -\cos \Omega \sin i & \cos i \end{bmatrix} \qquad 3.47$$

The position's radial magnitude (r) is computed from the semi-major axis (a), eccentricity (e), and true anomaly, using equation 2.130 for elliptical or hyperbolic trajectories and equation 2.116 for parabolic trajectories. Equation 3.48 expresses the position vector in the UVW coordinate system, recognizing it is directed purely in the $\widehat{U}$ Direction.

$$\bar{r}_{UVW} = \begin{bmatrix} r \\ 0 \\ 0 \end{bmatrix} \qquad 3.48$$

The velocity's radial (v_r) and intrack/intrack (v_θ) components are computed using equations 2.120 and 2.119, respectively. Equation 3.49 expresses the velocity vector in the UVW coordinate system.

$$\bar{v}_{UVW} = \begin{bmatrix} v_r \\ v_\theta \\ 0 \end{bmatrix} \qquad 3.49$$

Equation 3.50 transforms the position and velocity vectors to the ECI coordinate system, using the DCM in equation 3.47.

$$\bar{r}_{ECI} = (C_{ECI}^{UVW})^T \bar{r}_{UVW} \qquad \bar{v}_{ECI} (C_{ECI}^{UVW})^T \bar{v}_{UVW} \qquad 3.50$$

3.6.2   Spherical Elements

The spherical elements provide the means of expressing the trajectory's position and velocity vectors using spherical parameters. These elements use traditional astronomical parameters and terminology.

3.6.2.1   Spherical Elements [Level I – Descriptive]

In astronomical terms, *right ascension* (α) is the angle measured eastward in the equatorial plane, from the vernal equinox to the object referenced. The *declination* (δ) is the is the angle north or south of the equatorial plane. These angles and the position magnitude (r) are sufficient define the position vector ($\bar{v}$) in spherical coordinates.

The *azimuth* (Az) is the angle measured clockwise about local vertical (*zenith*) in the local horizontal plane from true north. Azimuth corresponds to a true heading. The azimuth, flight path angle (ϕ), and velocity magnitude (v) define the velocity vector ($\bar{v}$) in spherical coordinates.

➤ **Transition**: *You may continue this section with Spherical Elements at Level II, or you may skip to the beginning of the next section called Equinoctial Elements.*

3.6.2.2 Spherical Elements [Level II – Equations]

The position and velocity state vector may also be described in terms of spherical constituents, commonly called the ADBARV representation. The parameters in the acronym represent:
- A – right ascension (α)
- D – declination (δ)
- B – flight path angle (β – a.k.a. ϕ)
- A – azimuth
- R – radius (r)
- V – velocity (v)

The ADBARV right ascension, declination, and radius represent the ECI position vector. Equation 3.51 computes the position from the spherical elements; equation 3.52 computes the spherical elements from the ECI position vector. These correspond to equations 3.1 and 3.2.

$$\bar{r}_{ECI} = \begin{bmatrix} r_x \\ r_y \\ r_z \end{bmatrix} = r \begin{bmatrix} \cos\alpha\cos\delta \\ \sin\alpha\cos\delta \\ \sin\delta \end{bmatrix} \qquad 3.51$$

$$\tan\alpha = \frac{r_y}{r_z} \qquad \sin\delta = \frac{r_z}{r} \qquad r = |\bar{r}| \qquad 3.52$$

The velocity vector is described in the local topocentric coordinate system. Equation 3.53 transforms the velocity to *ENU* coordinates. Equation 3.54 computes the remaining spherical elements from the velocity vector.

$$\bar{v}_{ENU} = C_{ECI}^{ENU}\,\bar{v}_{ECI} \qquad 3.53$$

$$\sin\beta = \frac{v_u}{v} \qquad \tan Az = \frac{v_E}{v_N} \qquad v = |\bar{v}| \qquad 3.54$$

Since the right ascension (α) and azimuth (Az) both have ranges in all four quadrants, it is recommended they be computed using a two-argument inverse tangent function.

Equations 3.55 and 3.56 are used to compute the Cartesian velocity ($\bar{v}$) vector from ADBARV spherical elements.

$$\bar{v}_{ENU} = v \begin{bmatrix} \sin Az \cos \beta \\ \cos Az \cos \beta \\ \sin \beta \end{bmatrix} \qquad\qquad 3.55$$

$$\bar{v}_{ECI} = \left(C_{ECI}^{ENU} \right)^T \bar{v}_{ENU} \qquad\qquad 3.56$$

3.6.3 Equinoctial Elements

Equinoctial elements provide an alternative to Keplerian elements that avoid the numerical problems associated with low eccentricity and/or low inclination trajectories. When the eccentricity is zero, there is no unique periapsis. Thus, equations 3.42 and 3.43 would experience a divide by zero condition. Likewise, when the inclination is zero or π radians, there is no unique ascending node. The ascending node vector ($\bar{n}$) defined by equation 3.40 would be a zero vector as would the x- and y-components of the angular momentum vector ($\bar{h}$). Thus, an attempt to compute the RAAN in equation 3.41 would have a zero divided by zero condition.

Neither of these conditions are rare orbit types. Many low altitude orbits are nearly circular. Moreover, the geosynchronous orbit which is popular for both communications and meteorological spacecraft have nearly zero eccentricities and inclinations.

While it would seem reasonable to handle these exceptions in software by default values, there is a more subtle difficulty. The orbit determination process described in Chapter 9 performs a simultaneous fit of the orbital parameters to observed spacecraft locations or other tracking data. The poor definition of RAAN and/or argument of periapsis with low inclination or low eccentricity orbits tend to render the trajectory determination process unstable with traditional Keplerian elements. The equinoctial elements are particularly well suited for the trajectory determination process.

3.6.3.1 Equinoctial Elements Relationship to Keplerian Elements

The equinoctial elements avoid the numerical issues associated with Keplerian elements for trajectories with low eccentricity ($e \approx 0$) and low inclination ($i \approx 0$). Since they avoid numerical issues, equinoctial elements are advantageous as a stable solution set in statistical trajectory determination processes.

The equinoctial elements define two two-dimensional vectors. The *apsidal vector* has two components (h, k) that are functions of eccentricity, RAAN, argument of periapsis, and whether the trajectory is prograde $(i \leq 90°)$ or retrograde $(i > 90°)$. The planar vector has two components (p, q) that are functions of RAAN and inclination. The orbital phase is the mean longitude (λ), which is a function of RAAN, argument of periapsis, mean anomaly, and whether the trajectory is prograde or retrograde.

The parameter I indicates whether the trajectory is prograde or retrograde, though including the polar inclination $(i = 90°)$ as prograde, since the mathematics remains the same regardless of the choice made.

$$I = +1 \qquad \text{for } i \leq \pi/2$$
$$I = -1 \qquad \text{for } i > \pi/2$$

3.57

Equations 3.58 through 3.63 Define the equinoctial elements in terms of Keplerian Elements.

$$a = a \tag{3.58}$$

$$h = e \sin(\omega + I\Omega) \tag{3.59}$$

$$k = e \cos(\omega + I\Omega) \tag{3.60}$$

$$p = \left[\tan \left(\frac{i}{2} \right) \right]^{I} \sin \Omega \tag{3.61}$$

$$q = \left[\tan \left(\frac{i}{2} \right) \right]^{I} \cos \Omega \tag{3.62}$$

$$\lambda = M + \omega + I\Omega \tag{3.63}$$

The Keplerian elements may also be computed from the equinoctial elements.

$$e = \sqrt{h^2 + k^2} \tag{3.64}$$

$$i = \pi \left(\frac{1 - I}{2} \right) + 2I \tan^{-1} \sqrt{p^2 + q^2} \tag{3.65}$$

$$\tan \Omega = \frac{p}{q} \tag{3.66}$$

$$\tan(\omega + I\Omega) = \frac{h}{k} \qquad\qquad 3.67$$

$$M = \lambda - (\omega + I\Omega) \qquad\qquad 3.68$$

The Ω and $(\omega + I\Omega)$ angles should be computed with a two-argument inverse tangent function to resolve the correct quadrant.

Note that the use of the semi-major axis reciprocal ($\alpha = 1/a$) may be used with no loss of generality. This modification also avoids numerical issues for nearly parabolic eccentricities ($e \approx 1$), providing a universal trajectory set.

3.6.3.2   Equinoctial Elements Relationship to Cartesian Position and Velocity

The equinoctial elements may be computed directly from the Cartesian position ($\bar{r}$) and velocity ($\bar{v}$) vectors. The semi-major axis reciprocal (α) may be computed by the expression on the right of equation 2.125. The trajectory plane normal direction ($\hat{w}$) is computed by normalizing angular momentum.

$$\hat{w} = \frac{\bar{r} \times \bar{v}}{|\bar{r} \times \bar{v}|} \qquad\qquad 3.69$$

Equations 3.70 and 3.71 compute the equinoctial planar vector elements.

$$p = \frac{w_x}{1 + I w_z} \qquad\qquad 3.70$$

$$q = \frac{w_y}{1 + I w_z} \qquad\qquad 3.71$$

The eccentricity vector is computed by the expression on the right of equation 2.127. Equations 3.72 and 3.73 compute the in-plane equinoctial frame basis vectors ($\hat{f}$ and $\hat{g}$).

$$\hat{f} = \frac{1}{1 + p^2 + q^2} \begin{bmatrix} 1 - p^2 + q^2 \\ 2pq \\ -2Ip \end{bmatrix} \qquad\qquad 3.72$$

$$\hat{g} = \frac{1}{1 + p^2 + q^2} \begin{bmatrix} 2Ipq \\ (1 + p^2 - q^2)I \\ 2q \end{bmatrix} \qquad\qquad 3.73$$

Equations 3.74 and 3.75 compute the equinoctial apsidal vector components.

$$h = \bar{e} \cdot \hat{g} \qquad\qquad 3.74$$

$$k = \bar{e} \cdot \hat{f} \qquad\qquad 3.75$$

Equations 3.76 and 3.77 compute the position components in the equinoctial trajectory plane.

$$r_f = \bar{r} \cdot \hat{f} \qquad\qquad 3.76$$

$$r_g = \bar{r} \cdot \hat{g} \qquad\qquad 3.77$$

Equation 3.78 computes the b parameter, which is needed to determine the associated eccentric longitude (F) parameters.

$$b = \frac{1}{1 + \sqrt{1 - h^2 - k^2}} \qquad\qquad 3.78$$

Equations 3.79 and 3.80 compute the sine and cosine of the eccentric longitude.

$$\sin F = h + \frac{\alpha\left[(1 - bh^2)r_g - bhkr_f\right]}{\sqrt{1 - h^2 - k^2}} \qquad\qquad 3.79$$

$$\cos F = k + \frac{\alpha\left[(1 - bk^2)r_f - bhkr_g\right]}{\sqrt{1 - h^2 - k^2}} \qquad\qquad 3.80$$

Equation 3.81 computes the mean longitude (λ) using the equinoctial form of Kepler's equation.

$$\lambda = F + h \cos F - k \sin F \qquad\qquad 3.81$$

The Cartesian position $(\bar{r})$ and velocity $(\bar{v})$ may be computed directly from the equinoctial elements. The equinoctial in-plane $\hat{f}$ and $\hat{g}$ vectors may be computed from equations 3.72 and 3.73, using the equinoctial p, q, and I parameters. Equation 3.82 computes the $\hat{w}$ vector.

$$\hat{w} = \frac{1}{1 + p^2 + q^2} \begin{bmatrix} 2p \\ -2q \\ (1 - p^2 - q^2)I \end{bmatrix} \qquad\qquad 3.82$$

Equation 3.81 is the equinoctial form of Kepler's equation. The eccentric longitude (F) is obtained by a numerical solution.

1. Set $F_0 = \lambda$ as the preliminary estimate.

2. Initialize the iteration counter to zero.
3. Update the eccentric longitude estimate using Newton-Raphson:

$$\Delta F_i = \frac{F_i + h \cos F_i - k \sin F_i - \lambda}{1 - h \sin F_i - k \cos F_i}$$

$$F_{i+1} = F_i - \Delta F_i$$

4. Increment the iteration counter.
5. Perform an exit check:
   a. Converged if $|\Delta F_i| \leq \epsilon$ with ϵ representing a pre-configured convergence criteria greater than floating point precision.
   b. Exit criteria is met if converged of a pre-determined maximum iteration counter (20 iterations for example) is met.
6. Repeat steps 3 through 6 until exit criteria is met.

Equation 3.78 is used to compute b. Equation 3.83 computes the trajectory's mean motion (n).

$$n = \sqrt{\mu a^3} \qquad\qquad 3.83$$

Equations 3.84 and 3.85 compute the sine and cosine of the true longitude (L).

$$\sin L = \frac{(1 - bk^2) \sin F + bhk \cos F - h}{1 - h \sin F - k \cos F} \qquad\qquad 3.84$$

$$\cos L = \frac{(1 - bh^2) \cos F + bhk \sin F - k}{1 - h \sin F - k \cos F} \qquad\qquad 3.85$$

Equation 3.86 computes the magnitude of the position vector.

$$r = \frac{1 - h \sin F - k \cos F}{\alpha} \qquad\qquad 3.86$$

Equations 3.87 and 3.88 compute the equinoctial position components.

$$r_f = r \cos L \qquad\qquad 3.87$$

$$r_g = r \sin L \qquad\qquad 3.88$$

Equations 3.89 and 3.90 compute the equinoctial velocity components.

$$v_f = \frac{n(h + \sin L)}{\alpha \sqrt{1 - h^2 - k^2}} \qquad\qquad 3.89$$

$$v_g = \frac{n(k + \cos L)}{\alpha\sqrt{1 - h^2 - k^2}}$$

3.90

Equations 3.91 and 3.92 compute the position and velocity components.

$$\bar{r} = r_f\hat{f} + r_g\hat{g}$$

3.91

$$\bar{v} = v_f\hat{f} + v_g\hat{g}$$

3.92

3.7 Date and Time Representation

Accurate and unambiguous timekeeping is critical for space flight dynamics. While terrestrial timekeeping has it legacy from tracking the Sun throughout the day, a general-purpose time system is needed when expanding to space. In space, terrestrial time zones lose relevance. Likewise, a month-based calendar also loses relevance when leaving a terrestrial reference frame.

Ultimately, it is desirable to have a calendar and time reckoning that is readily added to or subtracted from for both relative times and past and future times. However, the new system must also be convertible to and from the terrestrial system.

3.7.1 Julian Dates [Level I – Descriptive]

Julian dates are a continuous count of days since 0.5 January 4713 Before Common Era (BCE). Julian dates have an astronomical origin and thus solar upper transit (i.e., local noon) was the observable against which astronomers could observe the length of a day. Hence the 0.5 fractional day (or noon) representing the beginning of the Julian day.

All times of day are represented as fractional days, offset from the noon datum. The beginning of the day is referenced to noon at the Greenwich meridian. Thus, all dates and times flow as linear increments of days and fractional days.

➢ **Transition**: *You may continue this section with Julian Dates at Level II, or you may skip to the beginning of the next section called Day of Week and Year.*

A Julian Day Number (JDN) is an integer date count that begins at noon of any given day. This may be converted to or from calendar (i.e., year, month, day) dates. The computation process uses a two-month offset to place February as the last month, since February has a variable number of days. Hence the use of the number 14 in the month computations. Equations 3.93 and 3.94 convert a year (Y), month (M), day (D) date format to a JDN. The $\lfloor \ \rfloor$ brackets indicate the use of a greatest integer function.

$$M' = \left\lfloor \frac{M - 14}{12} \right\rfloor \tag{3.93}$$

$$
\begin{aligned}
JDN = & \left\lfloor \frac{1461(Y + 4800 + M')}{4} \right\rfloor + \left\lfloor \frac{367(M - 2 - 12M')}{12} \right\rfloor \\
& + \left\lfloor \frac{3 \left\lfloor \frac{Y + 4900 + M'}{100} \right\rfloor}{4} \right\rfloor + D - 32075
\end{aligned}
\tag{3.94}
$$

The JDN for 1 January 2000 is 2451545. Equations 3.95 through 3.104 recover the calendar date from the JDN.

$$L = JDN + 68569 \tag{3.95}$$

$$N = \left\lfloor \frac{4 \times L}{146097} \right\rfloor \tag{3.96}$$

$$L = L - \left\lfloor \frac{146097 \times N + 3}{4} \right\rfloor \tag{3.97}$$

$$I = \left\lfloor \frac{4000(L + 1)}{1461001} \right\rfloor \tag{3.98}$$

$$L = L - \left\lfloor \frac{1461 \times I}{4} \right\rfloor + 31 \tag{3.99}$$

$$J = \left\lfloor \frac{80 \times L}{2447} \right\rfloor \tag{3.100}$$

$$D = L - \left\lfloor \frac{2447 \times J}{80} \right\rfloor \tag{3.101}$$

$$L = \left\lfloor \frac{J}{11} \right\rfloor \qquad\qquad 3.102$$

$$M = J + 2 - 12 \times L \qquad\qquad 3.103$$

$$Y = 100(N - 49) + I + L \qquad\qquad 3.104$$

3.7.3 Modified Julian Date (MJD)

The Modified Julian Date (MJD) is a continuous count of days since midnight on 17 November 1858. This date was chosen since it corresponds to a 2400000.5 Julian Date, allowing a reduction in the number of digits needed to express whole days. The net effect is more computer precision is preserved to express time as the fraction of the day. Equation 3.105 computes the MJD from the Julian Date (JD).

$$MJD = JD - 2400000.5 \qquad\qquad 3.105$$

3.7.4 Day of Week and Year

The sequential Day of Year (DOY) is a continuous integer day count within the year ranging from 1 to 365 (in a non-leap year) or to 366 (in a leap year). Day 1 corresponds to January 1 and day 365 is December 31 (in a non-leap year). Table 3-3 provides the number of days for each month in the modern Gregorian calendar, as well as the DOY *prior to* the beginning of the month (i.e., DOY_0). Thus, to compute the DOY corresponding to any date the day of the month is added to the DOY_0 value for the month.

Table 3-3 Day Counts by Month			
Month	Number of Days	DOY_0 (non-leap year)	DOY_0 (leap year)
January	31	0	0
February	28/29	31	31
March	31	59	60
April	30	90	91
May	31	120	121
June	30	151	152
July	31	181	182
August	31	212	213
September	30	243	244
October	31	273	274
November	30	304	305
December	31	334	335

The correct DOY_0 value depends on whether the year is a leap year since February has 29 days in a leap year. A *leap year* is any year that is evenly divisible by 4, *except* for centennial years that are *not* evenly divisible by 400.

The Day of Week (DOW) is an integer count of each day with Sunday designated as the first day (or day 1) and Saturday designated as the last day (or day 7). Equation 3.106 Computes the DOW from the MJD.

$$DOW = \lfloor \text{mod}(MJD - 4, 7) \rfloor \qquad\qquad 3.106$$

The DOW computation relies on the modulo (or remainder) and the greatest integer (or floor) functions.

Key Terms:

> **Day of Year**: is the sequential day count (1 to 366) within the year with day 1 corresponding to 1 January.
>
> **Leap Year**: is a year evenly divisible by 4 except for centennial years that are not evenly divisible by 400. A leap day is an extra day added to February (i.e., the 29th) during leap years.
>
> **Modified Julian Date**: is a continuous integer count of days since midnight on 17 November 1858 (i.e., Julian Date 2400000.5).
>
> **Julian Date**: is a continuous count of days since 0.5 January 4713 BCE.

References

1. Vallado, David A., *Fundamentals of Astrodynamics and Applications*, Second Edition, © 2001 by author, Kluwer Academic Publishers, ISBN 1-881883-12-4.

2. Bate, Roger R. et al., *Fundamentals of Astrodynamics*, © 1971 Dover Publications, Inc., ISBN 0-486-60061-0.

3. Curtis, Howard D., *Orbital Mechanics for Engineering Students*, Fourth Edition, © 2020 Elsevier Ltd., ISBN 978-0-08-102133-0.

4. Broucke, R.A. and Cefola, P.J., *On the Equinoctial Elements*, Celestial Mechanics 5 (1972) 303-310, © 1972 D. Reidel Publishing Company.

5. Danielson, D.A. et al., *Semianalytic Satellite Theory*, Mathematics Department, Naval Postgraduate School, Monterey, CA 93943.

6. Urban, Sean E., and Seidelmann, P. Kenneth, *Explanatory Supplement to the Astronomical Almanac*, 3rd Edition, 2012 University Science Books, ISBN 978-1-891389-85-6.

7. Gurfil, Pini and Seidelmann, P. Kenneth, *Celestial Mechanics and Astrodynamics: Theory and Practice*, © 2016 Springer-Verlag, ISBN 978-3-662-50368-3.

8. CCSDS, *Recommendation for Space Data Systems Time Code Formats*, CCSDS 301.0-B-4, November 2010 (Blue Book) Consultative Committee for Space Data Systems (CCSDS), https://public.ccsds.org.

4 Rocket Dynamics

Rocket propulsion is a momentum exchange that embodies Newton's action-reaction law. A rocket motor uses chemical energy to produce hot, high-pressure gas that is expanded and accelerated through a nozzle. The equal and opposite momentum exchange between the vehicle and the exhaust gas is represented as the action of force we know as *thrust*.

4.1 Thrust Equation

Thrust is generated by pressure in the combustion chamber and nozzle. The pressure causes an expulsion of combustion by-product gases as exhaust. The exhaust gas momentum is the product of the net mass of the expelled molecules and the exhaust velocity.

The throat is the small area between the combustion chamber and the nozzle. The small diameter acts as a choke, causing a supersonic jet in which the exhaust gas is dramatically accelerated. The nozzle pressure decreases as the nozzle exit area increases. Exhaust jet static pressure depends on the ratio of throat area to nozzle exit area. Perfect nozzle expansion occurs when nozzle exit pressure equals ambient (atmospheric) pressure. For this case, shown in figure 4-1, the exhaust momentum is directed only in the axial direction, resulting in the maximum possible thrust.

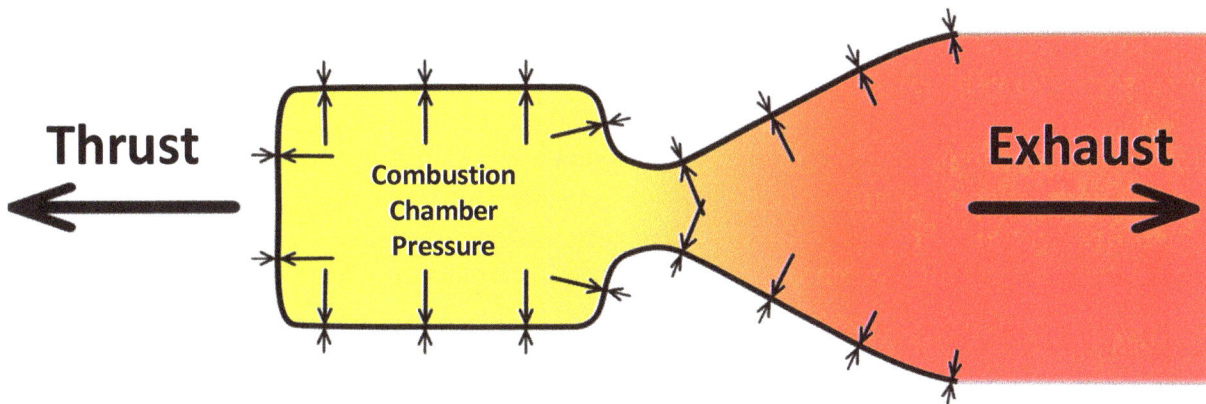

Figure 4-1 Rocket Combustion Chamber and Nozzle

Figure 4-2 illustrates exhaust nozzle pressure higher than ambient pressure on the left, and lower than ambient pressure on the right. When the pressure is higher than ambient, the exhaust plume expands at the exit, leading to an outward momentum component. Likewise, when the pressure is lower than ambient, the exhaust plume expands at the exit, leading to an inward momentum component.

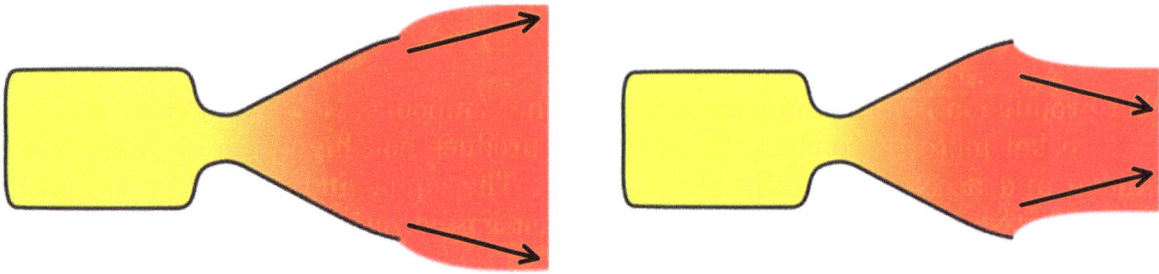

Figure 4-2 Exhaust Pressure Higher or Lower than Ambient Pressure

Given a circular nozzle, the net outward or inward momentum components cancel, leaving only the net axial component. For either case, the non-axial momentum components are wasted, and the remaining net axial component provides a lower thrust value.

4.1.1   Rocket Thrust [Level I – Descriptive]

Rocket thrust is the sum of two terms: *momentum thrust,* and *pressure thrust.* The momentum thrust is an application of Newton's Third Law, which embodies conservation of momentum. Momentum thrust is quantified as the product of the propellant mass flow rate and the exhaust gas velocity. Pressure thrust is the force generated by exhaust pressure against an atmosphere. Pressure thrust is the nozzle exit cross sectional area multiplied by the difference between the exhaust pressure and the ambient pressure. There is no pressure thrust in a vacuum.

Thrust is the time rate of change of the vehicle's momentum. A change made to the vehicle's momentum is called *impulse.* Thus, thrust applied over time imparts an impulse to the vehicle.

Momentum is the product of mass and velocity. The thrusting action results with a simultaneous decrease in the vehicle's mass and increase to its velocity, neither of which are trivial. This results in an increasing acceleration over time if thrust is constant. As a result, the change in velocity over time is not linear.

Key Terms:

> **Impulse**: is the change to the vehicle's momentum resulting from thrust applied over time.
> **Momentum Thrust**: is force generated from an action/reaction momentum exchange.
> **Pressure Thrust**: is force generated by exhaust pressure against an atmosphere.
> **Thrust**: is the force on a vehicle resulting from mass displacement. Thrust is the sum of momentum and pressure thrusts.

➢ **Transition**: *You may continue this section with Rocket Thrust at Level II, or you may skip to the beginning of the next section called Rocket Performance.*

4.1.2 Rocket Thrust [Level II – Equations]

Propellants are the reactive substances that produce pressure in the combustion chamber. A propellant's *specific impulse* (I_{sp}) is a measure of its impulse per unit weight; the weight being measured at Earth surface. As such, I_{sp} has units of seconds. Equation 4.1 relates I_{sp} to thrust (F_t), exhaust mass flow rate ($\dot{m}_e$), and Earth surface gravitational acceleration ($g_0 = 9.80665 \, m/s^2$).

$$I_{sp} = \frac{F_t}{\dot{m}_e g_0} \qquad\qquad 4.1$$

Equation 4.2 is the rocket equation which computes a vehicle's speed change (Δv) for a thrust event. The speed change is a function of I_{sp}, g_0, initial mass (m_0), and mass change (Δm_p) due to exhausted propulsion by-products.

The vehicle's mass following the thrust event (m_f) is the mass change subtracted from its initial mass ($m_f = m_0 - \Delta m_p$). Thus, Δm_p is also the propellant mass consumed in the thrust event.

$$\Delta v = I_{sp} g_0 ln\left(\frac{m_0}{m_0 - \Delta m_p}\right) = I_{sp} g_0 ln\left(\frac{m_0}{m_f}\right) \qquad\qquad 4.2$$

If the propellant mass flow rate is constant, the thrust event time may be computed as $\Delta t = \Delta m_p / \dot{m}_e$. Equation 4.3 computes the vehicle's post thrust event mass by algebraic manipulation of the rocket equation.

$$m_f = m_0 e^{-\left(\frac{\Delta v}{I_{sp} g_0}\right)} \qquad\qquad 4.3$$

Equation 4.4 computes the propellant consumed by the thrust event.

$$\Delta m_p = m_0 \left[1 - e^{-\left(\frac{\Delta v}{I_{sp} g_0}\right)}\right] \qquad\qquad 4.4$$

Equations 4.5 and 4.6 compute the thrust event time in terms of different available parameters.

$$\Delta t = \frac{m_0}{\dot{m}_e}\left[1 - e^{-\left(\frac{\Delta v}{I_{sp} g_0}\right)}\right] \qquad\qquad 4.5$$

119

$$\Delta t = \frac{I_{sp} g_0 m_0}{F_t} \left[1 - e^{-\left(\frac{\Delta v}{I_{sp} g_0}\right)} \right] \qquad\qquad 4.6$$

Key Term:

> **Specific Impulse**: is the ratio of a propellant's impulse to its equivalent weight at the Earth's surface.

> ➤ **Transition**: *You may continue this section with Rocket Thrust at Level II, or you may skip to the beginning of the next section called Rocket Performance.*

4.1.3   Rocket Thrust [Level III – Derivation]

Thrust generation can be evaluated using one-dimensional analysis. Ambient pressure (p_a) is the pressure of the environment surrounding the thruster. The nozzle exit pressure (p_e) is the exhaust gas pressure at the nozzle exit.

At any time (t), the vehicle has a mass m and a speed v. The process in the combustion chamber generates pressure, forcing the combustion by-products to exit. A small quantity of mass is expelled (δm) in a short time interval (δt) at absolute speed v_e, out of the nozzle with a cross-sectional exhaust area (A_e). This causes a net zero momentum exchange, resulting in a small speed change (δv) to the vehicle. The vehicle's momentum change is equal and opposite to the net external impulse. Equation 4.7 quantifies the momentum exchange in accordance with Newton's Second law.

$$[(m - \delta m)(v + \delta v) - \delta m v_e] - mv = (p_e - p_a) A_e \delta t \qquad\qquad 4.7$$

The mass flow rate $(\dot{m}_e)$ is the amount of mass crossing the exhaust nozzle area over time. The vehicle's time rate of change of mass has an equal and opposite magnitude to the mass flow rate.

$$\frac{dm}{dt} = -\dot{m}_e \qquad\qquad 4.8$$

Equation 4.9 provides the vehicle's mass as a function of time for a constant mass flow rate $(\dot{m}_e)$. Equation 4.10 quantifies the mass expelled in a short time interval.

$$m(t) = m_0 - \dot{m}_e t \qquad\qquad 4.9$$

$$\delta m = \dot{m}_e \delta t \qquad\qquad 4.10$$

Substituting equation 4.10 into equation 4.7 yields:

$$(m - \dot{m}_e \delta t)(v + \delta v) - \dot{m}_e \delta t v_e - mv = (p_e - p_a)A_e \delta t \qquad 4.11$$

$$m\delta v - \dot{m}_e \delta t(v + \delta v) - \dot{m}_e \delta t v_e = (p_e - p_a)A_e \delta t \qquad 4.12$$

Equation 4.13 results after dividing by δt and taking the limit as $\delta t \to 0$.

$$m\frac{dv}{dt} - \dot{m}_e(v + v_e) = (p_e - p_a)A_e \qquad 4.13$$

Equation 4.14 is the exhaust speed relative to the vehicle (c_a). (Note that the $\dot{m}_e \delta t \delta v$ term is negligible as $\delta t \to 0$.)

$$v_{rel} = v + v_e \qquad 4.14$$

$$m\frac{dv}{dt} - \dot{m}_e v_{rel} = (p_e - p_a)A_e \qquad 4.15$$

Equation 4.16 is the resulting sum of momentum and pressure thrust.

$$m\frac{dv}{dt} = \dot{m}_e v_{rel} + (p_e - p_a)A_e = F_t \qquad 4.16$$

After rearranging terms, equation 4.17 introduces the parameter the effective exhaust speed, denoted as v_{eff} in equation 4.18.

$$F_t = \dot{m}_e \left(v_{rel} + \frac{(p_e - p_a)A_e}{\dot{m}_e} \right) \qquad 4.17$$

$$F_t = v_{eff}\dot{m}_e \qquad 4.18$$

Using the I_{sp} definition in equation 4.1, the effective exhaust velocity is:

$$v_{eff} = \frac{F_t}{\dot{m}_e} = I_{sp}g_0 \qquad 4.19$$

Using the definition of acceleration from Newton's Second Law:

$$a = \frac{dv}{dt} = \frac{F_t}{m(t)} = \frac{I_{sp}g_0\dot{m}_e}{m(t)} \qquad 4.20$$

Integrating both sides by dt and recognizing that $\dot{m}_e = dm/dt$ provides:

$$\int_0^{\Delta t} \frac{dv}{dt} dt = I_{sp} g_0 \int_0^t \frac{dm}{dt} \frac{1}{m(t)} dt \qquad \text{4.21}$$

$$v(\Delta t) - v(0) = I_{sp} g_0 \int_0^{\Delta t} \frac{dm}{m(t)} \qquad \text{4.22}$$

$$\Delta v = c(\ln[m(\Delta t)] - \ln[m(0)]) \qquad \text{4.23}$$

The initial mass (at $t = 0$) is defined as m_0 and the final mass (at $t = \Delta t$) is defined as m_f. Therefore, equation 4.24 is the vehicle's velocity change which verifies equation 4.2.

$$\Delta v = I_{sp} g_0 \ln\left(\frac{m_0}{m_f}\right) \qquad \text{4.24}$$

4.2   Rocket Performance

The aspects of rocket performance to be evaluated are the traditional vertical booster ascent to orbit, practical considerations for mounting thrusters on spacecraft, propellant tank modeling, and thruster pulse modeling. These topics provide the insight needed for launching to orbit and changing the trajectory characteristics using the propulsion system.

4.2.1   Ascent to Orbit [Level I - Descriptive]

The traditional launch to orbit from a celestial body begins with what is initially a vertical ascent. The ascent profile, illustrated in figure 4-3 must climb to the orbital altitude while establishing the intrack (and possibly radial) velocity components at the end of the profile.

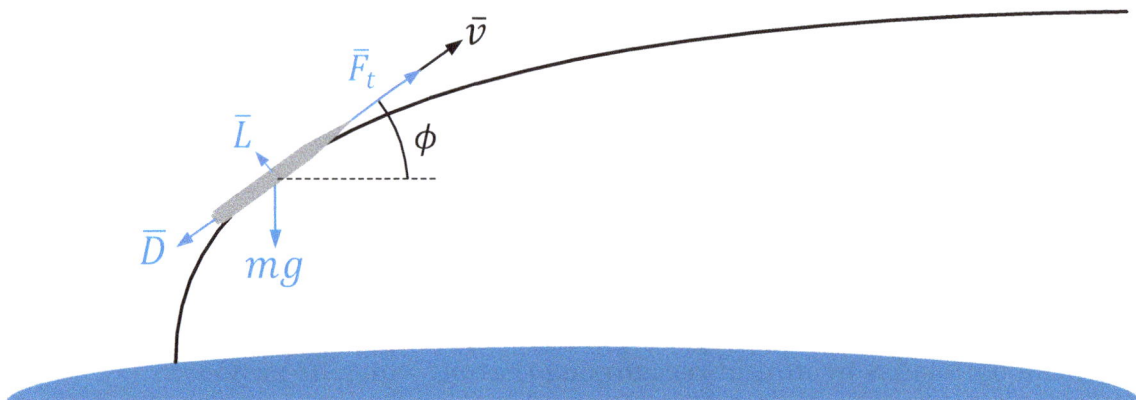

Figure 4-3 Vertical Launch Orbit Ascent Profile

Shortly after liftoff, the booster performs a pre-programmed roll maneuver and sets up a *pitching profile* that serves two purposes. The first purpose is to establish the launch azimuth that, coupled with the launch site declination, determines the trajectory's inclination. The second purpose is to allow the radially downward gravitational force (mg) to curve the ascent trajectory in the same way gravity curves a free flight trajectory. The steered pitch angle (ϕ) is selected to produce a *gravity turn* that establishes the desired intrack and radial velocity components at booster cutoff.

The axial forces are primarily along the launch booster's long (or longitudinal) axis. These include thrust (F_t) resulting from the momentum exchange that occurs when propellant is expelled. Opposing the thrust is *drag* (D), which occurs when launching from a celestial body with an atmosphere.

The drag force is associated with *dynamic pressure* (q), which varies rapidly during the early ascent timeline. The rapid change occurs because speed is increasing while atmospheric density is decreasing as the vehicle climbs in altitude. Since these factors change in opposite directions, the dynamic pressure reaches its maximum early in the ascent profile when the difference between its aerodynamic and ambient atmospheric pressure is maximum. Launch boosters may throttle back to ensure the maximum dynamic pressure is within their structural limits.

Aerodynamic *lift* (L) is force nearly perpendicular to the longitudinal axis that is also only applicable to launching within an atmosphere. This force is based on the booster's aerodynamic characteristics such as shape and the characteristics of the air flow around the vehicle.

Key Terms:

Drag: is a force imparted on a vehicle traveling through an atmosphere that opposes the direction of travel. Drag is the product of dynamic pressure, cross-sectional area along the flight direction, and a dimensionless drag coefficient that describes how streamlined is the vehicle.

Dynamic Pressure: is the pressure imparted on the booster due to travel through an atmosphere. Dynamic pressure is proportional to atmospheric density and the square of the booster's speed relative to the atmosphere.

Gravity Turn: is the curvature to an ascent profile caused by the downward gravitational acceleration on an off-vertical trajectory.

Lift: is a force imparted on a vehicle traveling through an atmosphere perpendicular to the direction of travel. Lift generally results from an asymmetric atmospheric gas flow around the vehicle.

Pitching Profile: is an initial change to the booster's pitch angle from vertical that establishes the launch azimuth and sets up the gravity turn to the launch profile.

➢ **Transition**: *You may continue this section with Ascent to Orbit at Level II, or you may skip to the beginning of the next section called Thrust Directive Effects.*

4.2.2   Ascent to Orbit [Level II – Equations]

The pitching profile begins shortly after liftoff, temporarily misaligning the thrust vector (F_t) with velocity (v). The misalignment angle between thrust and velocity is the *angle of attack* (α). Establishing a non-zero pitch angle sets the launch azimuth that determines the resulting trajectory's inclination angle. Equation 4.25 is the relationship between launch azimuth (Az), launch side declination (δ) and inclination (i).

$$\cos i = \cos \delta \sin Az \qquad\qquad 4.25$$

Except for a due east launch, there are two azimuth values that achieve the same inclination – one ascending and one descending. However, launch sites have restricted ranges of acceptable azimuths due to range safety constraints. The acceptable azimuths avoid populated areas and are typically over the ocean. For example, the azimuth limits for the Kennedy Space Center on Florida's east coast are in the 35° to 120° range.

It should be noted that performing the pitching off vertical early is important because the booster is easiest to turn while its speed is slow. This is particularly important when launching with an atmosphere, while the lateral drag components are within the booster's structural limits.

The forces are modeled with scalar equations, with directionality implied in figure 4-3 to facilitate conceptualization. Equation 4.26 provides the time rate of change of velocity.

$$\frac{dv}{dt} = \frac{F_t \cos \alpha - D}{m} - g \sin \phi \qquad\qquad 4.26$$

The first acceleration term is the net axial force (thrust minus drag), divided by the vehicle's instantaneous mass. Note that the thrust value is reduced by the cosine of the angle of attack, which is nominally zero after the initial pitching.

The second term is the gravitational acceleration which curves the ascent trajectory, inducing a gravity turn that persists throughout the ascent profile. Equation 4.27 is the product of velocity and the time rate of change of the pitch angle (ϕ).

$$v \frac{d\phi}{dt} = \frac{F_t \sin \alpha + L}{m} - \left(g - \frac{v^2}{r} \right) \cos \phi \qquad 4.27$$

Equation 4.28 is the time rate of change of *downrange distance* the vehicle flies during the ascent profile.

$$\frac{ds}{dt} = \frac{R}{r} v \cos \phi \qquad 4.28$$

Equation 4.29 is the time rate of change of radius, which is also the time rate of change of altitude.

$$\frac{dr}{dt} = \frac{dh}{dt} = v \sin \phi \qquad 4.29$$

Equation 4.30 recognizes that the vehicle mass in equations 4.26 and 4.27 is not constant, but instead is reduced during the powered flight phase by the propellant mass flow rate.

$$\frac{dm}{dt} = -\dot{m}(t) \qquad 4.30$$

Equation 4.31 shows the dynamic pressure (q) is a function of atmospheric density (ρ) and the square of the vehicle's speed (v^2) relative to the atmosphere. Equations 4.32 and 4.33 are the drag and lift forces, which directly depend on dynamic pressure.

$$q = \frac{1}{2} \rho v^2 \qquad 4.31$$

$$D = \frac{1}{2} \rho v^2 A C_D \qquad 4.32$$

$$L = \frac{1}{2} \rho v^2 A C_L \qquad 4.33$$

The forces in equations 4.32 and 4.33 are also proportional to the cross-sectional area (A) normal to the velocity and the two dimensionless force coefficients. The drag coefficient (C_D) reflects how streamlined is the portion of the vehicle exposed to the aerodynamic forces. The lift coefficient (C_L) similarly scales the lifting force.

As with the mass, the atmospheric density varies significantly over the full ascent profile. The atmospheric density decreases with altitude consistent with the *hydrostatic equation*, which states that the atmospheric pressure per unit area at any altitude is due to the weight of the column of air above that altitude. The pressure experiences an exponential density decrease with altitude.

Gravitational acceleration also decreases with altitude over the full ascent profile but changes far less than either the mass or atmospheric density. The variation nevertheless must be included to rigorously compute the ascent profile.

It should be noted that there is no closed-form analytic solution to determine a launch booster's ascent profile from the above equations. Instead, these equations are solved numerically by modeling them as a second order differential equation.

Key Terms:

Angle of Attack: is the misalignment angle between thrust and velocity, where the relative atmospheric gas flow opposes velocity.
Downrange Distance: is the surface distance from the launch site for the vehicle at any point in the ascent profile.
Hydrostatic Equation: states that the atmospheric pressure per unit area at any altitude is due to the weight of the column of atmospheric gas above that altitude.

4.2.3   Thrust Directive Effects

To this point, thrust has been presumed to have a purely translational effect on the vehicle. This is a common circumstance relying on the net thrust being directed through the vehicle's center-of-mass. When the net thrust direction is offset from the center-of-mass, a rotational force (or torque) is introduced. This section evaluates the net effect thruster placement has on vehicle motion.

4.2.3.1   Translational and Rotational Thrust Effects [Level I – Descriptive]

The effective thrust vector is the sum of all thrust forces, which may be distributed over multiple thrusters. The resultant force vector provides the direction for the instantaneous thrust acceleration, which provides an instantaneous translational (or linear) change to the velocity.

If the thrust vector is directed through the vehicle's center-of-mass, there are no thrust-induced steering forces. However, if the thrust vector has its direction offset from the path through the center-of-mass, there will be a thrust-induced torque on the vehicle. If there is no counter balancing torque applied, the vehicle will experience a thrust-induced rotational effect.

➤ **Transition**: *You may continue this section with Translational and Rotational Thrust Effects at Level II, or you may skip to the beginning of the next section called Propellant Tank Characteristics.*

4.2.3.2 Translational and Rotational Thrust Effects [Level II – Equations]

The net (or effective) thrust is the vector sum of all thrust forces over all thrusters. Equation 4.34 provides the net thrust sum of the individual thrust forces ($\bar{F}_i$). The vehicle's center-of-mass is the most convenient datum for locating the individual thrust sources.

$$\bar{F}_{net} = \sum_i \bar{F}_i \qquad\qquad 4.34$$

Equation 4.35 is the thrust-induced torque ($\bar{\tau}_{net}$), computed from the individual thrust values ($\bar{F}_i$) and thruster locations ($\bar{\rho}_i$) relative to the vehicle center-of-mass.

$$\bar{\tau}_{net} = \sum_i (\bar{\rho}_i \times \bar{F}_i) \qquad\qquad 4.35$$

Note that a single thruster with the thrust directed through the vehicle's center of mass will have the location vector colinear with the thrust. This causes a zero cross product and thus zero torque. Similarly, thruster pairs (or groups) that have locations and thrust values arranged symmetrically about the center-of-mass have cancellation and thus zero net torque imparted on the vehicle.

It is also possible to arrange thrusters in which there is a zero translational acceleration, with two (or more) thrusters contributing constructively to the net torque. Such a configuration requires the result of equation 4.34 to equal zero. This obviously requires thrusters in opposing directions. For equation 4.35 to have a non-zero result, the opposing thrusters need to be complemented by having locations symmetrically about the center-of-mass. Figure 4-4 illustrates a thruster arrangement that imparts a complementary torque with zero net translational acceleration.

A thruster system intended to impart torques for vehicle attitude control and/or translational accelerations is commonly called a *Reaction Control System* (RCS).

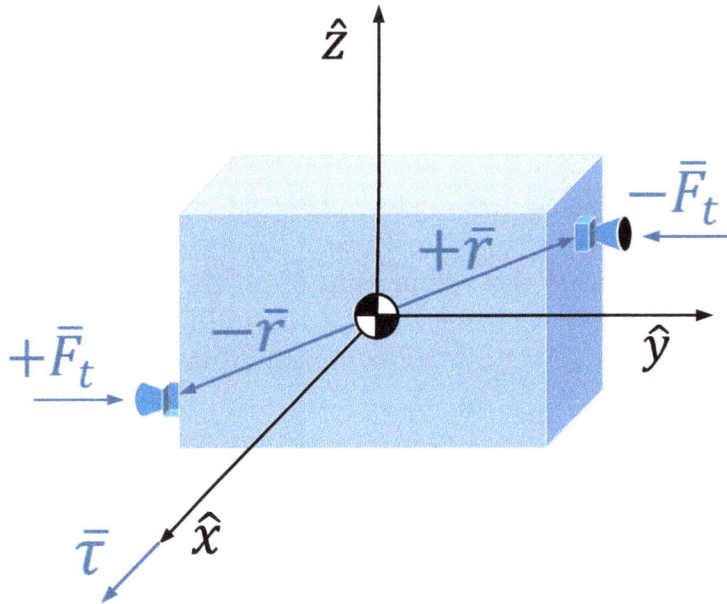

Figure 4-4 Thruster Induced Torque with Zero Translation

Key Term:

> **Reaction Control System (RCS)**: is a thruster system intended to impart torques on a vehicle for attitude control and/or impart translational accelerations.

4.2.4 Propellant Tank Characteristics

Propellant tank modeling allows the quantity of propellant remaining to be measured from telemetry parameters. Knowledge of the tank makeup, including size, shape, pressurant gas type, and method for inducing propellant flow to the thrusters permits modeling that facilitates thrust performance predictions.

4.2.4.1 Propellant Tank Modeling [Level I – Descriptive]

Propellant tanks are traditionally spherical or cylindrical with hemispherical endcaps as shown in figure 4-5. The tank radius and cylindrical length (if applicable) would be known characteristics. While a volume may be approximated from the geometry, the tank as manufactured will tend to deviate from the specified value. Moreover, there may be internal devices such as propellant bladders, diaphragms, pistons, or capillary vanes that direct the propellant out of the tank toward the thruster manifold(s) which will have known characteristics. Thus, a measured effective volume will also be needed as a tank characteristic.

Other known characteristics are the type of propellant and pressurant gas as well as the quantity of pressurant gas. A final characteristic is whether the tank is self-pressurized (i.e., a *blowdown system*) or whether it maintains a constant pressure from an external pressurant tank.

The self-pressurized (or blowdown system) has all the pressurant gas in the propellant tank. This is the simpler system with the penalty of tank pressure being highest at the beginning of the mission and decreasing as propellant is expended and the gas expands. While enough pressurant can be introduced to ensure workable pressure with a nearly empty tank, thruster performance (both thrust and specific impulse) varies with the feed pressure. Thus, thruster performance needs to be characterized over the full pressure range and thrust event planning (to get the desired impulse) becomes complicated.

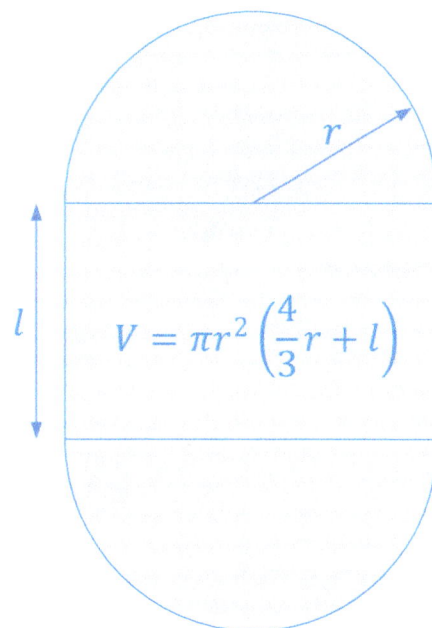

$$V = \pi r^2 \left(\frac{4}{3} r + l \right)$$

Figure 4-5 Cylindrical Tank with Hemispherical Endcaps

The *regulated system* is more convenient. This maintains a constant propellant tank pressure over the mission life. This is accomplished by an external tank that maintains propellant tank pressure through a regulator. While the hardware is more complex, the thrust event planning can rely on a steady thruster performance over the mission life.

Key Terms:

Blowdown System: is a self-pressurized propellant tank with pressure dropping as propellant is expelled and the pressurant gas expands.
Regulated System: is a propellant tank with constant pressure provided by an external tank supplying pressurant through a regulator.

➢ **Transition**: *You may continue this section with Propellant Tank Modeling at Level II, or you may skip to the beginning of the next section called Thrust Event Modeling.*

4.2.4.2   Propellant Tank Modeling [Level II – Equations]

The propellant quantity in a tank is measured indirectly using the ideal gas law, expressed in equation 4.36. The pressurant gas volume (V), or *ullage*, is a function of the number of mols (n) of pressurant gas, the universal gas constant (R), the pressurant absolute temperature (T) in Kelvins, and the gas pressure (P). The universal gas constant has a $R = 8.31446261815324\, J \cdot K^{-1}mol^{-1}$ value in SI units.

$$PV = nRT \qquad\qquad 4.36$$

The propellant volume is the difference between the measured effective tank volume and the pressurant volume. Equation 4.37 is the propellant mass (m_p) as a function of its volume and density (ρ).

$$m_p = \rho V \qquad\qquad 4.37$$

Propellant density can vary with temperature. For example, equation 4.38 is used to compute the density of Hydrazine (N_2H_4) as a function of temperature in degrees Celsius. Equation 4.39 computes the temperature in degrees Celsius from Kelvins.

$$\rho(T°\text{C}) = 1025.817 - 0.8742(T°\text{C}) - 0.0005(T°\text{C})^2 \; kg/m^3 \qquad 4.38$$

$$T°\text{C} = K - 273.15 \qquad\qquad 4.39$$

Table 4-1 lists the molecular masses of two common pressurant gases, namely Helium (He) and Nitrogen (N_2).

Table 4-1 Molecular Masses of Common Pressurant Gases	
Pressurant Gas	**Molecular Mass (kg/mol)**
Helium (He)	4.002602×10^{-3}
Nitrogen (N_2)	28.01340×10^{-3}

A significant propellant tank dynamic occurs when a thrust event has a high propellant mass flow rate. When propellant is expelled quickly from the tank, the pressurant gas expansion will be *isentropic* (i.e., taking place without an entropy change) and there will be a noticeable drop in pressurant temperature. This is an *adiabatic* expansion of the gas, which has the relationship expressed in equation

4.40 between the gas's absolute temperature (T), volume (V), and specific heat ratio (γ).

$$T \cdot V^{\gamma - 1} \equiv constant \qquad\qquad 4.40$$

Equation 4.41 computes the gas's final temperature (T_f) based on its initial temperature (T_i), initial volume (V_i), and final volume (V_f).

$$T_f = T_i \left(\frac{V_i}{V_f} \right)^{\gamma - 1} \qquad\qquad 4.41$$

Equation 4.41 indicates the expanding gas will experience isentropic cooling, which incidentally is the operating principle for refrigeration systems. Likewise, a gas being compressed will experience a temperature increase.

The specific heat ratio (γ) is the ratio of specific heat capacity (c_p) to molar heat capacity (c_v). The specific heat ratio is typically 1.67 for a monatomic gas such as Helium and approximately 1.40 for a diatomic gas such as Nitrogen. The specific heat ratios for these two gases are as follows:

- Helium: $\gamma = 1.667$
- Nitrogen: $\gamma = 1.400$

Key Term:

> **Adiabatic**: is a thermodynamic process that has no heat or mass is exchange.

4.2.5   Thrust Event Modeling

Thrust events are often treated as having discrete on and off demarcations in which full thrust is achieved instantaneously with the valve opening (i.e., on) and zero thrust occurring immediately with valve closing (i.e., off). This is obviously an approximation. There is a finite time interval in which pressure builds in the chamber as combustion reaches its peak level. The thrust value increases from zero toward the steady-state value as pressure builds. Similarly, the chamber pressure drops over a finite time interval after the valve closes and combustion erodes. The thrust level decays as chamber pressure drops.

4.2.5.1   Thruster Pulse Modeling [Level I – Descriptive]

A thrust event is modeled over a finite duration with three segments: rise, steady state, and decay. The rise and decay times generally have fixed durations, where the steady state is typically the major portion applying the thrusting impulse, at the rated thrust value. Figure 4-6 is a thruster pulse profile with the rise segment merging smoothly into the steady state segment. The steady state segment continues to thruster cutoff, which is caused by the valve closure. This results in an abrupt downward transition in the decay segment. The thrust value in the decay segment drops quickly until it begins to taper off toward zero. The area under the curve is the total impulse imparted to the vehicle.

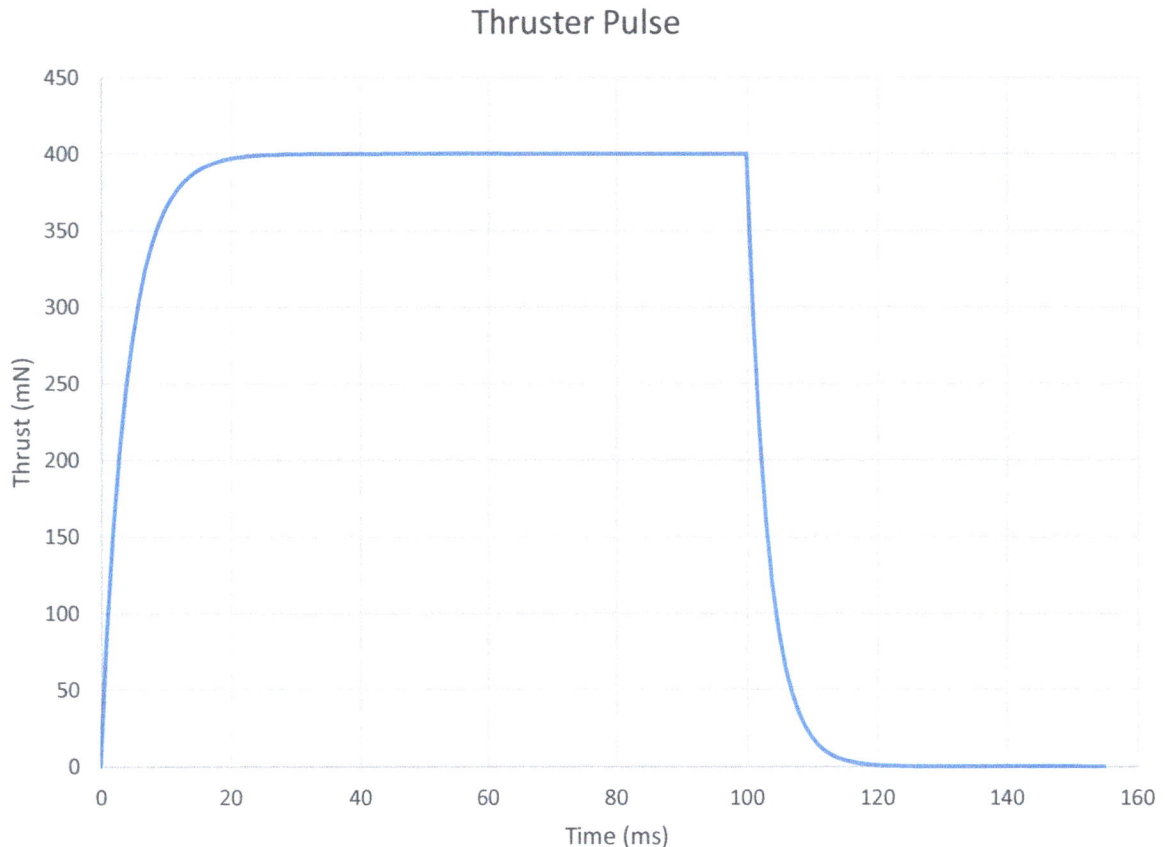

Figure 4-6 Thruster Pulse Profile

The rise and decay segments are only significant for short duration pulses. Long duration thrust events can be effectively approximated as a rectangular impulse.

With shorter thruster pulses the rise and decay segments are significant. An offset to the valve open and close times is chosen to center the impulse at the effective location in the trajectory where the velocity change should be applied. Characterization of the area under the rise and decay segments is needed to determine the necessary time offset.

➤ **Transition**: *You may continue this section with Thruster Pulse Modeling at Level II, or you may skip to the beginning of the next section called Chemical Propellant Properties.*

4.2.5.2 Thruster Pulse Modeling [Level II – Equations]

The thrust curve may be divided mathematically into the rise and decay segments. The steady state segment is the portion of the rise curve that asymptotically approaches the maximum thrust. Equation 4.42 characterizes the thrust versus time for the rise and steady state segments in terms of the maximum thrust (T_{max}), the rise curve time constant (τ_R), and the time since thruster valve opening (t_B), which is time since the beginning of burn.

$$T(t) = T_{max}\left(1 - e^{-\frac{t_B}{\tau_R}}\right) \qquad\qquad 4.42$$

Since the curve is smooth and asymptotically approaches the maximum thrust, the demarcation between the rise and steady state segments is subjective. The rise time constant (τ_R) characterizes the shape versus time.

When time equals the rise time constant, the thrust value has risen to approximately 63.2% of the maximum thrust. At twice the time constant, the thrust is approximately 86.5% of the maximum thrust. Table 4-2 characterizes the thrust value at various multiples of the rise time constant.

Table 4-2 Thruster Rise Time Fractions	
Time	**Fraction of Maximum Thrust**
$2\tau_R$	0.865
$2.2026\tau_R$	0.900
$2.9957\tau_R$	0.950
$4.6052\tau_R$	0.990
$6.9073\tau_R$	0.999

Thus, the steady state thrust segment is designated to begin somewhere between a factor of 2.2 and 6.9 times the rise time constant.

Equation 4.43 characterizes the thrust versus time for the decay segment in terms of the maximum thrust (T_{max}), the decay curve time constant (τ_D), and the time since thruster valve closure (t_B), which is also the burn time.

$$T(t) = T_{max}e^{-\frac{t_B}{\tau_D}} \qquad\qquad 4.43$$

Equation 4.44 is the total impulse imparted on the vehicle, computed as the integral of the thrust curve.

$$ I = T_{max} \left[t_B - \tau_R \left(1 - e^{-\frac{t_B}{\tau_R}} \right) + \tau_D \left(e^{-\frac{t_\infty - t_B}{\tau_D}} - 1 \right) \right] \qquad 4.44 $$

The t_∞ is the time the thruster reaches a zero thrust. Table 4-2 defines the practical limits, which should be designated as being between 2.2 and 6.9 times the decay time constant. Since the thrust at t_∞ is effectively zero, the total impulse is expressed more simply in equation 4.45.

$$ I = T_{max} \left[t_B - \tau_R \left(1 - e^{-\frac{t_B}{\tau_R}} \right) - \tau_D \left(1 - e^{-\frac{t_B}{\tau_R}} \right) \right] \qquad 4.45 $$

Equations 4.46 and 4.47 define the middle impulse times of the thrust curve rise and decay portions. The t_R parameter is the designated demarcation time between the rise and steady state segments.

$$ t_{Rmid} = \tau_R \, ln \left(\frac{1}{2} \right) \qquad 4.46 $$

$$ t_{Dmid} = t_B + \tau_D \, ln \left(\frac{1}{2} \right) \qquad 4.47 $$

Equation 4.48 is the time of the thrust impulse midpoint.

$$ t_{mid} = \tau_R \, ln \left(\frac{1}{2} \right) + \frac{1}{2} (t_B - t_R) + \tau_D \, ln \left(\frac{1}{2} \right) \qquad 4.48 $$

➤ **Transition**: *You may continue this section with Thruster Pulse Modeling at Level III, or you may skip to the beginning of the next section called Chemical Propellant Properties.*

4.2.5.3 Thruster Pulse Modeling [Level III – Derivation]

The total impulse for the thrust pulse is the integral of the thrust curve over the thrusting interval. Equations 4.49 through 4.51 integrate the thrust equations from zero to t_∞.

$$ I = T_n \left[\int_0^{t_B} \left(1 - e^{-t/\tau_R} \right) dt + \int_B^{t_\infty} e^{-(t - t_B)/\tau_D} dt \right] \qquad 4.49 $$

$$ I = T_n \left[t_B - \tau_R \left(1 - e^{-\frac{t_B}{\tau_R}} \right) + \tau_D \left(e^{-\frac{t_\infty - t_B}{\tau_D}} - 1 \right) \right] \qquad 4.50 $$

134

$$I = T_{max}\left[t_B - \tau_R\left(1 - e^{-\frac{t_B}{\tau_R}}\right) - \tau_D\left(1 - e^{-\frac{t_B}{\tau_R}}\right)\right] \qquad 4.51$$

The value for t_∞ is not particularly significant since it is the time where the thrust is zero. Thus, its effect is not present in equation 4.51.

The times for the half impulse for the rise and decay segments come from the integral of the exponential function in equation 4.52, which may be found in calculus textbooks or mathematical handbooks.

$$\int_0^\infty e^{-ax}dx = \frac{1}{a}e^{-ax} \qquad 4.52$$

The midpoint of the integral is based on the definition of a logarithm in equation 4.53. Equation 4.54 uses that definition to determine the exponent that produces the expression midpoint.

$$\log_e x = y \;\rightarrow\; e^y = x \qquad 4.53$$

$$\frac{1}{2a}e^{-ax} = \frac{1}{a}e^{-a\ln\left(\frac{1}{2}\right)} \qquad 4.54$$

4.2.6   Chemical Propellant Properties

There are numerous types of chemical fuels and oxidizers available. The choice of a fuel and oxidizer combination for a bipropellant system or just a fuel for a monopropellant system depends on the application. Applications vary from launch boosters to spacecraft maneuvering thrusters, RCS systems, and landing thrusters. Table 4-3 provides basic advantages and disadvantages of common propulsion system fuels.

Table 4-3 Liquid Fuel Characteristics		
Fuel	**Advantages**	**Disadvantages**
Hydrocarbon Fuels	• Good performance • Easy to handle	• Lower I_{sp} than liquid hydrogen
Liquid Hydrogen (LH$_2$)	• High performance (High I_{sp}) • Excellent regenerative coolant	• Low fuel density – requires bulky tanks • Requires well-insulated tanks • Not conducive to long-

		term storage
Hydrazine (N_2H_4)	• Reliable performance when compared with other fuels. • Storable – stable liquid • Useable as a monopropellant	• Highly reactive (store in high grades of stainless steel or aluminum • Highly toxic • Spontaneously combustible
Unsymmetrical Dimethyl Hydrazine (UDMH) $(CH_3)_2NNH_2$	• Storable • More stable than hydrazine • Lower freezing and higher boiling point than hydrazine	• Slightly lower I_{sp} than hydrazine • Same disadvantages as hydrazine
Monomethyl Hydrazine (MMH) CH_3NHNH_2	• Storable • Better shock resistance to blast waves • Better heat transfer properties • Lower freezing and higher boiling point than hydrazine • Small quantities (3 to 15%) added to hydrazine substantially quench explosive decomposition of N_2H_4	• lower I_{sp} than hydrazine (by 1% to 2%) • Same disadvantages as hydrazine

Monopropellant systems are simpler than bipropellant systems. Monopropellant systems only need a single fuel tank. However, they generally require a catalyst bed to cause the fuel to decompose to the heated exhaust gas. Catalyst beds usually operate most efficiently if they are warmed by electrical heaters or by a short sequence of small thruster pulses.

Bipropellant systems are more complicated because there is at least one fuel tank and at least one oxidizer tank, both with independent plumbing to the thrusters. The advantage to a bipropellant system is higher thrust and specific impulse than a corresponding monopropellant system. However, bipropellant systems require a selection and control of the fuel to oxidizer mixture ratio to get the optimal performance. Table 4-4 provides basic advantages and disadvantages of common propulsion system oxidizers.

Table 4-4 Liquid Oxidizer Characteristics		
Oxidizer	Advantages	Disadvantages
Liquid Oxygen (LO$_2$)	• Excellent performance with LH$_2$ and hydrocarbon and alcohol fuels • Safe handling and storage • Non-corrosive and non-toxic	• Not conducive to long-term storage
Liquid Fluorine	• Highest performance (High I$_{sp}$) • High specific gravity (1.5x)	• Extremely toxic • Extremely corrosive • Extremely reactive • Not conducive to long-term storage
Hydrogen Peroxide (H$_2$O$_2$)		• Storage stability problems • Severe skin burns
Nitric Acid (HNO$_3$)	• Spontaneous (hypergolic) ignition with many fuels	• Toxic fumes • Highly corrosive
Nitrogen Tetroxide (N$_2$O$_4$)	• Storable • High density / specific gravity • Hypergolic with many fuels	• Narrow temperature range

4.2.7 Electric Propulsion

Electric propulsion involves a class of thrusters (ion, arc jet, electro-thermal) that excite the propellant using electrical power. The electrical energy heats the exhaust gases and thus provides for extremely high exhaust velocities. The result is extremely high specific impulse (I_{SP}) due to the electrical energy boost.

In general, electric propulsion has low thrust. Thus, electric propulsion is typically characterized as long duration, low thrust events. The advantage of electric propulsion is the high specific impulse, resulting in low propellant mass requirements. The main disadvantage is the extremely high electrical power needed, thus limiting either the frequency in which thrust events may occur or other activities on the spacecraft requiring electrical power.

REFERENCES

1. Sutton, George P., *Rocket Propulsion Elements*, An Introduction to the Engineering of Rockets, Fifth Edition, © 1986 John Wiley & Sons, Inc., ISBN 0-471-80027-9.

2. Edberg, Don and Costa, Willie, *Design of Rockets and Space Launch Vehicles*, © 2020 by the authors, American Institute of Aeronautics and Astronautics (AIAA), ISBN 978-1-62410-593-7.

3. Curtis, Howard D., *Orbital Mechanics for Engineering Students*, Fourth Edition, © 2020 Elsevier Ltd., ISBN 978-0-08-102133-0.

4. Griffin, Michael D. and French, James R., *Space Vehicle Design*, Second Edition, © 2004 American Institute of Aeronautics and Astronautics (AIAA), ISBN 1-56347-539-1.

5. Brown, Charles D., *Elements of Spacecraft Design*, © 2002 American Institute of Aeronautics and Astronautics (AIAA), ISBN 1-56347-524-3.

6. Ley, Wilfried et al. (Editors), *Handbook of Space Technology*, © 2009 co-published by American Institute of Aeronautics and Astronautics (AIAA) and John Wiley & Sons, Ltd., ISBN 978-0-470-69739-9.

7. Gurfil, Pini and Seidelmann, P. Kenneth, *Celestial Mechanics and Astrodynamics: Theory and Practice*, © 2016 Springer-Verlag, ISBN 978-3-662-50368-3.

8. Sears, Francis W. and Zemansky, Mark W., *University Physics*, © 1949 Addison-Wessley Publishing Company, Inc., Library of Congress Catalog No. 55-5026.

9. Chemical Rubber Company, *CRC Handbook of Chemistry and Physics*, 64th Edition, © 1922 Chemical Rubber Company (CRC) Press, ISBN 0-8943-0464-4.

5 Orbital Maneuvering

Orbital maneuvering involves the use of translational thrust, imparting an impulse on a vehicle to change the trajectory characteristics in a controlled manner. The timing and thrust direction, plus the impulse changes the trajectory to achieve a desired result. Systematic methods are thus developed to achieve the desired trajectory outcome.

5.1 Impulsive versus Finite Duration Maneuvers

Maneuver planning involves the use of both *impulsive* and *finite duration maneuvers*. Impulsive maneuvers have an instantaneous application of impulse and thus have an instantaneous velocity change, referred to as *delta-V* (Δv). While no real thruster can impart an instantaneous Δv, the impulsive maneuver is an excellent approximation for maneuvers with a relatively high thrust and short burn duration. The impulsive maneuver is also a useful tool when planning finite duration maneuvers, since it provides a preliminary estimate using analytic methods.

Using finite duration maneuvers produces more realistic maneuver plans than ones that use impulsive maneuvers. Since the thrust events can be modeled in the same way the thruster accelerates the vehicle, the fidelity is higher than the impulsive maneuver's instantaneous Δv application. But this comes at the expense of lack of a closed-form solution. Finite duration maneuver plans are typically solved iteratively, with an impulsive maneuver used as a first approximation. Burn parameters are adjusted on each iteration until the trajectory targeting goals are achieved within the desired tolerances.

This chapter places emphasis on the analytics of the impulsive maneuver since these are solvable in closed form. This provides insight into what is needed to change the trajectory characteristics to the desired outcome. Higher fidelity maneuver planning involves numerical integration in which the maneuver thrust effects are modeled as an additional acceleration and the vehicle's mass is modeled as an additional state parameter.

Key Terms:

> **Delta-V**: is the effective velocity change made to a trajectory by a thrust impulse.
> **Finite Duration Maneuver**: is a maneuver event that applies impulse as a thrust-induced force over a discrete time interval.
> **Impulsive Maneuver**: is a maneuver approximation in which the impulse and thus the velocity change is imparted instantaneously.

5.1.1 Impulsive Maneuvers

Impulsive maneuvers are an idealized case. The position does not change during the impulse since the Δv is applied instantaneously. These conditions allow the delta-V to be determined analytically from the orbital geometry.

Since the impulse is applied instantaneously, the Δv can be computed using only two-body orbital equations. This provides a closed-form solution with the added benefit of allowing insight into how the desired orbital change is accomplished. This method is an accurate approximation for short duration, high thrust maneuvers as well as long duration very low thrust maneuvers.

5.1.2 Finite Duration Maneuvers

The finite duration maneuver is the realistic case since the thrust is applied as a force that is added to the environmental forces in the numerical integration process. Simultaneously, the mass is depleted at the appropriate rate, providing a high-fidelity acceleration profile. Thus, the spacecraft mass needs to be modeled as an additional state parameter in the integrator.

Maneuver plans are iterative processes in which the conditions are adjusted using an n-dimensional solver (such as a least-squares method for example). As previously indicated, the impulsive maneuver's $\Delta \bar{v}$ and effective time are used to develop the first approximation. The parameters that are typically adjusted include: ignition time, burn duration, and maneuver attitude.

Since the maneuver has a finite duration, its mid-Δv time should be centered on the impulsive maneuver's effective time. Equation 5.1 computes the time of the Δv midpoint (t_{mid}) using the rocket equation.

$$t_{mid} = \frac{m_0}{\dot{m}} \left[1 - e^{-\left(\frac{\Delta v}{2 I_{sp} \eta g_0} \right)} \right]$$

5.1

Equation 5-1 is a variant of equation 4-5 with half the exponent to determine the mid-Δv time. Equation 4-5 is used to compute the full burn time. The initial approximation for ignition time (t_{ig}) is half the burn time subtracted from the mid-Δv time.

5.2 Maneuver Coordinates

The maneuver velocity changes will be expressed in the UVW coordinate system, which is a.k.a. the RIC coordinate system. Referring to chapter 3, the $\hat{U}$ is the radial direction, $\hat{V}$ is in the intrack direction, and $\hat{W}$ is in the crosstrack direction. Three axis stabilized spacecraft generally accept thrust commands in the UVW or similar system such as velocity-normal-binormal (VNB) system. (The VNB aligns the V-axis with velocity and the N-Axis with angular momentum. The B-axis completes the orthonormal system.)

5.3 Intrack Maneuvers

Intrack maneuvers are the most common type of trajectory changes due to their versatility and efficiency. These maneuvers are generally intended to change the trajectory's semi-major axis with a consequential change to the eccentricity.

5.3.1 Intrack Orbital Maneuvers [Level I – Descriptive]

Figure 5-1 illustrates an example intrack maneuver that increases the semi-major axis of what is initially a circular orbit to an elliptical orbit. The thrust event (Δv) establishes what becomes the periapsis of the resulting elliptical orbit.

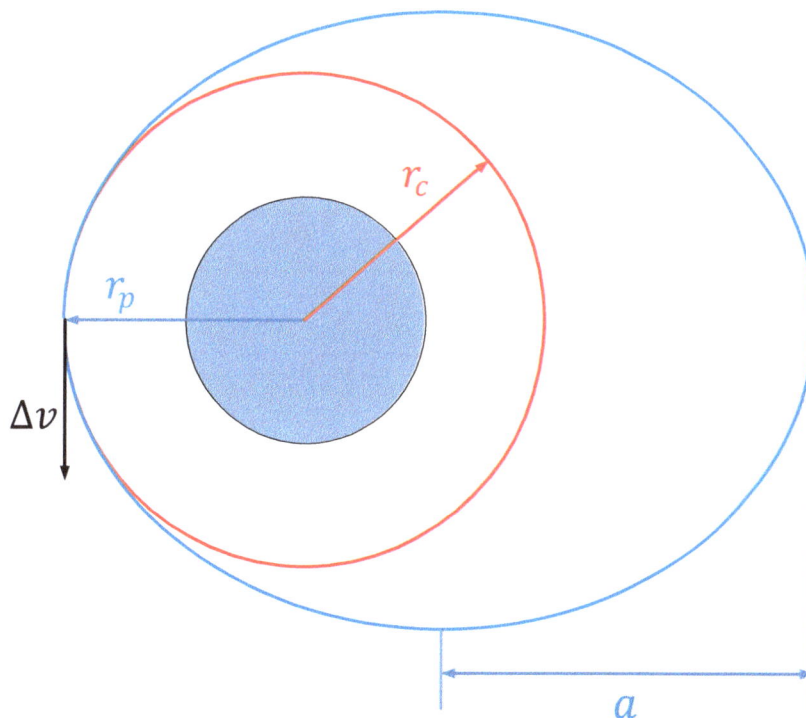

Figure 5-1 Semi-Major Axis Increase

➤ **Transition**: *You may continue this section with Intrack Orbital Maneuvers at Level II, or you may skip to the beginning of the next section called The Hohmann Transfer.*

For an intrack maneuver, the velocity change (Δv) is tangent to both the circular orbit and the resulting elliptical orbit's velocities. Hence it being in the intrack direction. Equation 5.2 is the Δv magnitude, which is the speed difference between the elliptical and circular orbits. It should be noted that the circular and periapsis radii are equal ($r_p = r_c$) and thus are simply designated as r.

$$\Delta v = \sqrt{\mu \left(\frac{2}{r} - \frac{1}{a} \right)} - \sqrt{\frac{\mu}{r}} \qquad\qquad 5.2$$

The intrack maneuver may also be used to circularize an elliptical orbit. Figure 5-2 illustrates an elliptical orbit being circularized at apoapsis. This is accomplished by an intrack speed increase, with both velocities in the intrack direction. The final speed must match what is needed for a circular orbit. Thus, the velocity change is the final circular speed minus the elliptical speed at apoapsis as provided in equation 5.3. Once again, since the circular and apoapsis radii are equal ($r_a = r_c$), they are designated simply as r.

$$\Delta v = \sqrt{\frac{\mu}{r}} - \sqrt{\mu \left(\frac{2}{r} - \frac{1}{a} \right)} \qquad\qquad 5.3$$

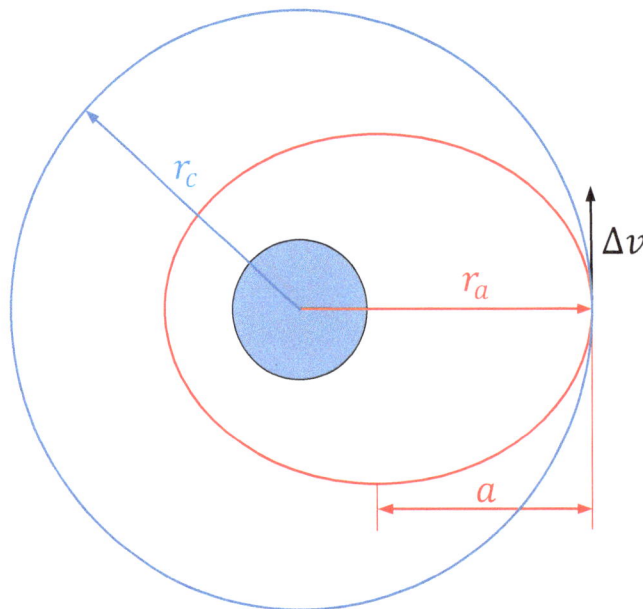

Figure 5-2 Elliptical Orbit Circularization

It should be noted that intrack maneuvers may also be performed in the negative intrack direction. In these instances, the Δv direction is reversed from what is shown in figures 5-1 and 5-2 and the identities of the initial and final orbits are reversed. Naturally, the signs on equations 5-2 and 5-3 would also be flipped.

5.4 The Hohmann Transfer

The Hohmann transfer, developed by German engineer Walter Hohmann (1880-1945) is a sequence of two intrack maneuvers separated by an odd multiple of 180-degree true anomaly intervals. The Hohmann transfer is used to transition between two concentric circular orbits.

5.4.1 Hohmann Transfers [Level I – Descriptive]

It is called a transfer because there are two distinct intrack maneuvers on either end of an intermediate cotangential transfer orbit. The three orbits involved are:
1. The initial circular or *parking orbit*.
2. The elliptical *transfer orbit*.
3. The target circular or *final orbit*.

Figure 5-3 illustrates a Hohmann transfer that begins with a low parking orbit and finishes with a high final orbit. The sequence pieces together the maneuver shown in figure 5-1 with the maneuver in figure 5-2. The first maneuver that transitions from the low circular to the elliptical transfer orbit is referred to as the periapsis boost. The second maneuver that transitions from the elliptical transfer orbit to the high circular orbit is referred to as apogee kick maneuver. The apogee kick maneuver occurs after an odd number of transfer orbit half revolutions.

Key Terms:

> **Final Orbit**: is the target orbit in an orbital transfer.
> **Hohmann Transfer**: is a maneuver sequence used to transition between circular orbits of different radii by use of a cotangential elliptical transfer orbit.
> **Parking Orbit**: is the initial circular orbit in an orbital transfer.
> **Transfer Orbit**: is an elliptical orbit tangent to circular orbits of different radii.

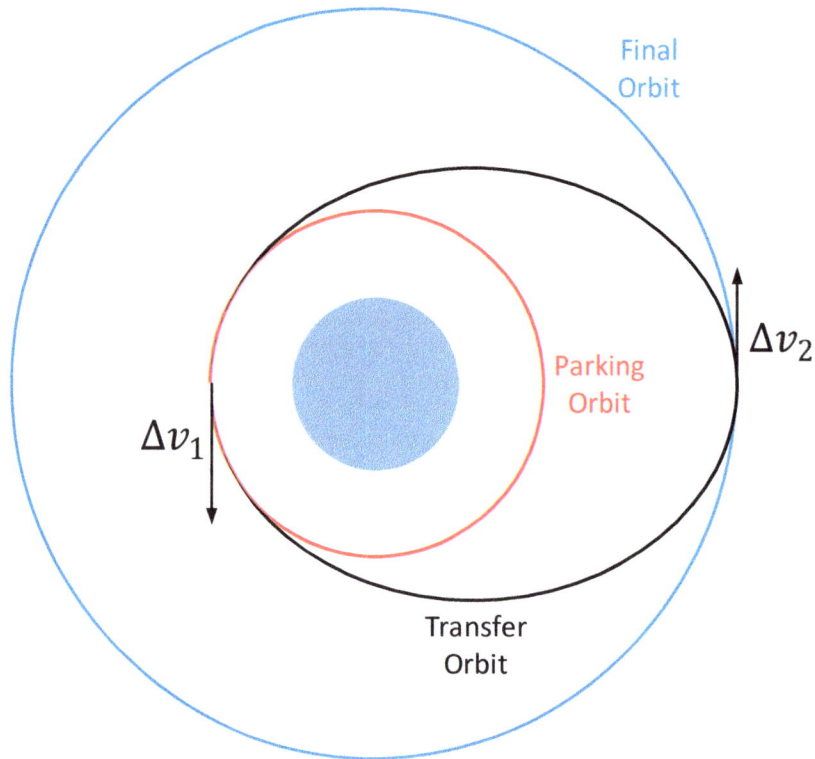

Figure 5-3 Hohmann Transfer

➢ **Transition**: *You may continue this section with The Hohmann Transfer at Level II, or you may skip to the beginning of the next section called The Bielliptic Transfer.*

5.4.2   Hohmann Transfers [Level II – Equations]

Equation 5.4 is the delta-V (Δv_1) for the first maneuver of a Hohmann transfer that increases orbital radius from an initial value (r_i) to a final orbital radius (r_f). Equation 5.5 is the delta-V (Δv_2) for the second burn of this Hohmann Transfer. Equation 5.6 provides he transfer orbit semi-major axis (a).

$$\Delta v_1 = \sqrt{\mu\left(\frac{2}{r_i} - \frac{1}{a}\right)} - \sqrt{\frac{\mu}{r_i}} \qquad\qquad 5.4$$

$$\Delta v_2 = \sqrt{\frac{\mu}{r_f}} - \sqrt{\mu\left(\frac{2}{r_f} - \frac{1}{a}\right)} \qquad\qquad 5.5$$

$$a = \frac{1}{2}\left(r_i + r_f\right) \qquad\qquad 5.6$$

144

As with any intrack maneuver sequence, the Hohmann Transfer may also be performed with retrograde (negative intrack) thrust events. This alternative transfers from a high circular orbit to one with a smaller radius.

5.5   The Bielliptic Transfer

A bielliptic transfer is an alternative method of changing from one circular radius to another. The traditional bielliptic transfer is more efficient than the Hohmann transfer when the ratio between the initial and final orbital radii is large (i.e., $r_{high} > 11.94\, r_{low}$).

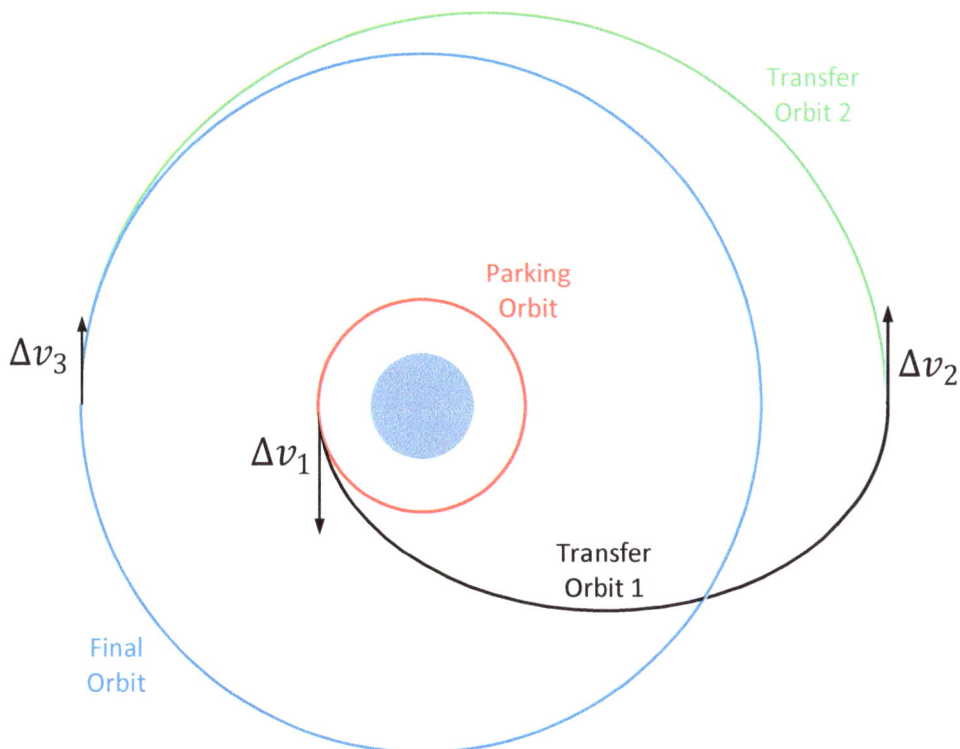

Figure 5-4 Bielliptic Transfer

5.5.1   Bielliptic Transfers [Level I – Descriptive]

The bielliptic transfer consists of two elliptical transfer orbits that are cotangential with the high and low circular orbits and themselves. The transition between the circular and elliptical orbits is made by three intrack maneuvers at the apses of the transfer orbits. Thus, the first maneuver causes a transition from the circular parking orbit to the first elliptical transfer orbit. The apoapsis of the first transfer orbit is higher than final circular orbit radius. The second maneuver occurs at the first transfer orbit's apoapsis. This raises the second transfer orbit's periapsis to

145

the final circular orbit radius. The third maneuver occurs at the second transfer orbit's periapsis. This is a retrograde burn that lowers the second transfer orbit's apoapsis altitude to the final circular radius.

Key Term:

> **Bielliptic Transfer**: is a maneuver sequence used to transition between circular orbits of different radii by use of two sequential cotangential elliptical transfer orbits.

> ➢ **Transition**: *You may continue this section with The Bielliptic Transfer at Level II, or you may skip to the beginning of the next section called The Bielliptic Transfer.*

5.5.2 Bielliptic Transfers [Level II – Equations]

Equation 5.7 is the delta-V (Δv_1) for the first maneuver of a Bielliptic Transfer that increases orbital radius from an initial value (r_i) to a final orbital radius (r_f). Equation 5.8 is the delta-V (Δv_2) for the second burn and equation 5.9 is the delta-V (Δv_3) of the third burn of this Bielliptic Transfer. Equations 5.10 and 5.11 provide the transfer orbit semi-major axes ($a_n, n = 1,2$).

$$\Delta v_1 = \sqrt{\mu \left(\frac{2}{r_i} - \frac{1}{a_1} \right)} - \sqrt{\frac{\mu}{r_i}} \qquad 5.7$$

$$\Delta v_2 = \sqrt{\mu \left(\frac{2}{r_a} - \frac{1}{a_2} \right)} - \sqrt{\mu \left(\frac{2}{r_a} - \frac{1}{a_1} \right)} \qquad 5.8$$

$$\Delta v_3 = \sqrt{\frac{\mu}{r_f}} - \sqrt{\mu \left(\frac{2}{r_f} - \frac{1}{a_2} \right)} \qquad 5.9$$

The apogee radius of the two transfer orbits is designated as r_a. Its value is typically determined by a numerical optimization process to achieve the minimum overall delta-V. Equation 5.10 is first transfer orbit's semi-major axis (a_1); equation 5.11 is the second transfer orbit's semi-major axis (a_2).

$$a_1 = \frac{1}{2} (r_i + r_a) \qquad 5.10$$

146

$$a_2 = \frac{1}{2}\left(r_a + r_f\right) \qquad\qquad 5.11$$

It should be noted that the delta-V savings from a bielliptic transfer are generally small compared with that required for a Hohmann transfer. However, the significant additional time needed for the bielliptic transfer often makes their use undesirable.

5.5.3 Step Up Bielliptic Transfers [Level I – Descriptive]

The bielliptic transfer may also be used to perform the equivalent of the Hohmann Transfer with the initial maneuver segmented into two smaller maneuvers. The sum of the two smaller maneuver delta-Vs is equivalent to the delta-V of the Hohmann Transfer's initial maneuver.

One application for the step up (or step down) bielliptic transfer is useful when the propulsion system lacks the impulse capability to accomplish the transfer. Another useful application is to reach the final orbit at a favorable time, such as when attempting to intercept and rendezvous with another spacecraft.

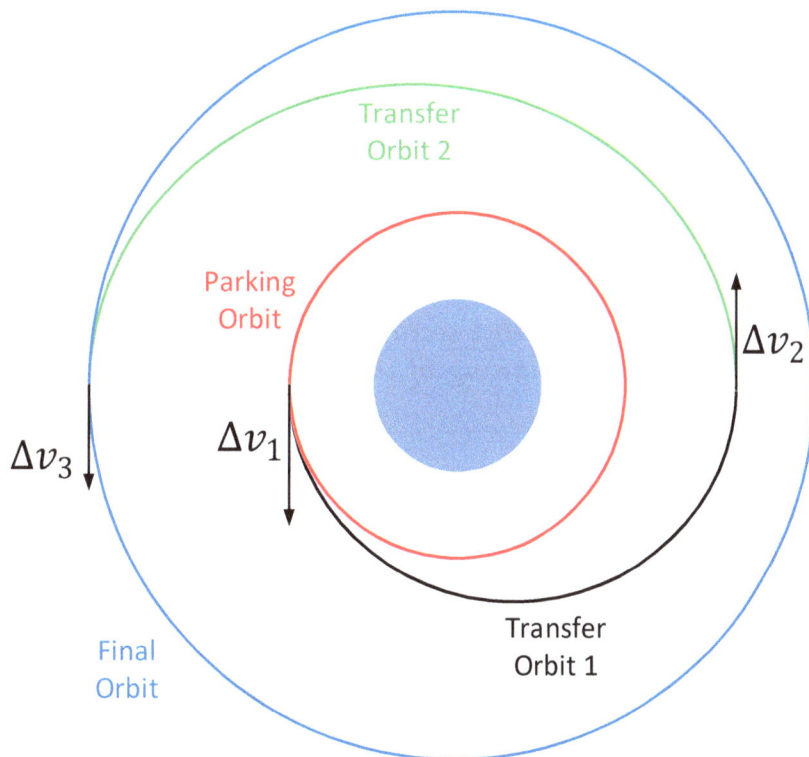

Figure 5-5 Step-Up Bielliptic Transfer

➢ **Transition**: *You may continue this section with The Step-Up Bielliptic Transfer at Level II, or you may skip to the beginning of the next section called The Eccentricity Change while Maintaining the Semi-Major Axis.*

5.5.4 Step Up Bielliptic Transfers [Level II – Equations]

The equations for the step up bielliptic transfer are like those for the traditional bielliptic transfer. But a significant difference is that while the traditional bielliptic transfer's third maneuver is retrograde, this maneuver is posigrade for the step up bielliptic transfer.

5.6 The Eccentricity Change while Maintaining Semi-Major Axis

As the name implies, an eccentricity chance alters the trajectory's eccentricity without changing its semi-major axis. There are several applications for this maneuver type, one of which provides a useful setup for rendezvous and proximity operations with another spacecraft.

5.6.1 Eccentricity Changes [Level I – Descriptive]

An eccentricity change that maintains the orbit's semi-major axis may only be performed at an orbital radius equal to the semi-major axis. The velocity direction is rotated, changing the flight path angle to that appropriate for the post maneuver eccentricity. Since the velocity magnitude does not change, the maneuver velocity change is the base of an isosceles triangle as seen in figure 5-6.

➢ **Transition**: *You may continue this section with The Eccentricity Change at Level II, or you may skip to the beginning of the next section called The Apsidal Line Rotation.*

5.6.1 Eccentricity Changes [Level II – Equations]

An eccentricity change maneuver requires knowledge of the target flight path angle (ϕ). Equations 5.11 and 5.12 provide a for the flight path angle when the orbital radius equals the semi-major axis.

$$\phi = \pm \cos^{-1}\left[\frac{|\bar{r} \times \bar{v}|}{rv}\right] = \pm \cos^{-1}\left[\frac{\sqrt{\mu a(1 - e^2)}}{rv}\right] \qquad 5.11$$

$$\phi = \pm \cos^{-1} \sqrt{1 - e^2} \qquad\qquad 5.12$$

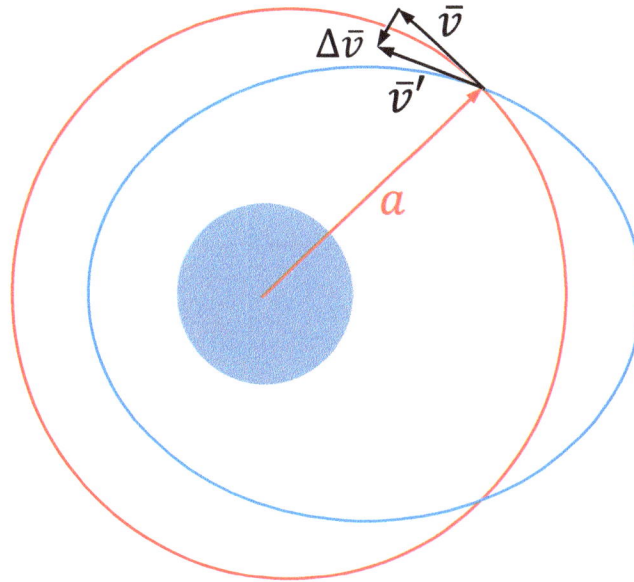

Figure 5-6 Eccentricity Change

Equation 5.13 expresses the orbital velocity ($\bar{v}$) in the UVW coordinate system. Equation 5.14 expresses the velocity change ($\Delta\bar{v}$) for the eccentricity change maneuver in the UVW coordinate system with ϕ and ϕ' as the pre-maneuver flight path angles, respectively.

$$\bar{v}_{UVW} = v \begin{bmatrix} \sin\phi \\ \cos\phi \\ 0 \end{bmatrix} \qquad\qquad 5.13$$

$$\Delta\bar{v}_{UVW} = v \begin{bmatrix} \sin\phi' - \sin\phi \\ \cos\phi' - \cos\phi \\ 0 \end{bmatrix} \qquad\qquad 5.14$$

➤ **Transition**: *You may continue this section with The Eccentricity Change at Level III, or you may skip to the beginning of the next section called The Apsidal Line Rotation.*

5.6.2 Eccentricity Change [Level III – Derivation]

Equation 5.11 derives from figure 1-8 and equations 1.33, 1.34, and 2.9. Equation 5.12 also substitutes the semi-major axis for the orbital radius with this maneuver type, plus equation 2-131 as an expression for the orbital speed. Thus, equation 5.15 is derived as follows:

$$\phi = \pm \cos^{-1} \left[\frac{\sqrt{\mu a(1 - e^2)}}{av} \right] \qquad\qquad 5.15$$

$$\phi = \pm \cos^{-1} \sqrt{\frac{\mu a(1 - e^2)}{a^2 v^2}} = \sqrt{\frac{\mu(1 - e^2)}{av^2}} \qquad\qquad 5.16$$

$$\phi = \pm \cos^{-1} \sqrt{\frac{\mu(1 - e^2)}{a\mu \left(\frac{2}{a} - \frac{1}{a} \right)}} \qquad\qquad 5.17$$

$$\phi = \pm \cos^{-1} \sqrt{1 - e^2} \qquad\qquad 5.18$$

5.7 Apsidal Line Rotation

An apsidal line rotation changes the orientation of the line of apsides without altering the trajectory's semi-major axis or eccentricity. The simplest apsidal line rotation is a single radial delta-V. The more complex (but more efficient) apsidal line rotation has two distinct delta-V events, with an intermediate trajectory that has an altered semi-major axis. However, the second of the two delta-V events restores the trajectory's original size.

5.7.1 Radial Apsidal Line Rotation [Level I – Descriptive]

The radial apsidal rotation employs a delta-V that flips/reverses the radial velocity component. This may be performed at any location in the trajectory through there will be zero effect when the radial velocity component is zero, such as at apoapsis or periapsis. The rotation angle for the apsidal line will be twice the true anomaly ($\Delta\omega = 2\theta$) of the delta-V location.

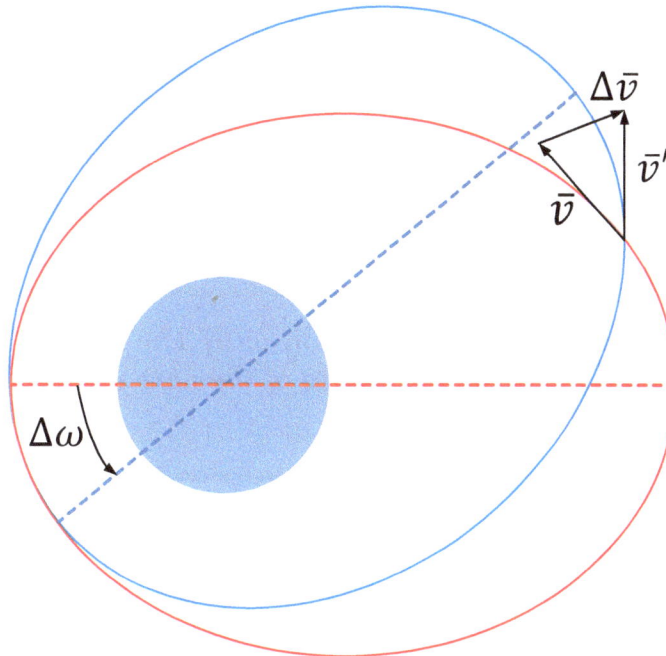

Figure 5-7 Radial Apsidal Line Rotation

➢ **Transition**: *You may continue this section with The Apsidal Line Rotation at Level II, or you may skip to the beginning of the next section called The Intrack Apsidal Line Rotation.*

5.7.2 Radial Apsidal Line Rotation [Level II – Equations]

Equation 5.19 expresses the velocity change for a radial apsidal line rotation maneuver. Equation 5.20 provides this delta-V in UVW coordinates and equation 5.21 provides the delta-V magnitude.

$$\Delta \bar{v} = -2v_r \hat{r} \qquad\qquad 5.19$$

$$\Delta \bar{v} = \begin{bmatrix} -2v \sin \phi \\ 0 \\ 0 \end{bmatrix} \qquad\qquad 5.20$$

$$\Delta v = 2e \sin \theta \sqrt{\frac{\mu}{a(1 - e^2)}} \qquad\qquad 5.21$$

Equations 5.22 and 5.23 relate the flight path angle (ϕ) to a given true anomaly (θ) as an alternative 2.121. Equation 5.24 is the relationship between the change to the argument of periapsis from the true anomaly of the maneuver location.

$$\sin \phi = \frac{e \sin \theta}{\sqrt{1 + e^2 + 2e \cos \theta}} \qquad\qquad 5.22$$

$$\cos \phi = \frac{1 + e \cos \theta}{\sqrt{1 + e^2 + 2e \cos \theta}} \qquad\qquad 5.23$$

$$\theta = -2\Delta\omega \qquad\qquad 5.24$$

➢ **Transition**: *You may continue this section with The Apsidal Line Rotation at Level III, or you may skip to the beginning of the next section called The Intrack Apsidal Line Rotation.*

5.7.3   Radial Apsidal Line Rotation [Level III – Derivation]

The elliptical orbital radius is described by equation 2.130 and the velocity is described by equation 2.131. Equation 2.123 provides an expression for the square of the angular momentum. These equations in combination, plus algebraic manipulation result in equation 5.28, which is an expression for speed as a function of eccentricity and true anomaly.

$$v^2 = \mu \left[\frac{2}{r} - \frac{1}{a}\right] \qquad\qquad 5.25$$

$$v^2 = \mu \left[\frac{2(1 + e \cos \theta)}{a(1 - e^2)} - \frac{1}{a}\right] \qquad\qquad 5.26$$

$$v^2 = \frac{\mu}{a(1 - e^2)}[1 + e^2 + 2e \cos \theta] \qquad\qquad 5.27$$

$$v = \frac{\mu}{h}\sqrt{1 + e^2 + 2e \cos \theta} \qquad\qquad 5.28$$

The radial velocity component in equation 2.120 is the product of speed and the sine of the flight path angle. Using the new velocity equation, equation 5.30 expresses the sine of the flight path angle in terms of eccentricity and true anomaly.

$$v_r = \frac{\mu}{h}e \sin \theta = v \sin \phi \qquad\qquad 5.29$$

$$\sin \phi = \frac{e \sin \theta}{\sqrt{1 + e^2 + 2e \cos \theta}} \qquad\qquad 5.30$$

Likewise, the intrack velocity component in equation 2.119 is the product of speed and the cosine of the flight path angle. Using the new velocity equation, equation

5.32 expresses the cosine of the flight path angle in terms of eccentricity and true anomaly.

$$v_\theta = \frac{\mu}{h}(1 + e \cos \theta) = v \cos \phi \qquad\qquad 5.31$$

$$\cos \phi = \frac{1 + e \cos \theta}{\sqrt{1 + e^2 + 2e \cos \theta}} \qquad\qquad 5.32$$

These results verify equations 5.22 and 5.23. Also note that equation 5.30 divided by equation 5.32 agrees with equation 2.121.

5.7.4   Intrack Apsidal Line Rotation [Level I – Descriptive]

The intrack apsidal line rotation is a sequence of two intrack delta-V events, performed at the same orbital radius. The first maneuver results in a transfer orbit with a different semi-major axis. The second maneuver occurs at the same orbital radius as the first maneuver. The second maneuver restores the original semi-major axis.

Low eccentricity orbits with relatively low delta-V values have the largest angular shift to the line of apsides when initiated near the 90° and 270° true anomalies. As either or both increases, the two optimum locations for these maneuvers shift toward apoapsis. The optimum maneuver true anomalies are the locations in which the desired intrack delta-V results in a change to the semi-major axis, but the eccentricity remains unchanged. But since the maneuver sequence intends to shift periapsis by a prescribed amount, there are two unknowns: the optimum maneuver true anomalies and the required delta-V.

Since the two unknown quantities are interdependent, the most obvious method to solve for the optimal solution is an iterative numerical process. However, a nearly optimal solution results from executing the maneuvers at the 90° and 270° true anomalies, with a total delta-V set to half of what is required to accomplish the equivalent apsidal rotation using the radial method. Thus, each of the two maneuvers uses approximately one quarter of the delta-V needed for a radial apsidal line rotation.

The intrack apsidal line rotation is particularly close to the optimum solution for nearly circular orbits and produces respectably optimal results for all but highly eccentric orbits. However, should an optimum solution be required, the nearly optimal solution provides an excellent starting point for an iterative numerical result.

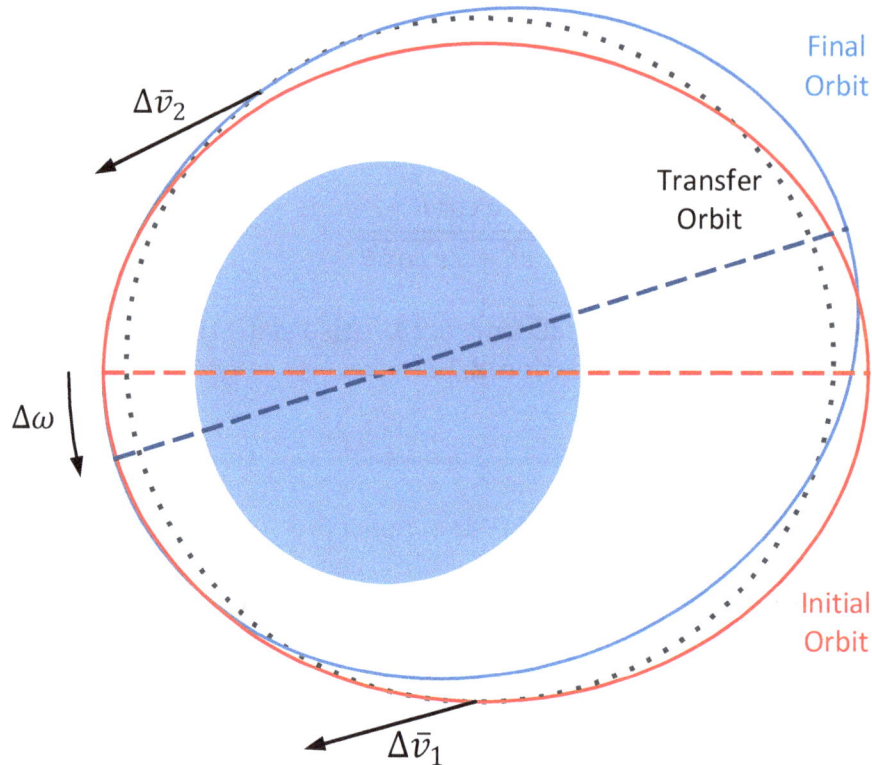

Figure 5-8 Intrack Apsidal Line Rotation

➢ **Transition**: *You may continue this section with The Intrack Apsidal Line Rotation at Level II, or you may skip to the beginning of the next section called Apsidal Line Rotation Method Comparison.*

5.7.5 Intrack Apsidal Line Rotation [Level II – Equations]

The nearly optimal intrack apsidal line rotation computes two velocity changes that are applied at 90° and 270° true anomalies associated with the initial and final orbits. The corresponding true anomalies for the transfer orbit are close in true anomaly to their 90° and 270° counterparts. The first maneuver may be at either the 90° or 270° true anomaly; the second maneuver is at the opposite true anomaly.

One of the maneuvers is posigrade and the other is retrograde. The posigrade or retrograde maneuver characteristics depend on whether the argument of periapsis increases (positive apsidal rotation: $\Delta\omega > 0$) or decreases (negative apsidal rotation: $\Delta\omega < 0$).

- For positive apsidal rotations, the first maneuver is retrograde, and the second maneuver is posigrade.
- For negative apsidal rotations, the first maneuver is posigrade, and the second maneuver is retrograde.

154

The first step is to evaluate the velocity change needed for the equivalent periapsis rotation ($\Delta\omega$), using the radial apsidal line rotation method. Equation 5.33 determines the radial burn true anomaly (θ_{rad}). Equation 5.34 is the delta-V needed for each of the intrack maneuvers.

$$\theta_{rad} = -2\Delta\omega \qquad\qquad 5.33$$

$$\Delta v = \pm \frac{\mu}{2h}(1 + e\cos\theta_{rad}) \qquad\qquad 5.34$$

The sign on the Δv is opposite the sign on $\Delta\omega$. The decision must be made whether the first maneuver is to occur at a $\theta = 90°$ or $\theta = 270°$ true anomaly. The first maneuver's position and velocity are computed using equations 2.130 and 2.131. The first maneuver's intrack and radial velocity components are also computed using equations 2.119 and 2.120. Equation 5.35 computes the magnitude of the pre-maneuver angular momentum (h).

$$h = rv\cos\phi \qquad\qquad 5.35$$

Equations 5.36 and 5.37 compute the post-maneuver (i.e., transfer orbit) in-track velocity and speed. Equation 5.38 is a relationship for the post-maneuver flight path angle.

$$v'_\theta = v_\theta + \Delta v \qquad\qquad 5.36$$

$$v' = \sqrt{\left(v'_\theta\right)^2 + v_r^2} \qquad\qquad 5.37$$

$$\cos\phi' = \frac{v'_\theta}{v'} \qquad\qquad 5.38$$

Equations 5.39, 5.40, and 5.41 compute the transfer orbit's semi-major axis, angular momentum magnitude, and eccentricity, respectively.

$$a' = \frac{1}{\left(\frac{2}{r} - \frac{(v')^2}{\mu}\right)} \qquad\qquad 5.39$$

$$h' = rv'_\theta \qquad\qquad 5.40$$

$$e' = \sqrt{1 - \frac{(h')^2}{\mu a'}} \qquad\qquad 5.41$$

Equation 5.42 is used to compute the transfer orbit's post maneuver true anomaly. The result is in the first two quadrants (i.e., $\theta' = \cos^{-1}(\cos\theta')$) if the flight path angle is positive and in the third or fourth quadrant (i.e., $\theta' = 2\pi - \cos^{-1}(\cos\theta')$) if the flight path angle is negative.

$$\cos\theta' = \frac{1}{e'}\left(\frac{a'[1-(e')^2]}{r} - 1\right) \qquad 5.42$$

The second maneuver occurs where the transfer orbit radius equals the orbital radius for the first maneuver. This will be near the remaining of the 90° or 270° true anomalies. The post maneuver (i.e., final) orbit will be exactly at the remaining of the 90° or 270° true anomalies. The applied delta-V is the same as was used for the first maneuver, but with the opposite sign.

5.7.6   Apsidal Line Rotation Method Comparison

The intrack track apsidal line rotation requires approximately half the delta-V to achieve the same rotation as its radial counterpart. Thus, the intrack option is the method of choice for orbital station keeping where a particular phase angle needs to be maintained against one or more other spacecraft or some other reference.

The radial option does have its place, however. This will be seen in the rendezvous and proximity operations section in the next chapter, where it is employed as a low-risk approach technique prior to terminal docking.

5.8   Intercept Targeting

Intercept is defined as starting from an initial position ($\bar{r}_0$) and meeting up with a target (or final) position ($\bar{r}_f$) at the same time the target is located at that position. This is commonly known as the Lambert problem (after Johann Heinrich Lambert who initially posed the problem).

Figure 5-9 illustrates the two positions with their respective times as well as the true anomaly difference between the positions. The Lambert problem determines the velocity ($\bar{v}_0$) needed at the initial position ($\bar{r}_0$) to establish a trajectory that intercepts the target position ($\bar{r}_f$) at the appropriate time.

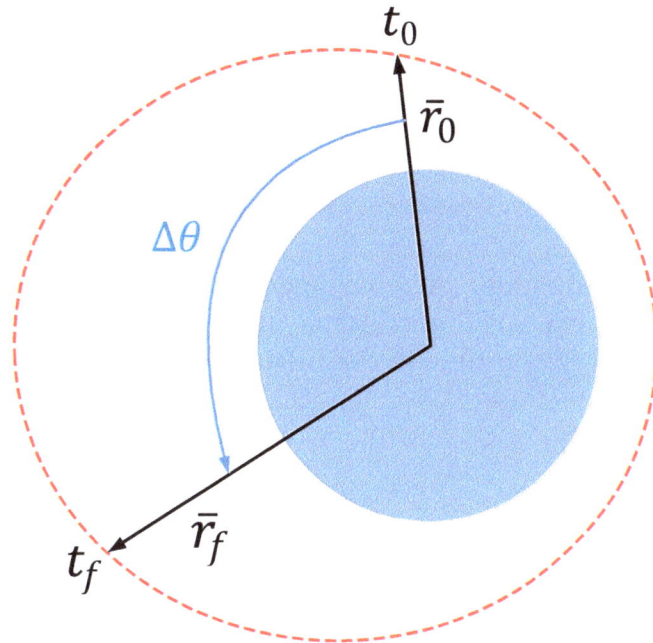

Figure 5-9 Lambert Intercept Targeting

Key Term:

Lambert Problem: determines the trajectory needed to transition from an initial position to a final position in a specified amount of time. This is also known as the Gauss problem.

5.8.1 Intercept using Universal Variables [Level I – Descriptive]

Most Lambert problem solutions use universal variables. This essentially involves an inversion of the orbital prediction for a time offset. The universal variables prediction problem requires the determination of three of the Lagrange coefficients. The intercept targeting takes the same approach, except the inputs are the initial position and the position after a time delay. The solution determines the velocity vector corresponding to the initial position. The velocity vector for the second position can also be computed from the determined Lagrange coefficients.

There are two solutions to the Lambert problem. They are the short path (i.e., true anomaly difference less than 180°) and the long path (i.e., true anomaly difference greater than 180°). Since the angular momentum direction is opposite for the two solutions, the preferred solution choice is usually dictated by the pre-existing orbital conditions.

> ➤ **Transition**: *You may continue this section with Intercept using Universal Variables at Level II, or you may skip to the beginning of the next section called Intercept by Hohmann or Bielliptic Timing.*

5.8.2 Intercept using Universal Variables [Level II – Equations]

This section provides a universal variables algorithm presented in Bate et al (1971). The intercept trajectory time-of-flight (Δt) is the difference between the time (t_0) of the target position (t_f) and the time of the initial position vector (t_0)

Equation 5.43 computes the cosine of the true anomaly difference between the two positions.

$$\cos \Delta\theta = \frac{\bar{r}_0 \cdot \bar{r}_f}{r_0 r_f} \qquad\qquad 5.43$$

Equation 5.44 computes the sine of the true anomaly difference. The sign is positive for the short path ($\Delta\theta < 180°$) and negative for the long path ($\Delta\theta > 180°$). Equation 5.45 is the relationship used to compute the true anomaly difference, using a two argument inverse tangent function.

$$\sin \Delta\theta = \pm \frac{\left| \bar{r}_0 \times \bar{r}_f \right|}{r_0 r_f} \qquad\qquad 5.44$$

$$\tan \Delta\theta = \frac{\left| \bar{r}_0 \times \bar{r}_f \right|}{\bar{r}_0 \cdot \bar{r}_f} \qquad\qquad 5.45$$

Equation 5.46 computes the A parameter. Equation 5.47 provides an initial guess for the z parameter, which is square of the eccentric anomaly.

$$A = \sqrt{\frac{r_0 r_f}{1 - \cos \Delta\theta}} \sin \Delta\theta \qquad\qquad 5.46$$

$$z = \Delta\theta^2 \qquad\qquad 5.47$$

Iterate in a loop until there are diminishing returns on the z value (or a maximum iteration count is reached). The maximum iteration count avoids an infinite loop, by ensuring loop exit.

1. Evaluate the Stumpff functions as infinite series using the current z estimate in equations 2.241 and 2.242 to estimate variables C and S. The infinite series representation is valid for all conic types.

2. Compute the y auxiliary variable.

$$y = r_0 + r_f - \frac{A(1 - zS)}{\sqrt{C}}$$

3. Compute the x variable.

$$x = \sqrt{\frac{y}{C}}$$

4. Compute the intercept trajectory time of flight (δt) corresponding to the current estimate.

$$\delta t = \frac{x^3 S + A\sqrt{y}}{\sqrt{\mu}}$$

5. Exit the loop if there is a small difference in absolute value between Δt and δt (i.e., convergence) or the loop counter reaches the maximum allowable iterations. Otherwise, refine the z estimate by Newton-Raphson step 6.

6. Compute the derivatives of C and S with respect to z using the derivatives of the equation 2.241 and 2.242 series.

$$C' = \frac{dC}{dz} = \frac{1}{4!} + \frac{2z}{6!} - \frac{3z^2}{8!} + \frac{4z^3}{10!} - \frac{5z^4}{12!} + \cdots$$

$$S' = \frac{dS}{dz} = \frac{1}{5!} + \frac{2z}{7!} - \frac{3z^2}{9!} + \frac{4z^3}{11!} - \frac{5z^4}{13!} + \cdots$$

7. Compute the derivative of the intercept time-of-flight with respect to z.

$$\frac{dt}{dz} = \frac{1}{\sqrt{\mu}}\left[x^3\left(S' - \frac{3SC'}{2C}\right) + \frac{A}{8}\left(\frac{3S\sqrt{y}}{C} + \frac{A}{x}\right)\right]$$

8. Refine the estimate for z.

$$z_{i+1} = z_i - \frac{(\delta t - \Delta t)}{\frac{dt}{dz}}$$

Equations 5.48 through 5.50 compute the applicable Lagrange coefficients.

$$f = 1 - \frac{y}{r_0}$$
5.48

$$g = A\sqrt{\frac{y}{\mu}}$$
5.49

$$\dot{g} = 1 - \frac{y}{r_f}$$
5.50

Equations 5.51 and 5.52 compute the velocities corresponding to the intercept trajectory's initial and final positions.

$$\bar{v}_0 = \frac{\bar{r}_f - f\bar{r}_0}{g}$$
5.51

$$\bar{v}_f = \frac{\dot{g}\bar{r}_f - \bar{r}_0}{g}$$
5.52

While the $\bar{r}_0$ and $\bar{v}_0$ are sufficient to define the intercept trajectory, knowledge of $\bar{v}_f$ is useful for comparison against a target's velocity at the $\bar{r}_f$ intercept position.

5.8.3   Intercept by Hohmann or Bielliptic Timing

The most efficient intercept method, in terms of minimal delta-V, involves cotangential ellipse transfers such as the Hohmann, step up bielliptic, or multi-elliptic transfers. These are excellent choices for rendezvous with a target in a nearly circular orbit since they facilitate setting up pre-docking rendezvous and proximity operations.

The Hohmann option is useful as an illustrative example but is often inconvenient when there is a time penalty needed to wait for favorable phasing. This is particularly acute for human spacecraft that must carry consumables for life support systems.

Fortunately, a series of multiple step-up cotangential maneuvers requires no more total delta-V than the equivalent Hohmann transfer. A well-planned sequence can change the phasing dynamics dramatically and thus significantly reduce the time to intercept.

Use of the Hohmann and step up bielliptic transfers to set up rendezvous and pre-docking proximity operations is addressed in the next chapter.

5.9   Plane Change Maneuvers

Plane change maneuvers tend to require the greatest amount of delta-V for a discernible change to the trajectory. This should not be surprising when one considers the speed required to orbit and that changing the trajectory plane requires a change to the angular momentum. As was seen from the first chapter, conservation of angular momentum is an extremely important physical law of orbital motion.

5.9.1   Plane Changes [Level I – Descriptive]

Changing a trajectory plane by even a small amount is a significant undertaking. Figure 5-10 illustrates a velocity change event in which the only change is the trajectory plane orientation. Changing only the planar orientation requires a redirection of the velocity and hence the angular momentum direction by the same amount.

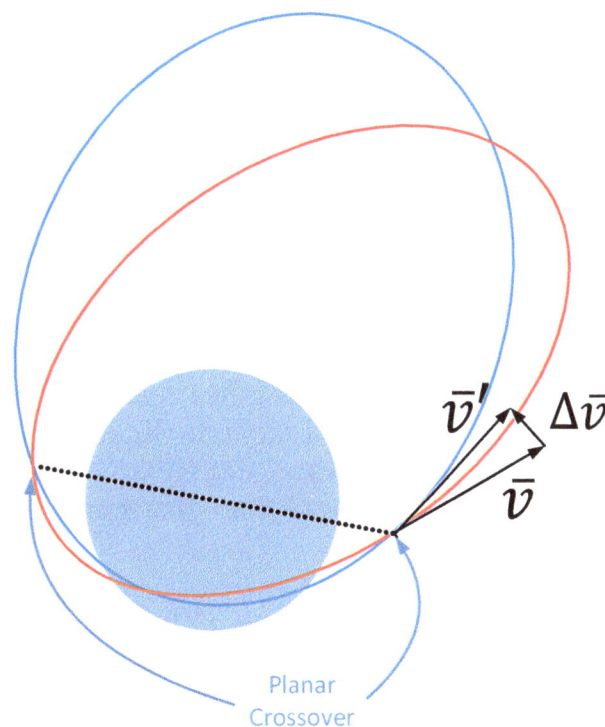

Figure 5-10 Plane Change Maneuver

Plane changes can only occur at the planar crossover "nodes" (i.e., intersections) of the pre- and post-maneuver trajectory planes. The Keplerian elements that orient the plane are the inclination (i) and RAAN (Ω). Maneuvers that change the inclination without affecting the RAAN may only be performed at the equator (i.e., ascending, or descending nodes). Maneuvers that change the RAAN by a small

161

angle without affecting the inclination may only be performed near the antinodes (i.e., the northern and southern latitude extremes). Maneuvers that change both the inclination and RAAN must consider the change to the angular momentum orientation that produces the desired values to both planar elements.

In operational practice, it is unusual to perform a plane change maneuver without making some other desired trajectory adjustments. The reason is because it is synergistic to combine in-plane and out-of-plane velocity changes into a single velocity change. Figure 5-11 illustrates the synergy with the combined velocity change as a resultant of the addition of two vectors.

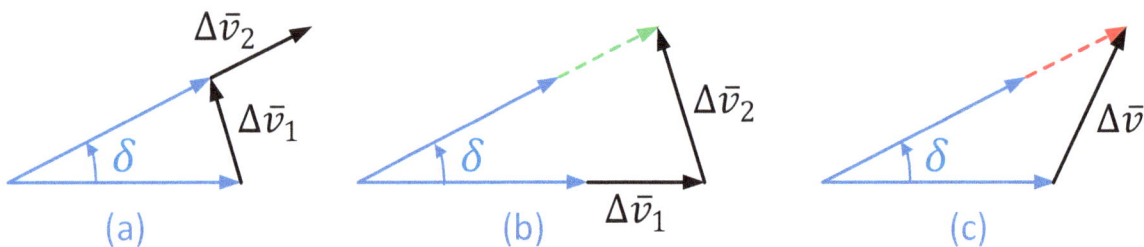

Figure 5-11 Combined Crosstrack and Intrack Maneuvers

While this principle can be proven mathematically, illustration in the figure by the three cases presented should be sufficient. Consider a combined maneuver consisting of an increase to the velocity magnitude as well as a plane change by the δ (dihedral) angle. For case (a) the plane change (velocity redirection) is performed first, followed by the increase in velocity. For case (b) the velocity magnitude increase is performed first, followed by the plane change. For case (c) the velocity magnitude increase is combined with the redirection into a single velocity change.

It should be apparent that the combined $\Delta\bar{v}$ in case (c) is less than the sum of $\Delta\bar{v}_1$ and $\Delta\bar{v}_2$ in either case (a) or case (b), if for no other reason than realizing that the shortest distance between two points in s a straight line. Nevertheless, it is also clear that adding the lengths of the separate maneuvers in cases (a) and (b) will inevitably require more of a velocity increase than the combined maneuver in case (c).

➢ **Transition**: *You may continue this section with Plane Changes at Level II, or you may skip to the beginning of the next chapter called Intercept, Rendezvous, Proximity Operations, and Docking.*

5.9.2   Plane Changes [Level II – Equations]

The mathematics of the combined velocity increase and redirection is a planar triangle solution. If v is the pre-maneuver speed, Δv_s is the magnitude of the speed change, and δ is the plane change dihedral angle, equation 5.53 computes the magnitude of the combined maneuver Δv using the law of cosines as is shown in figure 5-12.

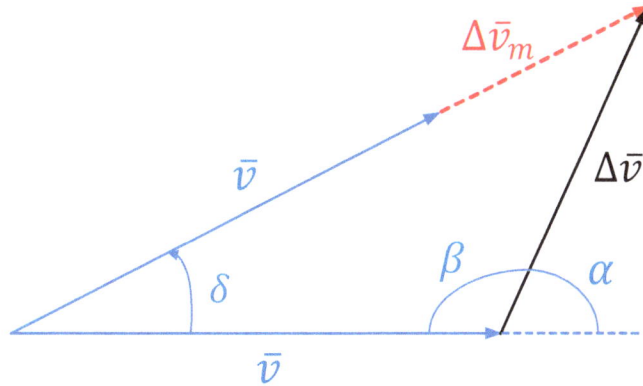

Figure 5-12 Combined Plane Change Maneuver Triangle

$$\Delta v = \sqrt{v^2 + (v + \Delta v_m)^2 - 2v(v + \Delta v_m) \cos \delta} \qquad 5.53$$

The angle (β) is computed using the law of sines in equation 5.54. The crosstrack steering angle (α) is provided in equation 5.55.

$$\sin \beta = \frac{\Delta v}{v + \Delta v_m} \sin \delta \qquad 5.54$$

$$\alpha = \pi - \beta \qquad 5.55$$

Equation 5.56 provides the combined intrack and crosstrack velocity change ($\Delta \bar{v}$) in Radial-Intrack-Crosstrack (RIC) coordinates.

$$\Delta \bar{v}_{RIC} = \Delta v \begin{bmatrix} 0 \\ \cos \alpha \\ \sin \alpha \end{bmatrix} \qquad 5.56$$

The locations where the plane change maneuver can be performed are illustrated in figure 5-13. The planar crossovers must be perpendicular to both trajectory planes' angular momentum vectors ($\bar{h}_1$, $\bar{h}_2$). Thus, the crossover node directions ($\hat{n}_1$, $\hat{n}_2$) may be computed by the vector cross product.

$$\hat{n}_1 = \frac{\bar{h}_1 \times \bar{h}_2}{|\bar{h}_1 \times \bar{h}_2|} \qquad 5.57$$

$$\hat{n}_2 = \frac{\bar{h}_2 \times \bar{h}_1}{|\bar{h}_2 \times \bar{h}_1|} \qquad\qquad 5.58$$

Equation 5.59 defines the celestocentric angular momentum direction in terms of the constituent Keplerian parameters. Naturally, angular momentum can also be determined as the cross product of position and velocity.

$$\hat{h} = \begin{bmatrix} \sin\Omega\sin i \\ -\cos\Omega\sin i \\ \cos i \end{bmatrix} \qquad\qquad 5.59$$

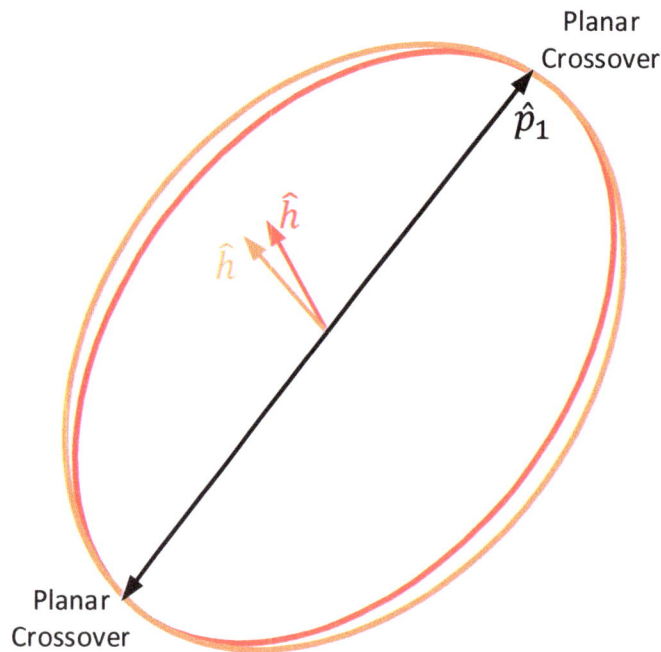

Figure 5-13 Planar Crossover Nodes

Executing a maneuver involving a plane change has the planar crossover constraint since the position vector needs to be in both the pre-and post-maneuver trajectory planes and thus perpendicular to both angular momentum directions. Likewise, the post-maneuver velocity vector must be in the post-maneuver trajectory plane.

Equation 5.60 relates the dihedral angle (δ) when the angular momentum of the pre- and post-maneuver trajectories ($\bar{h}, \bar{h}'$) are known in Cartesian form. Equation 5.61 uses the $\hat{w}$ row of the equation 3.28 transformation to relate the dihedral angle to the pre- (i, Ω) and post-maneuver (i', Ω') planar elements.

$$cos\delta = \frac{\bar{h} \cdot \bar{h}'}{hh'} \qquad\qquad 5.60$$

$$cos\delta = \cos(\Omega' - \Omega) \sin i \sin i' + \cos i \cos i'$$ 5.61

References

1. Prussing, John E. and Conway, Bruce A., *Orbital Mechanics*, © 1993 Oxford University Press, Inc., ISBN 0-19-507834-9.
2. Chobotov, Vladimir A (Editor), *Orbital Mechanics*, AIAA Education Series, © 1991 American Institute of Aeronautics and Astronautics (AIAA), ISBN 1-56734-007-1.
3. Vallado, David A., *Fundamentals of Astrodynamics and Applications*, Second Edition, © 2001 by author, Microcosm Press and Kluwer Academic Publishers, ISBN 1-881883-12-4.
4. Battin, Richard H., *An Introduction to The Mathematics and Methods of Astrodynamics*, © 1987 by author and the American Institute of Aeronautics and Astronautics (AIAA), ISBN 0-930403-25-8.
5. Curtis, Howard D, *Orbital Mechanics for Engineering Students*, Fourth Edition, © 2020 Elsevier, Ltd., ISBN 978-0-08-102133-0.
6. Gurfil, Pini and Seidelmann, P. Kenneth, *Celestial Mechanics and Astrodynamics: Theory and Practice*, © 2016 Springer-Verlag, ISBN 978-3-662-50368-3.
7. Bate, Roger R., et al., *Fundamentals of Astrodynamics*, © 1971 Dover Publications, Inc., ISBN 0-486-60061-0.

6 Intercept, Rendezvous, Proximity Operations, and Docking

Intercept and rendezvous are critical prerequisites for proximity operations and docking with another spacecraft. These are particularly important activities for human space missions where it is desirable to launch with ferrying craft and docking with space stations or for transfers to spacecraft used for landing on the lunar or other celestial body surfaces.

6.1 Intercept and Rendezvous

Intercept is defined as the solution to the Lambert problem: flying from the current trajectory to reach a position near the target vehicle at the same time the target vehicle is in the local proximity. Thus, the pursuing spacecraft must execute a maneuver (or series of maneuvers) to transition from its trajectory to a trajectory that causes it to intercept its intended target.

Rendezvous and Proximity Operations (RPO) starts at the intercept point, where the pursuing spacecraft performs a second maneuver to virtually match the target spacecraft velocity. Once rendezvous is accomplished, the spacecraft operate near each other, indicating proximity operations commencement.

Key Terms:

Intercept: occurs when a pursuing spacecraft executes a maneuver to match the near the location of a target spacecraft at a planned time.
Proximity Operations: occurs after rendezvous when two spacecraft operate in proximity with relatively small differences in velocity.
Rendezvous: occurs when an intercepting spacecraft virtually matches the target spacecraft velocity at or near the intercept point to cause the trajectories to match in relative proximity.

6.2 Rendezvous Mission Profile

The rendezvous mission profile seeks to perform a minimal energy transfer, typically from a lower circular orbit to that of the target orbit. This process uses a Hohmann transfer or a step up bielliptic (or multi-elliptic) transfer. The Hohmann transfer will be used for illustrative purposes as shown in figure 6-1.

The rendezvous mission profile goal is to maneuver to establish an orbit in proximity with the target spacecraft – the International Space Station (ISS) in this example. The intercept and rendezvous process are the coarse maneuver procedures by which proximity operations are established.

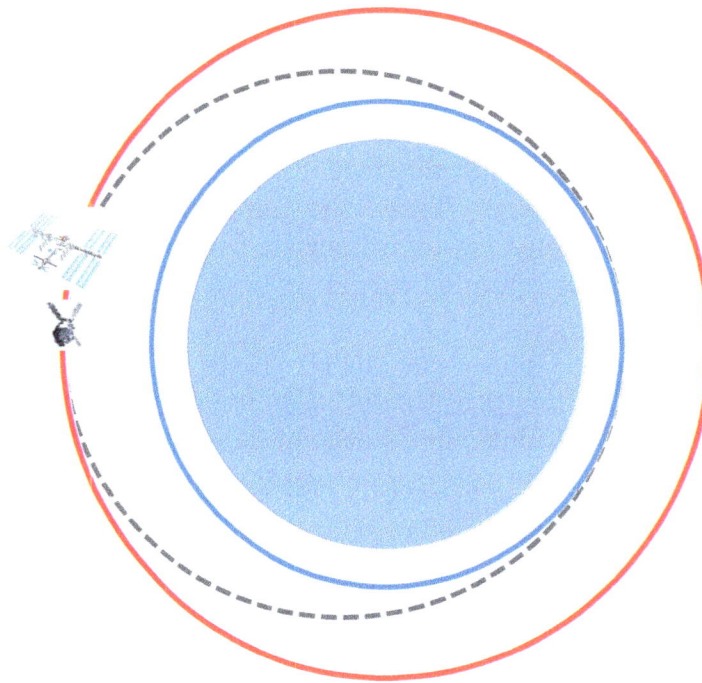

Figure 6-1 Hohmann Transfer Rendezvous Profile

The launch window is set up with the intention of inserting the pursuing ferry spacecraft in a parking orbit as close as practical to being coplanar with the target's orbit. If this is not adequately achieved, the first maneuver will be a plane change to match the target orbit's plane.

Launch opportunities inserting into the target plane will often have random in-plane phasing relative to the target. But since the ferry spacecraft will have a lower orbital radius, it will orbit faster than the target.

When using a Hohmann transfer for rendezvous, there might be an extended wait until a favorable opportunity exists to meet up with the target. The use of a step-up bielliptic transfer allows flexibility in choosing the apogee radius of the first transfer ellipse, thus adding a degree of freedom that facilitates a significant reduction in the wait time. In the figure 6-1 illustration, the rendezvous proximity places the ferry spacecraft in the same orbit, but slight ahead of the ISS target spacecraft.

The ferry spacecraft will normally use the Hohmann or bielliptic transfer to step up to a circular orbit of slightly lower radius than the target and use linearized maneuver planning equations to plan and execute the transition to proximity operations more dynamically.

> **Transition**: *You may continue this section with Rendezvous Mission Profile at Level II, or you may skip to the beginning of the next chapter called Clohessy-Wiltshire Equations.*

6.2.2 Rendezvous Mission Profile [Level II – Equations]

Intercept and rendezvous by a Hohmann transfer may be planned systematically. Since the ferry spacecraft and the target are in circular orbits at different radii, the relative phasing between the spacecraft changes at a constant rate in the range from $0°$ *to* $360°$ after which the cycle repeats. The time after which the cycle repeats is called the *synodic period* between the orbits. Thus, the synodic period is the maximum wait for a favorable intercept phasing. Equation 6-1 computes the synodic period (T_{syn}) as a function of either the two orbital periods (T_1, T_2) or the mean motions (n_1, n_2) of the two orbits.

$$T_{syn} = \frac{T_1 T_2}{|T_1 - T_2|} = \frac{2\pi}{|n_1 - n_2|} \qquad \text{6-1}$$

The location to begin the Hohmann transfer sets a relative phase angle between the pursuing spacecraft and the target. Given the typical case of the pursuing spacecraft in the lower orbit the target vehicle needs to lead the pursuing spacecraft's orbital phase such that the two vehicles meet up in the planned arrangement of the transfer orbit. Equation 6-2 is the semi-major axis of the transfer orbit (a_x). Equation 6-3 is the transfer orbit's period (T_T) with r_1 being the pursuing spacecraft's orbital radius and r_2 being the target orbital radius.

$$a_T = \frac{1}{2}(r_1 + r_2) \qquad \text{6-2}$$

$$T_T = 2\pi \sqrt{\frac{a_T^3}{\mu}} \qquad \text{6-3}$$

Mean motion (n) is its angular orbital rate in radians per second. Equation 6-4 is the target's orbital phase change during the transfer orbit. Since the pursuing spacecraft travels π radians in that time, equation 6-5 provides the target orbit's lead angle ($\Delta\theta_T$) for a half revolution Hohmann transfer.

$$\Delta\theta_2 = n_2 T_T \qquad\qquad 6\text{-}4$$

$$\Delta\theta_T = \pi - n_2 T_T \qquad\qquad 6\text{-}5$$

The pursuing spacecraft may intercept the target at apoapsis after any odd number of transfer orbit half revolutions. Thus, the target's lead angle may be represented in discrete increments as expressed in equation 6-6, where ($k = 1, 2, \cdots$) represents the number of full or half transfer orbits.

$$\Delta\theta_T = \pi - (2k - 1)n_2 T_T \qquad\qquad 6\text{-}6$$

If the time to reach the available phase angles, plus $(2k - 1)$ transfer orbit periods is unacceptable (or undesirably) high, the intercept and rendezvous should be planned with a step up bielliptic transfer. This may be planned numerically, by an iterative process in which the initial maneuver time and the first transfer orbit apogee radius are adjusted to achieve a desirable wait time.

Key Term:

> **Synodic Period**: the time in which the relative phase of two orbits of different radii changes by a full circle (i.e., 2π radians).

6.3   Clohessy-Wiltshire Equations

The relative motion of two spacecraft in circular orbits of equal or nearly equal orbital radius can be accurately modeled using a linearized relationship known as the Clohessy-Wiltshire (or Hills) equations. Because the equations are linearized, the relative motion between the spacecraft can be readily modeled and adjusted by straight forward velocity changes.

The scope for using the Clohessy-Wiltshire equations is to provide an alternative to the Hohmann or bielliptic intercept and rendezvous process, in the form of what is called a co-elliptic approach. The Hohmann or bielliptic approach can be used preliminary to raise the pursuing spacecraft orbit to an altitude close to that of the target spacecraft. Doing so can set up more favorable conditions for the co-elliptic rendezvous approach.

6.3.1   Clohessy-Wiltshire Equations [Level I – Descriptive]

The linearized equations are expressed relative to the local vertical, local horizontal (LVLH) coordinate system standardly used by NASA spacecraft. Thus, the x-axis is aligned with forward orbital motion (V-Bar), the y-axis is aligned opposite the orbital angular momentum (H-Bar), and the z-axis is aligned with geocentric nadir (R-Bar) as depicted in figure 6-2.

The equations are periodic in nature, with the angular rate equaling the target spacecraft orbital rate (i.e., mean motion). The target spacecraft is located at the origin where the x and z axes cross.

Position in the orbital plane follows a "drifting" relative motion ellipse as shown in figure 6-3. The ellipse's semi-major axis (a) is twice that of its semi-minor axis (b). The ellipse's center (x_c, z_c) drifts in the positive x-direction if the ellipse center is below the target spacecraft ($z_c > 0$) and in the negative direction if the ellipse center is above the target spacecraft ($z_c < 0$). Thus, when the pursuing spacecraft is at the same radius as the target ($z_c = 0$), the ellipse's center does not drift ($v_c = 0$) relative to the target spacecraft.

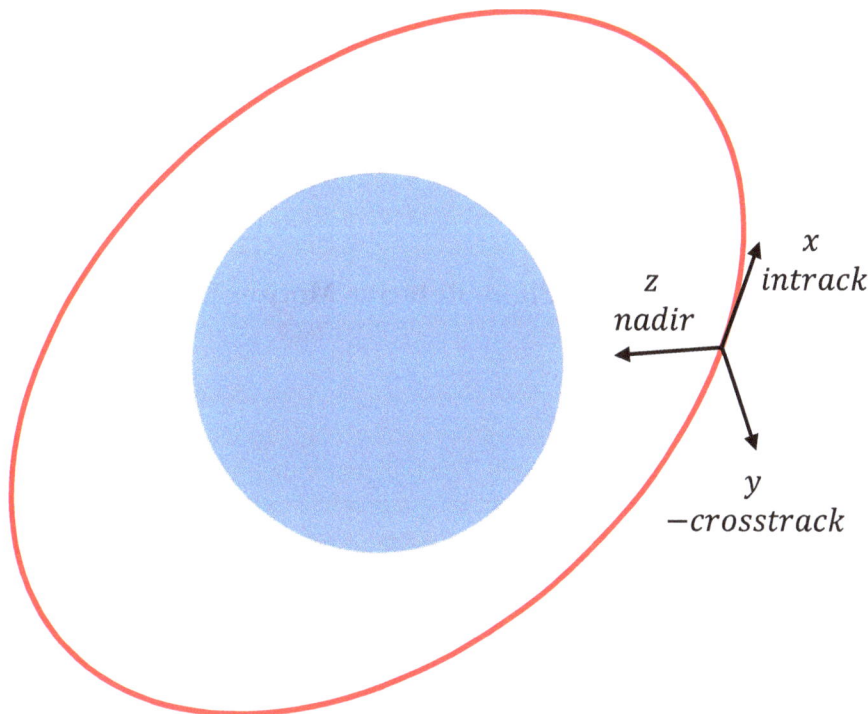

Figure 6-2 Local Vertical, Local Horizontal (LVLH) Coordinate Frame

The pursuing spacecraft's position relative to the target follows the [dashed line] relative motion ellipse. Motion in the x-direction is toward the target when the pursuing spacecraft is above the ellipse center and away from the target when the pursuing spacecraft is below the ellipse center. Thus, for the depicted axes orientations, the pursuer spacecraft motion along the ellipse is clockwise.

171

Figure 6-4 shows an example pursuing spacecraft's motion relative to the target spacecraft. This example illustrates the general case of a drifting relative motion ellipse.

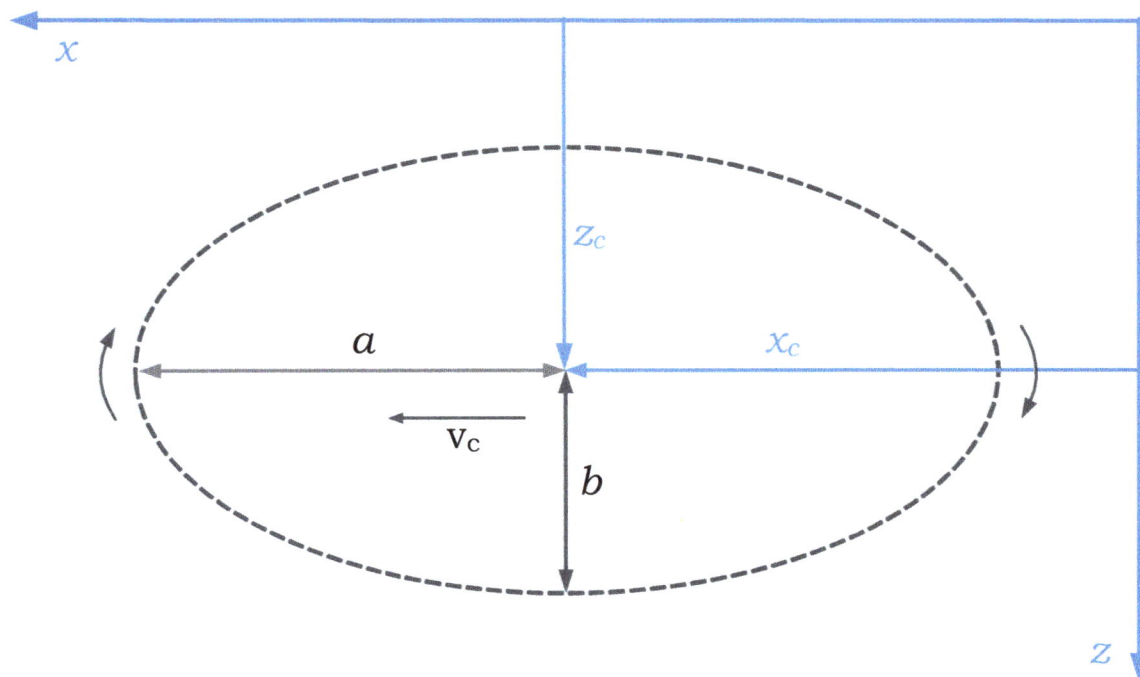

Figure 6-3 In Plane Relative Motion Ellipse

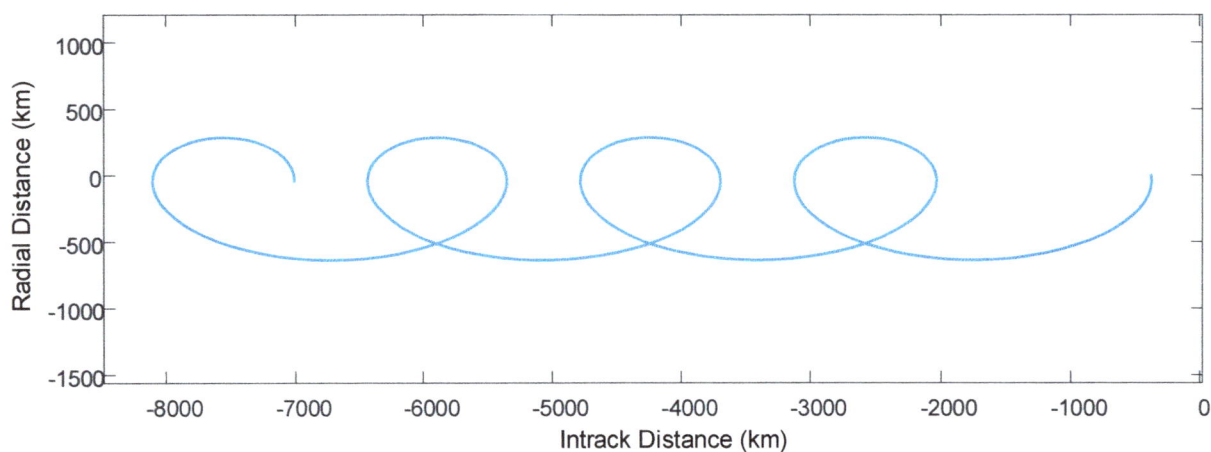

Figure 6-4 Relative Motion with a Drifting Ellipse Center

Key Terms:

H-BAR: opposes the crosstrack/angular momentum direction and is along the LVLH y-axis.
LVLH: is the Local Vertical, Local Horizontal coordinate system with the x-axis in the intrack direction, the y-axis opposing the angular momentum direction, and the z-axis in the local geocentric nadir direction.
R-Bar: is oriented in the local geocentric nadir direction (i.e., opposing the radial direction) and is along the LVLH z-axis.
V-Bar: is oriented in the local geocentric horizontal direction in the intrack direction and is along the LVLH x-axis.

➤ **Transition**: *You may continue this section with Clohessy-Wiltshire Equations at Level II, or you may skip to the beginning of the next section called Proximity Operations.*

6.3.2 Clohessy-Wiltshire Equations [Level II – Equations]

The Clohessy-Wiltshire equations provide an excellent closed-form approximation to the motion of a pursuing spacecraft relative to its target. The pursuer's relative position $\bar{r}(t)$ may be determined at any desired time, given an initial position ($\bar{r}_0$) and velocity ($\bar{v}_0$) offset relative to the target spacecraft as expressed in equations 6.7 through 6.11 in the LVLH coordinates. In these equations, a_T is the target spacecraft's orbital semi-major axis and ω is the LVLH coordinate system rotation rate (which is also the target spacecraft's orbital mean motion).

$$\omega = \sqrt{\frac{\mu}{a_T^3}} \qquad\qquad 6.7$$

$$\bar{r}(t) = \begin{bmatrix} x(t) \\ y(t) \\ z(t) \end{bmatrix} \qquad \bar{r}_0 = \begin{bmatrix} x_0 \\ y_0 \\ z_0 \end{bmatrix} \qquad \bar{v}_0 = \begin{bmatrix} \dot{x}_0 \\ \dot{y}_0 \\ \dot{z}_0 \end{bmatrix} \qquad\qquad 6.8$$

$$x(t) = \left(\frac{4\dot{x}_0}{\omega} - 6z_0\right)\sin \omega t - \frac{2\dot{z}_0}{\omega}\cos \omega t + (6\omega z_0 - 3\dot{x}_0)t + \left(x_0 + \frac{2\dot{z}_0}{\omega}\right) \qquad 6.9$$

$$y(t) = y_0 \cos \omega t + \frac{\dot{y}_0}{\omega}\sin \omega t \qquad\qquad 6.10$$

$$z(t) = \left(\frac{2\dot{x}_0}{\omega} - 3z_0\right)\cos \omega t + \frac{\dot{z}_0}{\omega}\sin \omega t + \left(4z_0 - \frac{2\dot{x}_0}{\omega}\right) \qquad 6.11$$

The relative velocity $v(t)$ is the time derivative of the relative position $\bar{r}(t)$ as expressed in equations 6.12 through 6.15.

$$v(t) = \begin{bmatrix} \dot{x}(t) \\ \dot{y}(t) \\ \dot{z}(t) \end{bmatrix} \qquad 6.12$$

$$\dot{x}(t) = \left(\frac{4\dot{x}_0}{\omega} - 6z_0\right)\omega \cos \omega t + 2\dot{z}_0 \sin \omega t + (6\omega z_0 - 3\dot{x}_0) \qquad 6.13$$

$$\dot{y}(t) = -\omega y_0 \sin \omega t + \dot{y}_0 \cos \omega t \qquad 6.14$$

$$\dot{z}(t) = \left(3z_0 - \frac{2\dot{x}_0}{\omega}\right)\omega \sin \omega t + \dot{z}_0 \cos \omega t \qquad 6.15$$

Equations 6.16 and 6.17 provide the coordinates for the center of the relative motion ellipse for the initial conditions (i.e., at $t = 0$).

$$x_c = x_0 + \frac{2\dot{z}_0}{\omega} \qquad 6.16$$

$$z_c = \frac{2\dot{x}_0}{\omega} - 4z_0 \qquad 6.17$$

Equation 6.18 provides the velocity (i.e., drift rate) of the relative motion ellipse.

$$v_c = (6\omega z_0 - 3\dot{x}_0)\hat{x} \qquad 6.18$$

Equation 6.19 provides the semi-minor and semi-major axis dimensions for the relative motion ellipse.

$$b = \sqrt{\left(\frac{z_0}{\omega}\right)^2 + \left(\frac{2\dot{x}_0}{\omega} - 3z_0\right)^2} \qquad a = 2b \qquad 6.19$$

The equations for relative position $\bar{r}(t)$ and relative velocity can be factored into matrix form as shown in equations 6.20 through 6.24.

$$\begin{bmatrix} \bar{r}(t) \\ v(t) \end{bmatrix} = \begin{bmatrix} \mathbf{\Phi}_{rr} & \mathbf{\Phi}_{rv} \\ \mathbf{\Phi}_{vr} & \mathbf{\Phi}_{vv} \end{bmatrix} \begin{bmatrix} \bar{r}_0 \\ \bar{v}_0 \end{bmatrix} \qquad 6.20$$

$$\mathbf{\Phi}_{rr} = \begin{bmatrix} 1 & 0 & 6(\omega t - \sin \omega t) \\ 0 & -\cos \omega t & 0 \\ 0 & 0 & 4 - 3\cos \omega t \end{bmatrix} \qquad 6.21$$

$$\mathbf{\Phi}_{rv} = \begin{bmatrix} 4(\sin \omega t - 3t)/\omega & 0 & 2(1 - \cos \omega t)/\omega \\ 0 & -\sin \omega t/\omega & 0 \\ 2(\cos \omega t - 1)/\omega & 0 & \sin \omega t/\omega \end{bmatrix} \qquad 6.22$$

$$\mathbf{\Phi}_{vr} = \begin{bmatrix} 0 & 0 & 6\omega(\cos \omega t - 1) \\ 0 & \omega \sin \omega t & 0 \\ 0 & 0 & 3\omega \sin \omega t \end{bmatrix} \qquad 6.23$$

$$\mathbf{\Phi}_{vv} = \begin{bmatrix} 4\cos \omega t - 3 & 0 & 2 \sin \omega t \\ 0 & -\cos \omega t & 0 \\ -2 \sin \omega t & 0 & \cos \omega t \end{bmatrix} \qquad 6.24$$

Equations 6.21 through 6.24 are (3x3) submatrices of the full (6x6) state transition matrix ($\mathbf{\Phi}$), which is a function of time.

$$\mathbf{\Phi} = \begin{bmatrix} \mathbf{\Phi}_{rr} & \mathbf{\Phi}_{rv} \\ \mathbf{\Phi}_{vr} & \mathbf{\Phi}_{vv} \end{bmatrix} \qquad 6.25$$

Since the state transition provides relative position and velocity as a function of time, the velocity needed at any given time to intercept a desired relative position at some desired (future) time can be solved analytically. This is commonly known as Clohessy-Wiltshire targeting.

6.3.2.1 Clohessy-Wiltshire Targeting

The general process only requires that the pursuer meet up with the target at a specific time. The targeting process only needs to constrain the position parameters in equation 6.20. Thus, the general targeting constraint is expressed in equation 6.26.

$$\bar{r}(t) = \mathbf{\Phi}_{rr}\bar{r}_0 + \mathbf{\Phi}_{rv}(\bar{v}_0 + \Delta \bar{v}) \qquad 6.26$$

Equation 6.27 algebraically solves for the $\Delta \bar{v}$ needed to match the desired $\bar{r}(t)$, which would be a zero vector the desire is exactly meet up with the target.

$$\Delta \bar{v} = \mathbf{\Phi}_{rv}^{-1}[\bar{r}(t) - \mathbf{\Phi}_{rr}\bar{r}_0] - \bar{v}_0 \qquad 6.27$$

The $\bar{r}(t)$ is the desired standoff position from the target at intercept. The rendezvous is completed by establishing the desired rendezvous conditions at the intercept (see next section for rendezvous conditions). Standoff positions with a zero z and $\dot{x}$ values will maintain the target proximity with no drift to the relative motion ellipse.

The inverse matrix $\mathbf{\Phi}_{rv}^{-1}$ may be computed numerically. However, since it is a (3x3) matrix, and analytic inversion may be defined. The analytic inverse is:

$$d = 8(1 - \cos\omega t) - 3\omega t \sin\omega t \qquad\qquad 6.28$$

$$\mathbf{\Phi}_{rv}^{-1} = \frac{\omega}{d}\begin{bmatrix} -\sin\omega t & 0 & 2(\cos\omega t - 1) \\ 0 & -d/\sin\omega t & 0 \\ -2(\cos\omega t - 1) & 0 & 4(\sin\omega t - 3t) \end{bmatrix} \qquad\qquad 6.29$$

6.3.3   Trigger Angle Constraint

Equation 6.29 provides a general solution to the Clohessy-Wiltshire targeting problem. However, it is desirable to impose additional operational constraints. The new constraints presume that there will be imperfect knowledge of the pursuer's relative position to the target. The goal of trigger angle targeting is to add constraints that provide natural feedback in the intercept process and thus correct small knowledge errors. The additional constraints are:

- The pursuer must have line-of-sight (LOS) contact with the target (visible or radio).
- The direction of the $\Delta\bar{v}$ causing the intercept must align with the LOS. Mathematically, this second constraint requires that $\Delta\bar{v} \times \bar{r}_0 = 0$.

These constraints provide a co-elliptic approach for orbital intercept as illustrated in figure 6-5. Mathematically, they impose a coupling between the *orbital transfer angle* (θ) and the elevation *trigger angle* (e) that the pursuer points to establish LOS with the target. The transfer angle is the change in true anomaly from the Terminal Phase Initiation (TPI) maneuver to the intercept point. The TPI initiates the process by placing the pursuing spacecraft on the intercept trajectory, with intercept occurring forward of the TPI location by the transfer angle.

6.3.3.1   Trigger Angle Targeting [Level I – Descriptive]

Trigger angle targeting was evaluated by Woffinden et. al as an optimization problem with direct comparison to the parameters used in the NASA Gemini and Apollo programs. In doing so, they validated the NASA parameters (130° orbital transfer angle and approximately 27° elevation angle) as having the best overall solution in terms of the following desirable characteristics:

- TPI is possible within LOS constraints (i.e., $\Delta\bar{v}$ aligned to the LOS).
- TPI maneuver propellant consumption is within a feasible and minimal range.
- The *dispersion* on the intercept error is insensitive to small errors in LOS pointing and/or thrust duration.

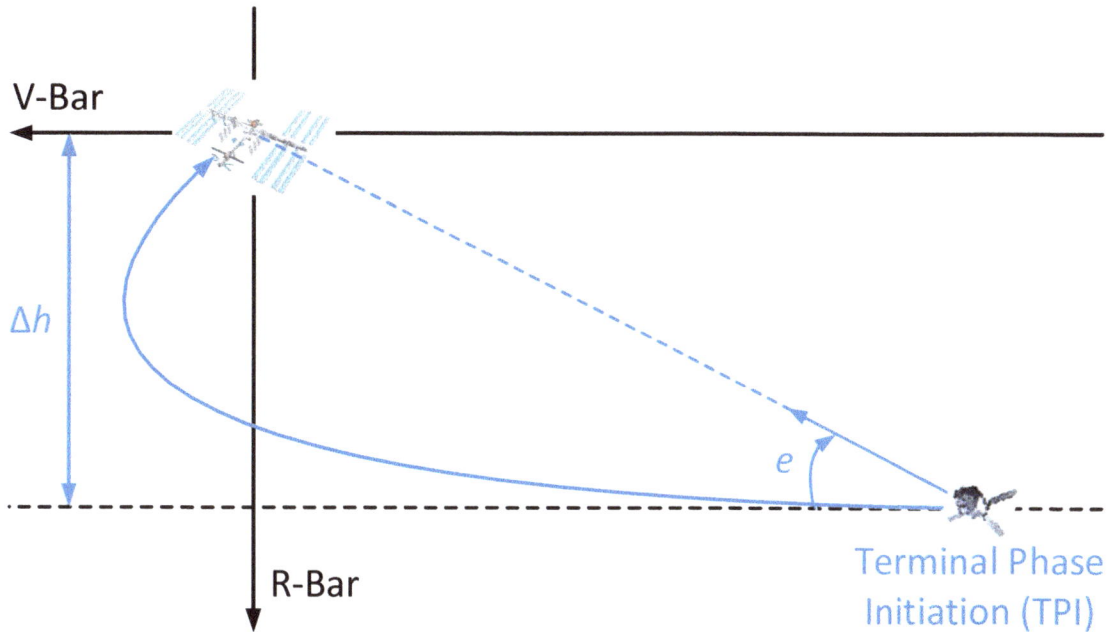

Figure 6-5 Co-Elliptic Approach for Orbital Intercept

Table 6-1 provides their general conclusions regarding the co-elliptical approach intercept characteristics for various approach angles.

Table 6-1 Intercept Characteristics for Transfer Angle Ranges		
Intercept Characteristics	**Transfer Angle Range**	**Trigger Angle Elevation Range**
TPI is <u>not</u> possible with $\Delta\bar{v}$ aligned to the LOS (i.e., the primary constraint is <u>not</u> satisfied).	195° to 286°	N/A
TPI is possible with reasonable propellant consumption.	70° to 170°	27° to 35°
TPI is possible with: • Minimal propellant consumption. • Low intercept error dispersion due to TPI execution errors.	125° to 145°	26.8° to 27.8°

➤ **Transition**: *You may continue this section with Trigger Angle Targeting at Level II, or you may skip to the beginning of the next section called Rendezvous Conditions.*

6.3.3.2   Trigger Angle Targeting [Level II – Equations]

This section provides the trigger angle targeting algorithm as derived by Woffinden et. al. Per table 6-1, the orbital transfer angle (θ) should be constrained between 70° and 170°, with a strong preference for the range from 125° to 145°. Transfer angles in the range from approximately 195° to 286° will provide a mathematically imaginary result (i.e., square root of a negative number) since trigger angle targeting is not possible in this range. The orbital transfer angle is defined as the difference in true anomaly (±180° range) between the target (θ_T) and the pursuing spacecraft (θ_P).

$$\theta = mod(\theta_T - \theta_P + 180°, 360°) - 180° \qquad\qquad 6.30$$

Equations 6.31 through 6.34 are the preliminary quantities computed for the Clohessy-Wiltshire equations with the trigger angle LOS constraint.

$$\alpha = 6\sin\theta(\sin\theta - \theta) + 2(1 - \cos\theta)(4 - 3\cos\theta) \qquad\qquad 6.31$$

$$\beta = \sin\theta - 12(\cos\theta - 1)(\sin\theta - \theta) - (4 - 3\cos\theta)(4\sin\theta - 3\theta) \qquad\qquad 6.32$$

$$\gamma = 2(1 - \cos\theta) \qquad\qquad 6.33$$

$$d = 4\sin^2\theta - 3\theta\sin\theta + 4(1 - \cos\theta)^2 \qquad\qquad 6.34$$

Equation 6.35 is the rendezvous ratio (κ), which is zero for a direct target intercept. However, when intercepting a position at a standoff location from the target, x_f is the intrack (V-bar) standoff component and Δh is the altitude change from the initial orbit to the standoff location.

$$\kappa = x_f/\Delta h \qquad\qquad 6.35$$

Equations 6.36 through 6.39 are the coefficients of and solution to the quadratic equation used to find the trigger angle.

$$a = \gamma \qquad\qquad 6.36$$

$$b = \beta - \kappa\gamma \qquad\qquad 6.37$$

$$c = \alpha - \frac{3}{2}d - \kappa\sin\theta \qquad\qquad 6.38$$

$$\xi = \frac{-b \pm \sqrt{b^2 - 4ac}}{2a}$$

6.39

There are two trigger angle solutions corresponding to the inverse cotangent of the quadratic solution. The cotangent also has two possible solutions. Compliant trigger angles are the cotangent solution directed toward, rather than away from the target. The two trigger angles will have different delta-V magnitudes needed to intercept the target. The trigger angles with the lower delta-V tend to have lower maneuver dispersions. Solutions outside the nominal NASA ranges need to be evaluated on a case-by-case basis.

Key Terms:

Dispersion: is the variation in achieved versus planned target intercept due to maneuver input or execution errors.

Orbital Transfer Angle: is the downrange difference in true anomaly from TPI to target intercept.

Terminal Phase Initiation (TPI): is the maneuver that initiates a target intercept, directed along the trigger angle.

Trigger Angle: is the target elevation angle in which the intercept delta velocity for a chosen orbital transfer angle is aligned with the LOS from the pursuing spacecraft to the target.

6.3.4 Rendezvous Conditions

Orbital rendezvous occurs at the intercept position, which is typically near the target and at the same radius/altitude. The positional offset is often only in the intrack direction. The relative motion ellipse equations (6.16 through 6.19) provide insight into the proximity operations conditions that will ensue. Thus, these equations are useful for establishing the conditions desired for operating in proximity with the target.

Equations 6.16 and 6.17 establish the relative motion ellipse's *instantaneous* center. It is sometimes desirable to have the ellipse's center remain stationary relative to the target (i.e., zero intrack drift), and having some intrack offset. Per equation 6.18, this can only occur when the relative intrack velocity component ($\dot{x}_0$) is properly coordinated with the radial position offset (z_0), specifically when $\dot{x}_0 = 2\omega z_0$.

It is sometimes desirable to hold a at a constant position relative to the target (see the later section on safety zones and holding). This may be established by neutralizing (i.e., zeroing out) relative velocity in the both the x- and z-directions, when the pursuing vehicle is at the same radius as the target (i.e., $\dot{x}_0 = 0$, $\dot{z}_0 = 0$, when $z_0 = 0$).

The relative motion conditions become more complex when setting up an approach for docking with the target spacecraft. The specific approach strategies proved a means to approach along the intrack or radial paths, both of which are covered in subsequent sections.

6.4 Proximity Operations

Proximity operations encompass all relative motion between the pursuing and target spacecraft, from rendezvous to the commencement of the docking final approach. Proximity operations are well-modeled using the Clohessy-Wiltshire equations, which may be effectively used to predict, monitor, and modify the relative motion between the two spacecraft.

6.4.1 Safety Zones and Holding

The International Space Station (ISS) has two defined zones, facilitating safety during proximity operations. Docking, which results in physical contact between the two spacecraft, maneuvering needs to be done in a safe and controlled manner to join the two docking fixtures. It is otherwise necessary to ensure the pursuer's relative motion does not pose a collision hazard.

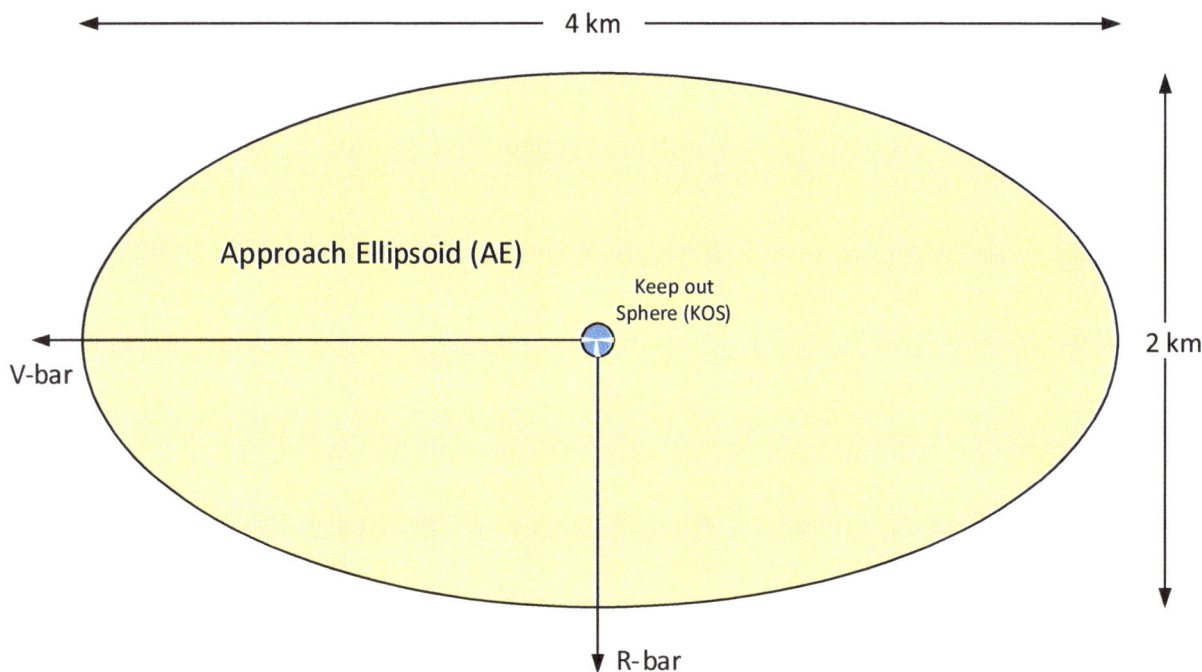

Figure 6-6 ISS Approach Ellipsoid

The ISS safety zones are the Approach Ellipsoid (AE) and the Keep Out Sphere (KOS), illustrated in figures 6-5 and 6-6. The AE has a 2 km maximum radial dimension and a 4 km horizontal dimension and has the ISS at its center. The AE horizontal extent is in both the intrack and crosstrack directions.

A spacecraft's entry into the AE is only permitted after clearance is given by the ISS ground control center. Such permission is only granted after confirmation that the pursuing spacecraft is established on a predictable trajectory for 90 minutes (approximately one orbital revolution) before entry. Entry to the AE is called Approach Initiation (AI).

The KOS is a 200-meter spherical volume centered on the ISS. The pursuing spacecraft performs a final hold outside the KOS, while aligned with the ±10° conical final approach corridor. The hold provides the opportunity to perform final checks of the pursuer spacecraft, ensuring it can safely maneuver and dock while not posing a collision hazard. This includes the ability of aborting the approach and withdrawing to a safe holding position at the request of the ISS. All spacecraft within the KOS must remain within the final approach corridor for the intended docking port.

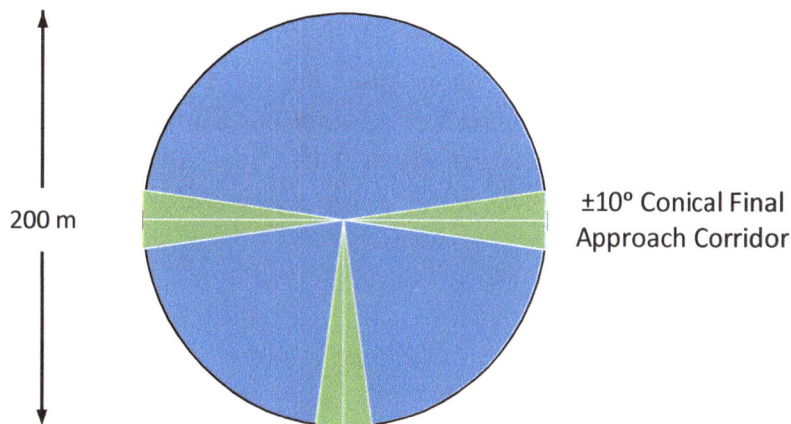

±10° Conical Final Approach Corridor

200 m

Figure 6-7 ISS Keep Out Sphere

Two approach strategies exist that leverage the relative motion expressed by the Clohessy-Wiltshire equations: *V-Bar Hopping* is the horizontal approach strategy and *R-Bar Hopping* is the vertical approach strategy.

Key Terms:

Approach Ellipsoid (AE): is the safety volume centered on the ISS requiring ground controller permission prior to entry. Its maximum dimensions are 4 km along the V-Bar and H-Bar directions and 2 km along the R-Bar direction.
Keep Out Sphere (KOS): is the 200 m safety volume centered on the ISS with 10° conical final approach volumes aligned with the docking ports.

6.4.2 V-Bar Hopping

V-Bar Hopping provides a controlled horizontal approach with the pursuing spacecraft in proximity operations forward of the ISS. The pursuing vehicle matches the ISS orbital semi-major axis with a zero radial (z) position offset and a positive intrack (x) position offset and a slight orbit eccentricity induced by a $\dot{z}$ velocity component. Each hop is bounded by a maneuver that reverses the radial velocity component ($\dot{z}$) at the ISS orbital radius (i.e., radial apsidal line rotation maneuvers).

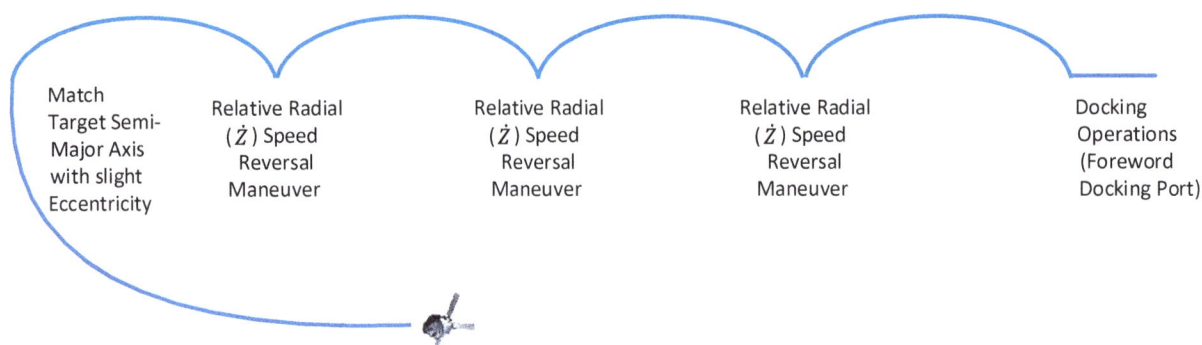

Figure 6-8 V-Bar Hopping Approach

When the pursuing spacecraft is positioned ahead of the target (i.e., +V-Bar) the effect is to create a 180° flip of the line of apsides, by flipping the true anomaly from 270° to 90°. As a result, the pursuing spacecraft flies through apoapsis during each hop, slowing down relative to the ISS. This allows the ISS to close the intrack position gap with the pursuer by a distance equal to the relative motion ellipse major axis.

The V-Bar Hopping approach has inherent collision avoidance safety since the pursuing spacecraft has a relative motion ellipse with zero drift. Thus, failure to execute a radial speed reversal maneuver results in the pursuer continuing in a relative motion ellipse that is stationary with respect to the ISS. The next opportunity to perform the radial maneuver is deferred to one revolution later.

Key Term:

V-Bar Hopping: is a sequence of radial apsidal line rotation maneuvers performed at the target altitude. The pursuing spacecraft has the same orbital semi-major axis as the target orbit but has slight eccentricity whereas the target is in a circular orbit.

R-Bar Hopping provides a controlled vertical approach with the pursuing spacecraft in proximity operations below the ISS. The pursuing spacecraft successively executes maneuvers while crossing the ISS nadir (R-Bar) direction, targeting a position directly above and in the R-Bar direction.

The maneuver delta-V has components in both intrack (V-Bar) and radial (R-bar) directions, with the intrack having the larger magnitude as computed by the Clohessy-Wiltshire targeting process. These maneuvers are like small Hohmann transfers, with the addition of a radial component that counteracts the relative motion ellipse's drift.

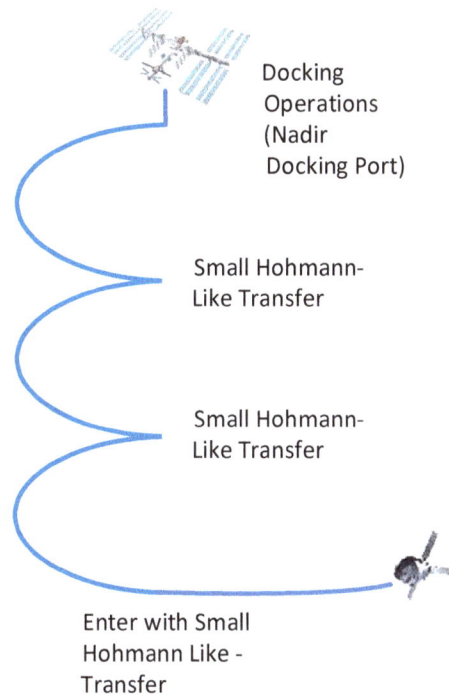

Docking
Operations
(Nadir
Docking Port)

Small Hohmann-
Like Transfer

Small Hohmann-
Like Transfer

Enter with Small
Hohmann Like -
Transfer

Figure 6-9 R-Bar Hopping Approach

Key Term:

R-Bar Hopping: is a sequence of Clohessy-Wiltshire targeting maneuvers performed in line with the R-Bar direction. The effect is changing altitude toward the target with each successive hop.

Docking operations occur with the pursuer and target in very close proximity. Since they are in virtually the same orbit, the approach is in nearly a straight line. However, there will be tendencies to drift from a straight approach that require small corrections.

The best practice for manual docking is to align the pursuing spacecraft attitude parallel to that of the target spacecraft along each axis. The rotational controller sets rates along one or more spacecraft axes to correct any angular errors. Since there is no resistance, a rotational maneuver must be counteracted with an opposing rotational thruster firing to stop the rotation.

- A horizontal approach (along the V-Bar) will have the pursuer at the same orbital radius as the target. Since the pursuer will have a slightly smaller speed than the target, it will tend to decrease in radius (drift in the R-Bar direction).
- A vertical approach (along the R-bar) will have the pursuer with a slightly larger intrack speed than the target since it is at a lower orbital radius, resulting in a drift in the V-bar direction.

Translational maneuvers should align the pursuing spacecraft docking port with that of the target spacecraft. Once aligned, a maneuver should be performed to establish a slow straight line closure rate to the docking port. Since there will be small drifts during closure, small translational corrections should be applied to maintain alignment between the docking ports.

REFERENCES
1. Ley, Wilfried et al. (Editors), *Handbook of Space Technology*, © 2009 co-published by American Institute of Aeronautics and Astronautics (AIAA) and John Wiley & Sons, Ltd., ISBN 978-0-470-69739-9.
2. Chamitoff, Gregory E. and Vadali, Srinivas Rao (editors), *Human Spaceflight Operations: Lessons Learned from 60 Years in Space*, © 2021 by editors, American Institute of Aeronautics and Astronautics (AIAA), ISBN 978-1-62410-399-5.
3. Aldrin, Edwin E., *Line-of-Sight Guidance Techniques for Manned Orbital Rendezvous*, Submitted in Partial Fulfilment of the Degree Doctor of Science, Massachusetts Institute of Technology, January 1963.
4. Lunney, Glynn S., *Summary of Gemini Rendezvous Experience*, AIAA Paper 67-272, AIAA Flight Test, Simulation and Support Conference, Cocoa Beach, FL, February 6-8, 1967.

5. Parten, Richard P. and Mayer, John P., *Development of the Gemini Operational Rendezvous Plan*, Journal of Spacecraft and Rockets, Volume 5, No. 9, September 1968.

6. Woffinden, David C., et al., *Trigger Angle Targeting for Orbital Rendezvous*, The Journal of the Astronautical Sciences, Volume 56, No. 4, October-December 2008.

7. Clohessy, W.H. and Wiltshire, R.S., *Terminal Guidance for Satellite Rendezvous*, Journal of the Aerospace Sciences, September 1960, pp. 653-658 and 674.

8. Townsend Jr., G.E. and Russak, S.L., *Space Flight Handbooks, Orbital Flight Handbook*, NASA SP 33, National Aeronautics and Space Administration, 1963.

7 Perturbational Effects on Trajectories

Conic trajectories result from the gravitational acceleration between two bodies. Orbital perturbations are deviations from two body conic section trajectories, resulting from additional forces introduced. In most cases the additional forces are small in comparison to the gravity of the major attracting celestial body and thus the perturbations result in small disturbances from a trajectory that is well approximated by a conic section. Nevertheless, precise, and/or long-term trajectory predictions require that any significant perturbational forces be modeled to maintain prediction accuracy.

7.1 Trajectory Propagation

A trajectory's position and velocity state may be propagated forward or backward to determine the corresponding position and velocity state at a desired time. The Trajectory Prediction Using Universal Variables section in chapter 2 provides an example method of this process. A shortfall of this method is that it does not have a provision to include perturbational accelerations. Thus, its primary usefulness is for either preliminary or short-term predictions.

The next sections provide trajectory propagation methods that permit the inclusion of perturbational effects. The two main categories differ in their approach to integrating the perturbational accelerations, since one is analytic and the other uses numerical techniques. It should be noted that hybrid (semi-analytic) techniques also exist but given the high speed of modern computers the necessity for their application is limited in scope.

7.1.1 Variation of Parameters and General Perturbations Techniques

Variation of parameters is a method by which predictions are performed analytically and perturbational effects are expressed as equations that represent variations to the orbital parameters as a function of time. Given Keplerian elements as an example, the in-plane position and velocity are determined by a solution to Kepler's equation to map the desired time to true anomaly, using equations in chapter 2. The angular elements are then used to orient the orbital plane to the desired inertial coordinate system as illustrated in chapter 3.

The equations representing variations to the orbital parameters are integrated using analytic techniques. As a result, application of perturbational effects tends to be efficient, requiring low computational overhead. However, the efficiency gained is not without inherent drawbacks.

To analytically integrate perturbational accelerations, the level of modeling sophistication must be selected prior to the integration being performed. Whether or not to include various accelerations and the degree to which effects are modeled (such as when the mathematics involves approximations by infinite series) is determined by the selections made before the integration. A particular set of equations expressing the variations to orbital parameters therefore limits the applicable scope of the model.

Because of the limitations, only limited examples of equations varying orbital parameters will be presented. Readers interested in a rigorous treatment of this topic should refer to Curtis (2020).

7.1.1.1 Earth Oblateness Evaluation

The Earth's oblate shape due to its equatorial bulge creates perturbations to its gravitational acceleration. These effects were measured by observing the orbits of artificial Earth-orbiting satellites during International Geophysical Year (IGY), which spanned from 01 July 1957 to 31 December 1958.

The first order effects of Earth oblateness can be expressed in equations that provide a drift rate to the Keplerian RAAN (Ω) and argument of periapsis (ω) orbital parameters. Equations 7.1 and 7.2 provide the drift rates as derivatives with respect to time. The rates depend mainly on the orbit's semi-major axis (a) and inclination (i) with a second order contribution from eccentricity (e).

$$\dot{\Omega} = -\frac{3}{2}J_2 \frac{R_e^2\sqrt{\mu}}{(1-e^2)^2 a^{7/2}} \cos i \qquad 7.1$$

$$\dot{\omega} = \frac{3}{4}J_2 \frac{R_e^2\sqrt{\mu}}{(1-e^2)^2 a^{7/2}} (4 - 5\sin^2 i) \qquad 7.2$$

The J_2 parameter is a coefficient specific to the Earth's equatorial bulge mass concentration, R_e is the Earth's equatorial radius, and μ is the Earth's gravitational parameter.

Inspection of these equations provides general insights as to the nature of perturbational effects, based on orbital parameters. One obvious conclusion is these disturbances progressively decrease as semi-major axis increases. This should be intuitively obvious since the Earth should begin to appear gravitationally as a point mass as distance increases.

Conclusions about the rates can also be made from the orbital inclination. The RAAN drift rate ($\dot{\Omega}$) is maximum for low inclinations and zero for polar ($i = 90°$)

inclinations. The nodes drift westward for prograde $(i < 90°)$ inclinations and eastward for retrograde $(i > 90°)$ inclinations.

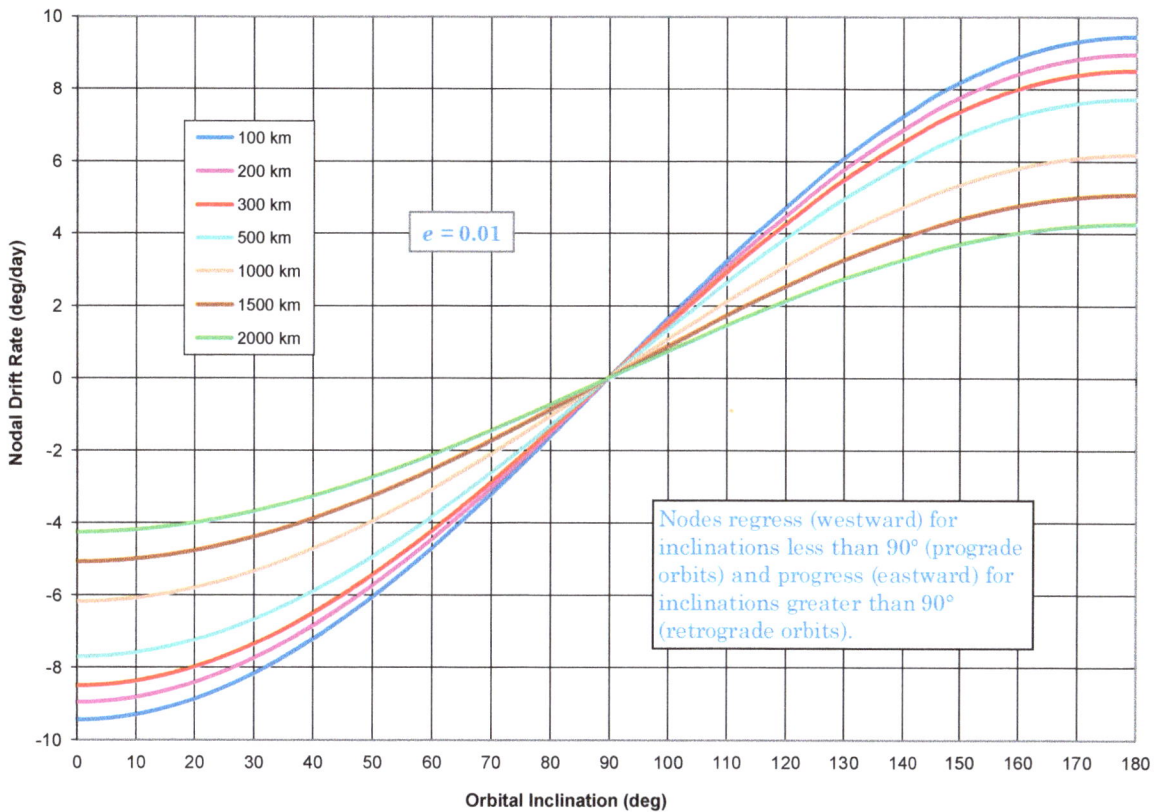

Figure 7-1 Nodal Drift Rate due to Earth Oblateness

Drift in the argument of periapsis is slightly more complicated. The apsidal drift rate $(\dot{\omega})$ is zero at the two critical inclinations $(i \approx 63.4°, 116.6°)$ resulting from the relationship $\sin i = \sqrt{4/5}$ derived from equation 7.2. Figures 7-1 and 7-2 Illustrate the nodal and apsidal drift rates for various altitudes and inclinations.

7.1.1.2   The Simplified General Perturbations version 4 (SGP4) Propagator

The North American Aerospace Defense Command (NORAD) developed the SGP4 orbit propagator during the Cold War as a method to accurately and efficiently propagate orbits. Since NORAD was tasked to detect, track, identify, and maintain surveillance on all artificial Earth-orbiting objects, there was a need to propagate the orbits of large numbers of satellites using the slow computers available at the time. The Project Space Track report provides details of the model and FORTRAN source code. Source code is also available for the C++, C#, FORTRAN, Java,

MATLAB, and Pascal programming languages on the Celestrak Web site (https://celestrak.org/publications/AIAA/2006-6753/).

SGP4 and its related methods are required for correct propagation of NORAD Two-Line Element (TLE) sets that are available on the Space Track and Celestrak Web sites. To expedite the processing, the TLEs are what are known as *mean elements* (as opposed to *osculating elements*). Osculating elements are instantaneous orbital parameters that would be obtained from the direct Cartesian position and velocity conversion in chapter 3. The elements in the TLEs are mean values determined by removing periodic perturbational variations in a manner compatible with the SGP4 and related propagators.

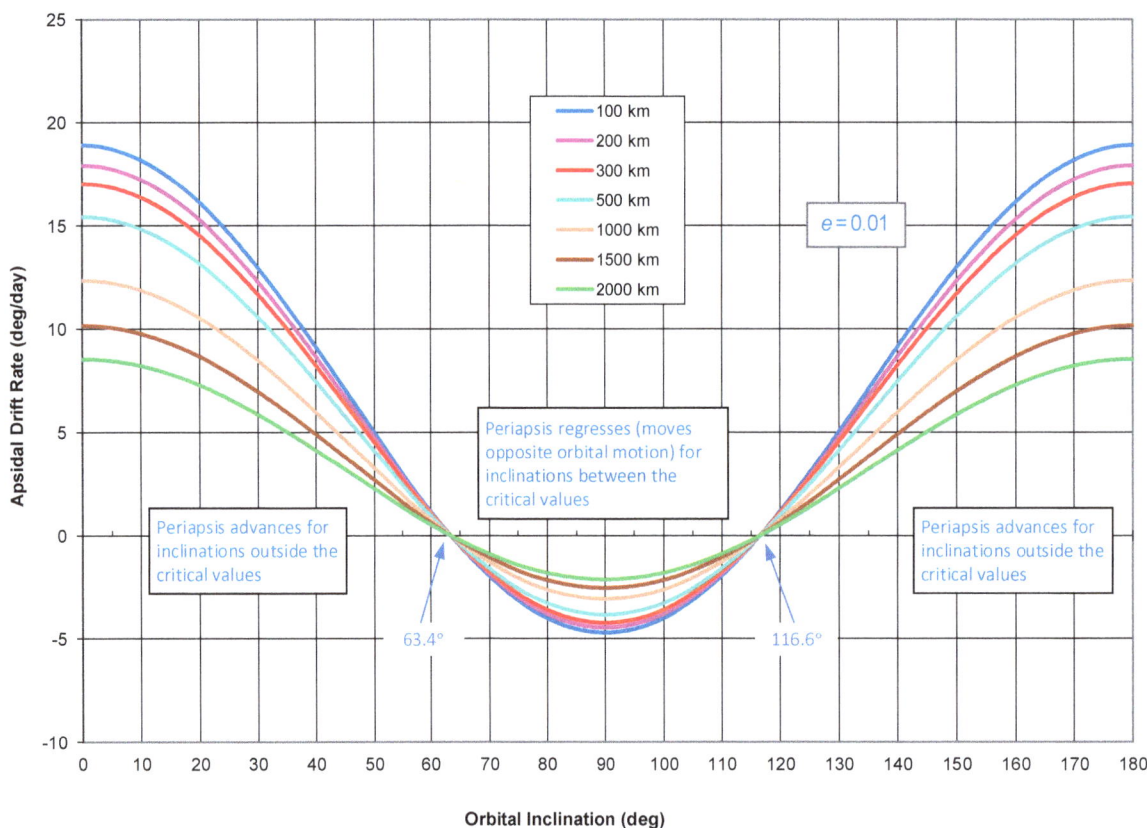

Figure 7-2 Apsidal Drift Rate due to Earth Oblateness

Two-Line Element sets were formerly known as Two-Card Element sets since the obscure format was created to fit on two Hollerith punch cards. This allowed the elements to be easily sent to NORAD's world-wide network of tracking sites, using the electronic data transmission methods available during the Cold War.

It should also be noted that that the position and velocity vectors output by SGP4 are in what is known as the True Equator, Mean Equinox (TEME) coordinate system. This is an ECI system that approximates the TOD system to within a few hundred meters per Earth radii of the position magnitude. Vallado provides a method to convert TEME vectors to or from the MOD coordinate system.

Key Terms:

> **Mean Elements**: are orbital parameters with the average perturbational effects applied.
> **Osculating Elements**: are the instantaneous orbital elements at any given time.

7.1.2 High Fidelity Perturbation Modeling Using Numerical Integration

Numerical integration provides the most accurate and flexible methods for orbit propagation. A significant advantage of these methods is the ability to select the perturbational force set to include at run time. This advantage includes the ability to specify the number of terms to include for perturbative forces computed by infinite series.

Trajectory propagation using numerical integration begins with the equation 2.21 as a second order Ordinary Differential Equation (ODE). Equation 2.21 considered the central attracting body's gravity as the only acceleration. Adding a net perturbative acceleration ($\bar{a}_p$) as is done in equation 7.3 complicates the integration process.

$$\ddot{\bar{r}}_{net} = -\frac{\mu}{r^3}\bar{r} + \bar{a}_p \qquad\qquad 7.3$$

The analytic integration seen in general perturbative formulations is inherently limited to a pre-defined set of perturbative acceleration terms. The numerical formulations permit an open-ended choice of perturbative acceleration terms that can be tailored to the problem at hand, by the user at run time. This offers the greatest flexibility.

7.1.2.1 Encke's Method [Level I – Descriptive]

Encke's method performs an analytic two-body propagation of an osculating orbit. The perturbational forces (i.e., all forces other than the central two-body gravity) are integrated and that difference is applied to the two-body propagation result. Periodically, the two-body osculating orbit is rectified to the current perturbative correction, usually when the position difference reaches a pre-defined threshold.

Encke's method is presented primarily for historic significance, since given the speed of modern computing, Cowell's method is extremely effective.

➤ **Transition**: *You may continue this section with Encke's Method at Level II, or you may skip to the beginning of the next section called Cowell's Method.*

7.1.2.2 Encke's Method [Level II – Equations]

Encke's formulation keeps the two-body central gravitational acceleration ($\ddot{\bar{r}}$) separated from the perturbative acceleration as presented in equation 7.3. The conic trajectory is propagated using an analytic method such as universal variables or any equivalent process to update the conic trajectory position ($\bar{r}$) and velocity ($\dot{\bar{r}}$) vectors to the current time.

Numerical integration is limited to the perturbational accelerations, which will include any central body gravitational deviations from that of a homogeneous sphere, plus any other perturbative accelerations modeled. The process computes deviations from the conic trajectory in position ($\delta\bar{r}$) and velocity ($\delta\dot{\bar{r}}$) by numerically integrating the deviation to acceleration ($\delta\ddot{\bar{r}}$). Equations 7.4 and 7.5 represent the perturbed position ($\bar{r}_p$) and velocity ($\dot{\bar{r}}_p$) at any given time (i.e., the actual state) plus the sums of the conic trajectory parameters and their corresponding perturbed deviations.

$$\bar{r}_p = \bar{r} + \delta\bar{r} \qquad\qquad 7.4$$

$$\dot{\bar{r}}_p = \dot{\bar{r}} + \delta\dot{\bar{r}} \qquad\qquad 7.5$$

Equations 7.6 and 7.7 provide intermediate quantities that are numerically well-conditioned equations for determining the deviation to the acceleration ($\delta\ddot{\bar{r}}$), which is numerically integrated to update the perturbed velocity deviation ($\delta\dot{\bar{r}}$) and position deviation ($\delta\bar{r}$). (Note that $\delta\bar{r}$ and $\delta\dot{\bar{r}}$ are zero vectors for the initial and post rectification states.)

$$\varepsilon = \frac{\bar{r} \cdot \delta\bar{r}}{r^2} \qquad\qquad 7.6$$

$$f = 1 - \frac{1}{(1 - 2\varepsilon)^{3/2}} \qquad\qquad 7.7$$

Equation 7.8 is an expression for the deviation to the acceleration that is provided to the numerical integrator.

$$\delta\ddot{\bar{r}} = \bar{a}_p + \frac{\mu}{r^3}\left[f\bar{r}_p - \delta\bar{r}\right] \qquad\qquad 7.8$$

192

When the ratio of $\delta \bar{r} / \bar{r}$ exceeds the pre-defined threshold, rectification is performed to align the current osculating position and velocity ($\bar{r}$ and $\dot{\bar{r}}$) to the current perturbed state and the perturbed vectors ($\delta \bar{r}$ and $\delta \dot{\bar{r}}$) are reset as zero vectors.

$$\bar{r} = \bar{r} + \delta \bar{r} \tag{7.9}$$

$$\dot{\bar{r}} = \dot{\bar{r}} + \delta \dot{\bar{r}} \tag{7.10}$$

➢ **Transition**: *You may continue this section with Encke's Method at Level III, or you may skip to the beginning of the next section called Cowell's Method.*

7.1.2.3 Encke's Method [Level III – Derivation]

Equation 7.11 is the total acceleration ($\ddot{\bar{r}}_p$) on the perturbed state, which is the sum of two body acceleration on the perturbed state and the net perturbative acceleration.

$$\ddot{\bar{r}}_p = -\frac{\mu}{r_p^3} \bar{r}_p + \bar{a}_p \tag{7.11}$$

Equation 7.12 is the deviation to acceleration, which is the difference between the total acceleration on the perturbed state and the two-body acceleration. Equation 7.13 combines equation 7.6 with the two-body acceleration (i.e., using equations 2.21 and 7.4) algebraically.

$$\delta \ddot{\bar{r}} = \ddot{\bar{r}}_p - \ddot{\bar{r}} \tag{7.12}$$

$$\delta \ddot{\bar{r}} = \bar{a}_p + \frac{\mu}{r^3} \left[\left(1 - \frac{r^3}{r_p^3} \right) \bar{r}_p - \delta \bar{r} \right] \tag{7.13}$$

The quantity in parentheses in equation 7.13 is nearly zero, which can create some numerical conditioning issues. Kaplan (1976) introduces a process that eliminates this numerical concern. Equation 7.14 defines the small quantity ε and provides for the relationship in equation 7.15.

$$2\varepsilon = 1 - \frac{r_p^2}{r^2} \tag{7.14}$$

$$\frac{r_p^2}{r^2} = \frac{1}{[1 - 2\varepsilon]^{3/2}} \tag{7.15}$$

Equation 7.16 substitutes this relationship into equation 7.13. Equation 7.17 simplifies the form of equation 7.16 by defining a parameter f expressed in equation 7.18.

$$\delta\ddot{\bar{r}} = \bar{a}_p + \frac{\mu}{r^3}\left[\left(1 - \frac{1}{[1-2\varepsilon]^{3/2}}\right)\bar{r}_p - \delta\bar{r}\right] \qquad 7.16$$

$$\delta\ddot{\bar{r}} = \bar{a}_p + \frac{\mu}{r^3}\left[f\bar{r}_p - \delta\bar{r}\right] \qquad 7.17$$

$$f = 1 - \frac{1}{[1-2\varepsilon]^{3/2}} \qquad 7.18$$

This result provides the origin for the expression in equation 7.7.

7.1.2.4 Cowell's Method [Level I – Descriptive]

Like Encke's method, Cowell's method augments the two-body ODE with the net perturbational acceleration as expressed in equation 7.3. The fundamental difference is that Cowell's method integrates the complete net acceleration without any presumption of an underlying conic trajectory. Thus, the Cowell method directly integrates the net acceleration.

➤ **Transition**: *You may continue this section with Cowell's Method at Level II, or you may skip to the beginning of the next section called Sources of Perturbative Accelerations.*

7.1.2.5 Cowell's Method [Level II – Equations]

Cowell's method formulates the position and velocity vectors into a consolidated state ($\bar{y}$). Equation 7.19 provides the arrangement for $\bar{y}$ and equation 7.20 provides the arrangement for its first time derivative ($\dot{\bar{y}}$).

$$\bar{y} = \begin{bmatrix} \bar{r} \\ \dot{\bar{r}} \end{bmatrix} = \begin{bmatrix} r_i \\ r_j \\ r_k \\ \dot{r}_i \\ \dot{r}_j \\ \dot{r}_k \end{bmatrix} \qquad 7.19$$

$$\dot{\bar{y}} = \begin{bmatrix} \dot{\bar{r}} \\ \ddot{\bar{r}} \end{bmatrix} = \begin{bmatrix} \dot{r}_i \\ \dot{r}_j \\ \dot{r}_k \\ \ddot{r}_i \\ \ddot{r}_j \\ \ddot{r}_k \end{bmatrix} \qquad\qquad 7.20$$

This formulation is compatible for direct integration by numerical methods, including those provided in Appendix F. Equation 7.21 indicates the net acceleration ($\ddot{\bar{r}}$) includes the gravity of the central attracting body and all modeled perturbative accelerations.

$$\ddot{\bar{r}} = -\frac{\mu}{r^3}\bar{r} + \bar{a}_p \qquad\qquad 7.21$$

There is significant flexibility in selecting perturbative acceleration models and consolidating their specification in what is commonly known as the *force model*. The individual models are typically sub-functions called by integrator's force function.

It should be noted that thrust events have mass depletion, which is a unique feature for this perturbative acceleration. When thrust events are included, it is a good practice to augment equation 7.19 with the spacecraft total mass (m) and 7.20 with the propellant mass expulsion rate ($\dot{m}$). Thus, the $\bar{y}$ *and* $\dot{\bar{y}}$ state vectors would have seven instead of six elements. Propulsion parameters provide the thrust to instantaneous mass ratio as an acceleration term and the mass expulsion rate as part of the thruster model.

7.2 Sources of Perturbative Accelerations

Perturbative accelerations are commonly modeled from forces in four categories: gravitational, atmospheric drag, radiation pressure, and those induced by thrust events. Table 7-1 summarizes the perturbations applicable to an Earth-orbiting spacecraft.

Table 7-1 Perturbations Applicable to Earth-Orbiting Spacecraft		
Category	**Description**	**Affected Orbits**
Gravitational	Acceleration magnitude and direction variations due to the central body's non-spherical mass distribution.	All, but variations are strongest for low-altitude orbits.
	Point mass attraction by other celestial bodies (Sun, Moon, and	All, but effects are more dominant perturbations for

	planets).	higher altitude orbits.
Atmospheric Drag	Upper atmosphere gas molecule impacts (momentum transfer).	Primarily lower altitude orbits. Effects are measurable below a 1000 km altitude and negligible above 2500 km.
Radiation Pressure	Solar – photon impact (momentum transfer) from direct sunlight: at 1 AU ~4.56 x 10^{-6} N/m^2 (or ~1367 W/m^2).	All, but effects are more dominant perturbations for higher altitude orbits.
	Body Albedo – visible light photon momentum transfer from surface reflection (for Earth ~459 W/m^2 or 10-35% of sunlight reflected on the daytime side.	All, but effects are strongest for low-altitude orbits with an inverse square drop off versus altitude.
	Body Thermal – infrared photon momentum transfer (for Earth ~230 W/m^2).	
Thrust	Acceleration due to mass expulsion (momentum transfer) during thrusting events.	All trajectories.

7.2.1 Gravitational Perturbations [Level I – Descriptive]

Gravitational perturbations result from any gravitational accelerations other than the two-body attraction that provides a conic section trajectory. These disturbing gravitational accelerations fall into two categories: a major attracting celestial body with a non-spherical mass distribution and gravitational attraction from other celestial bodies.

When the major attracting body has a mass distribution other than that of a uniform sphere, the gravitational acceleration is offset in magnitude and direction from the body's center of mass. A realistic celestial body typically has a shape that deviates from being perfectly spherical and/or has mass and density concentrations specific to that body. Fortunately, most significant celestial bodies are nearly spherical in shape. This characteristic allows for a general mathematical form to which a celestial body's particular mass distribution may be precisely characterized through a series of coefficients, defined at increasingly small resolution levels. This model, which uses *spherical harmonics*, will be presented in detail.

Gravitational attraction from other celestial bodies provides an additional acceleration that perturbs the net acceleration on a trajectory. Whether the effect of gravitational attraction from one or more other celestial bodies should be considered depends on whether the disturbance is strong enough to affect the trajectory at the desired accuracy level. The effects of a celestial body's gravity can be neglected if the ratio of its acceleration to that of the primary attracting body is below the floating-point precision of the computational platform. Often a comparison is made between including and excluding the body's gravity to determine whether the effect is enough to accept the additional computational burden.

Key Term:

Spherical Harmonics: are a mathematically method for modeling the effects of mass concentrations for a nearly spherical body. The models resolution may be increased to any level by using a greater degree and order of the infinite series.

> ➢ **Transition**: *You may continue this section with Third Body Gravity at Level II, or you may skip to the beginning of the next section called Sources of Perturbative Accelerations.*

7.2.2   Third Body Gravity [Level II – Equations]

This section provides the process for computing a gravitational acceleration from a celestial object other than the central attracting body. Spherical harmonics modeling of the central attracting body is covered in the subsequent section.

The typical presumption is that the celestial object is at a distance that the perturbation may be accurately computed by treating the object as a point mass.

The perturbation resulting from gravitational attraction from a third body ($\ddot{\bar{r}}_B$) recognizes that the additional celestial body exerts an attraction on both the central attracting celestial body and the spacecraft. Equation 7.22 provides the resulting perturbational acceleration relative to the spacecraft.

$$\ddot{\bar{r}}_B = \mu_B \left(\frac{\bar{r}_B - \bar{r}}{|\bar{r}_B - \bar{r}|^3} - \frac{\bar{r}_B}{r_B^3} \right) \qquad 7.22$$

In this equation, $\bar{r}$ is the spacecraft position relative to the central attracting body and $\bar{r}_B$ is the position of the disturbing celestial body and μ_B is the gravitational parameter for the disturbing celestial body.

7.2.3   Spherical Harmonics Perturbations [Level II – Equations]

A celestial body's gravitational field may be empirically fit to a mathematically convenient representation. Since many celestial bodies are nearly spherical shape, spherical coordinates are the chosen system. In the Earth-fixed system, theta (θ)

represents longitude and phi (ϕ) represents co-declination. The spherical coordinates are illustrated in figure 7-3.

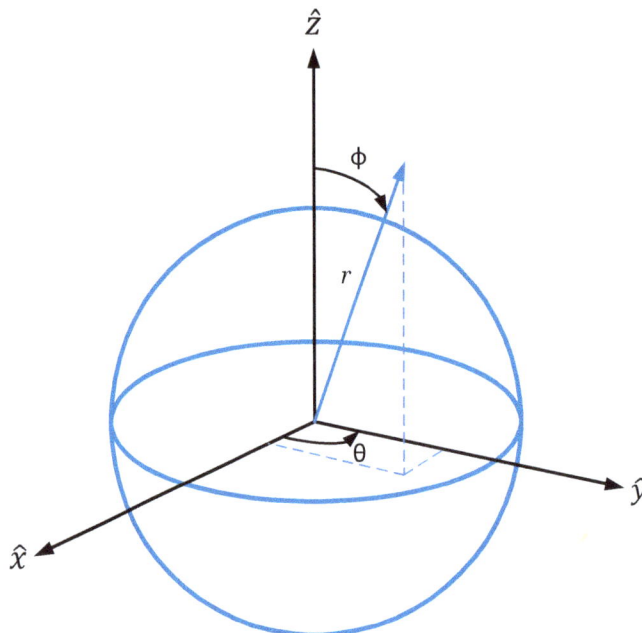

Figure 7-3 Spherical Coordinate System

Note that co-declination is selected over the ellipsoidal latitude counterpart to declination since this is easily mathematically related to Cartesian coordinates in closed form in both the forward and inverse transformations. The inverse will allow the Cartesian orbital position to be easily related to the spherical gravity field. The conversion between spherical and Cartesian coordinates is as follows:

$$\bar{r} = \begin{bmatrix} r \cos\theta \sin\phi \\ r \sin\theta \sin\phi \\ r \cos\phi \end{bmatrix} \qquad 7.23$$

The spherical elements may be recovered by:

$$r = |\bar{r}| \qquad \tan\theta = \frac{r_y}{r_x} \qquad \phi = \sin^{-1}\frac{r_z}{r} \qquad 7.24$$

The θ angle spans $0 \le \theta \le 2\pi$ radians, requiring quadrant resolution for which a two-argument inverse tangent function is recommended.

The next step is developing a functional form to relate gravitational potential to position in spherical coordinates. This form also needs to provide for coefficients that may be fit to the body's actual mass distribution such that the gravitational potential may be modeled to any desired accuracy level. This method is directly applicable to solid celestial bodies that are approximately spherical in shape. The mathematical form remains the same, with the only differences being the unique coefficients that describe a particular body's mass distribution.

The process begins with the first principles of Laplace's equation represented in spherical coordinates and produces the gravitational potential in spherical harmonics form. Certain derivations are omitted as they deviate too far from the central intent presented here. These are the conversion of Laplace's equation from Cartesian to Spherical and the solution of Legendre's differential equation. The presentation of Laplace's equation in spherical coordinates is sufficiently ubiquitous in many mathematical and scientific references and textbooks, that it may be considered a given. Likewise, the solution of Legendre's differential equation is well known. Nevertheless, the interested reader may refer to Kaula (2000) for both derivations.

Many of the assumptions made in the mathematical manipulations are not at all apparent and are not at all obvious why they are chosen. They are typically the result of numerous cases of laborious attempts until a desired result is achieved. As such, the intent herein is to show that given the assumptions, the desired result is achieved by a legitimate mathematical process.

The process begins with Laplace's Equation for scalar potential (V) outside the mass distribution:

$$\nabla^2 V = 0 \qquad\qquad 7.25$$

Solutions of Laplace's Equation in spherical coordinates are known as *spherical harmonics functions*. The representation of $\nabla^2 V$ in spherical coordinates is available in many standard references:

$$\nabla^2 V = \frac{1}{r^2}\frac{\partial}{\partial r}\left(r^2\frac{\partial V}{\partial r}\right) + \frac{1}{r^2 \sin\phi}\frac{\partial}{\partial \phi}\left(\sin\phi\frac{\partial V}{\partial \phi}\right) + \frac{1}{r^2 \sin^2\phi}\frac{\partial^2 V}{\partial \theta^2} = 0 \qquad 7.26$$

Substituting declination for co-declination ($\delta = \pi/2 - \phi$) and longitude for theta ($\lambda = \theta$) slightly alters the form to accommodate these convenient parameters:

$$\nabla^2 V = \frac{1}{r^2}\frac{\partial}{\partial r}\left(r^2\frac{\partial V}{\partial r}\right) + \frac{1}{r^2 \cos\delta}\frac{\partial}{\partial \delta}\left(\cos\delta\frac{\partial V}{\partial \delta}\right) + \frac{1}{r^2\cos^2\delta}\frac{\partial^2 V}{\partial \lambda^2} = 0 \qquad 7.27$$

Multiply both sides by r^2 to eliminate the common $1/r^2$ term:

$$r^2\nabla^2 V = \frac{\partial}{\partial r}\left(r^2\frac{\partial V}{\partial r}\right) + \frac{1}{\cos\delta}\frac{\partial}{\partial \delta}\left(\cos\delta\frac{\partial V}{\partial \delta}\right) + \frac{1}{\cos^2\delta}\frac{\partial^2 V}{\partial \lambda^2} = 0 \qquad 7.28$$

The solution is constrained to be the product of three independent functions of radius (R), declination (Δ), and longitude (Λ) for convenience:

$$V = R(r)\Delta(\delta)\Lambda(\lambda) \qquad\qquad 7.29$$

Equation 7.29 is then substituted into equation 7.28 and divided by $R\Delta\Lambda$ to get:

$$\frac{1}{R}\frac{d}{dr}\left(r^2\frac{dR}{dr}\right) + \frac{1}{\Delta\cos\delta}\frac{d}{d\delta}\left(\cos\delta\frac{d\Delta}{d\delta}\right) + \frac{1}{\Lambda\cos^2\delta}\frac{d^2\Lambda}{d\lambda^2} = 0 \qquad\qquad 7.30$$

Note the following derivative substitutions below. Note also that partial derivatives are no longer needed since the variables are separated.

$$\frac{\partial V}{\partial R} = \Delta\Gamma\frac{dR}{dr} \qquad\qquad \frac{\partial V}{\partial\Delta} = R\Gamma\frac{dV}{d\Delta} \qquad\qquad \frac{\partial V}{\partial\Gamma} = R\Delta\frac{dV}{d\Gamma} \qquad\qquad 7.31$$

The leftmost term of equation 7.30 is only a function of r. Because of that versus the form of the full equation, this term must equate to a constant that equates to: $n(n+1)$; a form that will be convenient later.

$$\frac{1}{R}\frac{d}{dr}\left(r^2\frac{dR}{dr}\right) = n(n+1) \qquad\qquad 7.32$$

Multiplying this expression by R and carrying out the differentiation yields:

$$r^2\frac{d^2R}{dr^2} + 2r\frac{dR}{dr} - n(n+1)R = 0 \qquad\qquad 7.33$$

By inspecting this form, it is noted that R and each of its derivatives are multiplied by an equivalent power of r. This suggests R has the form: r^k as evaluated below. Taking the derivatives:

$$r^2\frac{d^2r^k}{dr^2} + 2r\frac{dr^k}{dr} = n(n+1)r^k \qquad\qquad 7.34$$

$$k(k+1)r^2r^{k-2} + 2krr^{k-1} = n(n+1)r^k \qquad\qquad 7.35$$

Equation 7.35 reduces algebraically to:

$$k(k+1) = n(n+1) \qquad\qquad 7.35$$

The most obvious solution to the above expression is: $k = n$. However, there is also a second solution of: $k = -(n+1)$. The general solution for the differential equation is the sum of the individual solutions:

200

$$R = \alpha r^n + \beta r^{-(n+1)} \qquad 7.36$$

In this expression, α and β are arbitrary constants. Since this is solving for a gravity field in which R must go to zero at an infinite distance, α must equal zero. Thus:

$$R = \beta r^{-(n+1)} \qquad 7.37$$

The angular solutions are determined by back substituting the constant for R into our most recent form of Laplace's equation and multiplying through by $cos^2\delta$ to get:

$$n(n+1)cos^2\delta + \frac{cos\,\delta}{\Delta}\frac{d}{d\delta}\left(cos\,\delta\frac{d\Delta}{d\delta}\right) + \frac{1}{\Lambda}\frac{d^2\Lambda}{d\lambda^2} = 0 \qquad 7.38$$

Since the rightmost term is only a function of λ, it must also equate to a constant designated as $-m^2$, again for convenience.

$$\frac{1}{\Lambda}\frac{d^2\Lambda}{d\lambda^2} = -m^2 \qquad 7.39$$

With this form, there is a general solution for Λ that fits the differential equation 7.39, in which the c and s coefficients are arbitrary constants:

$$\Lambda = c\cos m\lambda + s\sin m\lambda \qquad 7.40$$

Substituting $-m^2$ for the rightmost term of Laplace's equation 7.38 and multiplying through by $\Delta/cos^2\delta$, the result is an equation that is only a function of δ:

$$\frac{1}{cos\,\delta}\frac{d}{d\delta}\left[cos\,\delta\frac{d\Delta}{d\delta}\right] + \left[n(n+1) - \frac{m^2}{cos^2\delta}\right]\Delta = 0 \qquad 7.41$$

Now substituting $x = \sin\delta$ and noting consequentially that: $dx = \cos\delta\,d\delta$, produces:

$$\frac{d}{dx}\left[(1-x^2)\frac{d\Delta}{dx}\right] + \left[n(n+1) - \frac{m^2}{1-x^2}\right]\Delta = 0 \qquad 7.42$$

After differentiating, equation 7.42 reduces to *Legendre's Associated Differential Equation*:

$$(1-x^2)\frac{\Delta}{dx^2} - 2x\frac{d\Delta}{dx} + \left[n(n+1) - \frac{m^2}{1-x^2}\right]\Delta = 0 \qquad 7.43$$

The solutions of Legendre's associated differential equation are the associated Legendre functions, which are recursive:

$$\Delta = P_n^m(x) \qquad\qquad 7.44$$

$$P_n^m(x) = (-1)^m (1 - x^2)^{m/2} \frac{d^m}{dx^m} P_n(x) \qquad\qquad 7.45$$

The $P_n(x)$ functions $(m = 0)$ are Legendre polynomials which may be generated using the Rodrigues' recursion formula:

$$P_n(x) = \frac{1}{2^n n!} \frac{d^n}{dx^n} [(x^2 - 1)^n] \qquad\qquad 7.46$$

The first five Legendre polynomials (i.e., $n = 0 \cdots 4$) are listed below:

$$P_0(x) = 1$$

$$P_1(x) = x$$

$$P_2(x) = \frac{1}{2}(3x^2 - 1) \qquad\qquad 7.47$$

$$P_3(x) = \frac{1}{2}(5x^2 - 3x)$$

$$P_4(x) = \frac{1}{8}(3x^5 - 30x^2 + 3)$$

Recalling that: $x = \sin \delta$:

$$P_0(\sin \delta) = 1$$

$$P_1(\sin \delta) = \sin \delta$$

$$P_2(\sin \delta) = \frac{3}{2} sin^2 \delta - \frac{1}{2} \qquad\qquad 7.48$$

$$P_3(\sin \delta) = \frac{5}{2} sin^2 \delta - \frac{3}{2} \sin \delta$$

$$P_4(\sin \delta) = \frac{3}{8} sin^5 \delta - \frac{15}{4} sin^2 \delta + \frac{3}{8}$$

The solutions of Δ for $m = 0$ (i.e., Legendre Polynomials) represent the axially (or rotationally) symmetric components of the scalar potential. The contributions by these terms, which vary only with latitude, are called *zonal harmonics*.

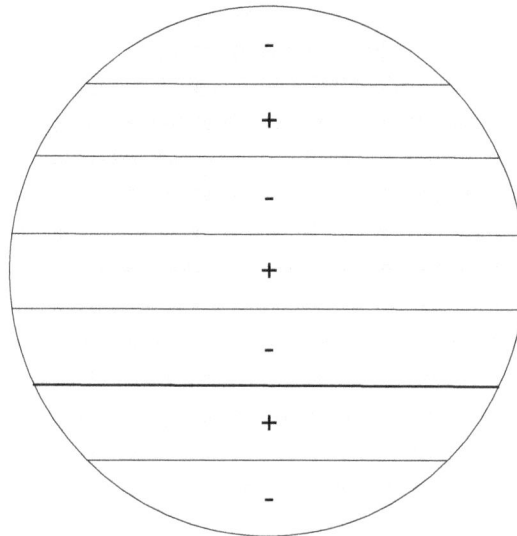

Figure 7-4 Zonal Spherical Harmonics

The solutions involving associated Legendre functions where $m = n$, known as *sectoral harmonics*, vary only with longitude.

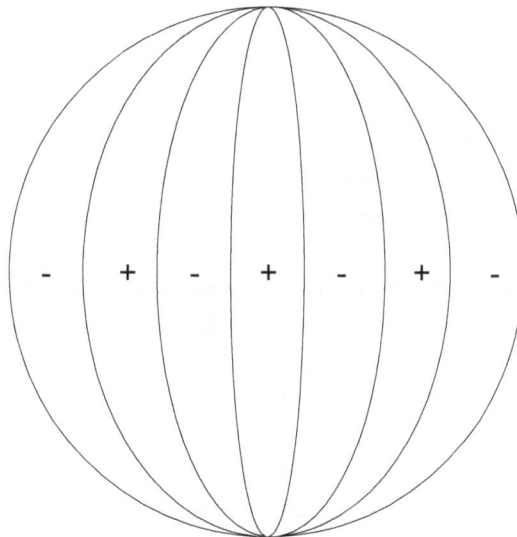

Figure 7-5 Sectoral Spherical Harmonics

The solutions involving associated Legendre functions where $0 < m \leq n$, known as *tesseral harmonics*, vary with both longitude and latitude.

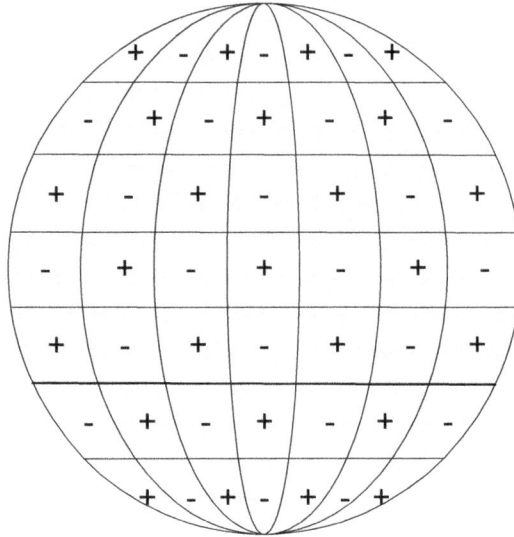

Figure 7-6 Tesseral Spherical Harmonics

The general solution of the scalar potential is the product of the solutions for $R, \Delta,$ and Λ.

$$V = \sum_{n=1}^{\infty} \sum_{m=0}^{n} \frac{1}{r^{n+1}} [c_n^m \cos(m\lambda) + s_n^m \sin(m\lambda)] P_n^m(\sin \delta) \qquad 7.49$$

In this solution, the cosine (c_n^m) and sine (s_n^m) related coefficients are an infinitely growing set of empirically fit values. They are specific to the body being modeled and represent the arrangement of density and mass concentrations (or deviations from a homogeneous sphere). (Note that the β constant in the solution for R was absorbed as factors in the c_n^m and s_n^m coefficients.)

Equation 7.50 is the perturbed acceleration due to the celestial body's mass distribution is the gradient of the scalar potential in equation 7.49 ($\ddot{\vec{r}} = \nabla V$).

$$\ddot{\vec{r}} = \begin{bmatrix} \sum_{n,m} \ddot{x}_{n,m} \\ \sum_{n,m} \ddot{y}_{n,m} \\ \sum_{n,m} \ddot{z}_{n,m} \end{bmatrix} \qquad 7.50$$

The interested reader should refer to Montenbruck and Gill (2000) and Cunningham (1970) for implementation details, including the Legendre function recursions.

Key Terms:

Sectoral Harmonics: are spherical harmonics that depend only on longitude variations.
Tesseral Harmonics: are spherical harmonics that depend on both latitude and longitude variations.
Zonal Harmonics: are spherical harmonics that depend only on latitude variations.

7.2.1 Atmospheric Drag Acceleration [Level I – Descriptive]

Atmospheric drag accelerations result from the spacecraft impacting upper atmospheric gas atoms and molecules, resulting in a momentum transfer. While the effect of drag highly depends on the spacecraft's cross-sectional density (i.e., the ratio of area perpendicular to the velocity, divided by the mass), some general characterizations can be made for a typical spacecraft makeup versus the atmospheric density as a function of altitude.

The perturbational effect of Earth atmospheric drag is observable at altitudes below 1000 km and effectively unmeasurable above 2500 km altitudes. Drag effects are particularly significant below a 600 km altitude and particularly acute below a 300 km altitude. A 100 km altitude (i.e., the Kármán line) is the threshold below which atmospheric drag makes orbital flight unsustainable.

Drag is a potential perturbation to trajectories when the major attracting celestial body has a significant atmosphere.

➢ **Transition**: *You may continue this section with Atmospheric Drag Acceleration at Level II, or you may skip to the beginning of the next section called Radiation Pressure Acceleration.*

7.2.2 Atmospheric Drag Acceleration [Level II Equations]

Atmospheric drag acceleration depends highly on atmospheric density (ρ), the square of the velocity at which the spacecraft is moving through the atmosphere, the spacecraft shape-induced drag coefficient (C_D), and the spacecraft's cross-sectional density normal to its velocity. Equation 7.51 is the perturbative drag acceleration.

$$\ddot{\bar{r}}_D = -\frac{1}{2} C_D \frac{A}{m} \rho v_e^2 \hat{v}_e \qquad\qquad 7.51$$

The drag coefficient (C_D) is a dimensionless parameter indicating the resistance to atmospheric gas flow. A sphere or hemisphere has a C_D value equal to 1.0 while a more streamlined shape would have a value between 0 and 1. A flat plate normal to the flow would have a C_D value of 2. Typical C_D values for a spacecraft are approximately 2.1.

The cross-sectional density is the ratio of the cross-sectional area (A) to mass (m). The applicable area is measured normal to the velocity at which the spacecraft is moving relative to the atmosphere ($\bar{v}_e$). Since the atmosphere is presumed to rotate with the Earth, the applicable velocity is an Earth-fixed velocity. Equation 7.52 Computes a relative velocity when only an inertial velocity ($\bar{v}$) is available. The Earth rotational velocity is the Earth rotation rate in the $\hat{k}$ direction ($\bar{\omega} = \omega\hat{k}$). The relative velocity is computed by subtracting the cross product of the Earth rotational velocity and the inertial position ($\bar{r}$) from the inertial velocity.

$$\bar{v}_e = \bar{v} - \bar{\omega} \times \bar{r} \qquad\qquad 7.52$$

The atmospheric density (ρ) decreases with altitude consistent with the hydrostatic equation for the static models. Dynamic atmospheric density models tend to have more empirical elements. These model a different density versus altitude profile at a diurnal bulge caused by atmospheric solar heating. They are also adaptive to current solar activity to the degree that the density profiles are adapt based on measured 10.7 cm (F10.7) solar flux values and the planetary geomagnetic indices (a_p or k_p) that are available from the National Oceanic and Atmospheric Administration (NOAA). Commonly-used dynamic Earth atmospheric density algorithms include the Harris-Priester, Jacchia-Roberts, and the Naval Research Laboratory (NRL) Mass Spectrometer Incoherent Scatter (MSIS) models.

7.2.3   Radiation Pressure Acceleration [Level I – Descriptive]

Radiation pressure accelerations result from the spacecraft absorbing or reflecting incident photons, resulting in a momentum transfer. Radiation pressure effects are readily observable for photons in the visible and infrared regions of the electromagnetic spectrum.

Radiation pressure perturbations require a direct path from the source to the spacecraft. Solar radiation effects occur fully when the spacecraft is fully illuminated and have no effect when the Sun is fully occulted (i.e., eclipsed) by another celestial body such as the Earth or Moon. Effective modeling also applies partial effects when the Sun is partially occulted.

Radiation that occurs from sunlight reflecting off the surface of another celestial body also perturbs trajectories. *Albedo* is a physical parameter indicating the fraction of incident sunlight a celestial body reflects.

The effect of reflected visible radiation also depends on the fraction of the celestial body's illuminated surface that is directly visible to the spacecraft. With infrared radiation, however, the body itself is the radiation source. The entire surface emits radiation, with strength as a function of absolute temperature, in accordance with the Stefan-Boltzmann law.

Radiation pressure always decreases by the inverse square of the distance between the source and the spacecraft.

> ➢ **Transition**: *You may continue this section with Radiation Pressure Acceleration at Level II, or you may skip to the beginning of the next section called Thrust Acceleration.*

7.2.4   Radiation Pressure Acceleration [Level II – Equations]

A simplified solar radiation pressure algorithm is used to model the average effect on the spacecraft. The solar radiation flux at one Astronomical Unit (AU = 149597870700 m) is $\Phi = 1367\ W/m^2$. The corresponding solar radiation pressure ($P_\odot$) is obtained by dividing the flux by the speed of light ($c = 299792458\ m/s$) to obtain $P_\odot \approx 4.56 \times 10^{-6}\ N/m^2$. Equation 7.53 is the perturbative acceleration ($\ddot{\vec{r}}_R$) due to solar radiation pressure.

$$\ddot{\vec{r}}_R = -P_\odot C_R \frac{A}{m} \frac{\vec{r}_\odot}{r_\odot^3} \text{AU}^2 \qquad\qquad 7.53$$

The area to mass cross-sectional density (A/m) is like what was used in the drag computation. The area value may however be different if a different profile is consistently sun-facing. The Sun's position relative to the spacecraft ($\vec{r}_\odot$) is expressed in units of meters, with the AU^2 term converting the square of the distance to inverse an inverse square pressure drop off in AU units.

Equation 7.54 relates the dimensionless radiation pressure coefficient (C_R) to the spacecraft's average reflectivity (ε).

$$C_R = 1 + \varepsilon \qquad\qquad 7.54$$

207

A zero reflectivity ($\varepsilon = 0$) would indicate complete absorption, while $\varepsilon = 1$ would correspond to complete specular reflectivity like a highly polished mirror. Thus, valid the range for the radiation pressure coefficient is $1 \leq C_R \leq 2$ with 1.4 being a typical value for a spacecraft.

The Earth has an average albedo of approximately 0.34, providing a surface radiance of 459 W/m^2 with an equivalent $1.53 \times 10^{-6} N/m^2$ radiation pressure in the visible spectrum. The surface infrared emission is 230 W/m^2 with an equivalent $7.67 \times 10^{-7} N/m^2$ radiation pressure. Modeling the acceleration is more difficult than solar radiation pressure, particularly for the contribution from visible light portion. Since the effect is smaller and drops off quickly with altitude, these perturbative accelerations are often neglected.

7.2.5 Thrust Acceleration

Thrust accelerations may be included as net acceleration when the trajectory is predicted using numerical integration techniques. Acceleration is the ratio of the thrust force, divided by the instantaneous spacecraft mass. Since acceleration is also a function of mass, rigorous modeling dictates that the mass depletion due to exhaust gas expulsion also be modeled. Chapter 4 provides guidance for modeling propulsion system thrust and mass depletion rates for inclusion in a numerical integration model.

7.3 Special Orbit Types

Special orbit types presented in this section are those that have useful operational applications in certain Earth-orbiting contexts. These orbits leverage standard orbital properties, plus Earth oblateness-induced perturbations in two instances to produce orbits with desirable features.

7.3.1 Geostationary Earth Orbit (GEO)

The *Geostationary Earth Orbit (GEO)* is circular with a period that matches the Earth's inertial rotation rate and an equatorial inclination. The GEO was first proposed in 1945 by Arthur C. Clarke and is thus often referred to as a Clarke orbit. The combination of the orbital revolution rate matching Earth rotation and the equatorial inclination gives the illusion of a stationary spacecraft, allowing direct communication from an antenna on the Earth's surface pointed in a fixed direction.

While a mean *solar day* is 24 hours in length, the Earth does a 360-degree rotation in a *sidereal day* which is approximately 23 hours, 56 minutes, 4.09 seconds in length. The difference between a solar and sidereal day stems from a solar day being the interval between successive solar upper transits (i.e., noon events) and the combined motion of Earth rotation and the revolution around the Sun as illustrated in figure 7-7. Since the Earth orbits the Sun in approximately 365.25 days, its orbital rate is approximately 0.9856 degrees per day. Thus, in the interval of a solar day, the Earth rotates nearly 361 degrees.

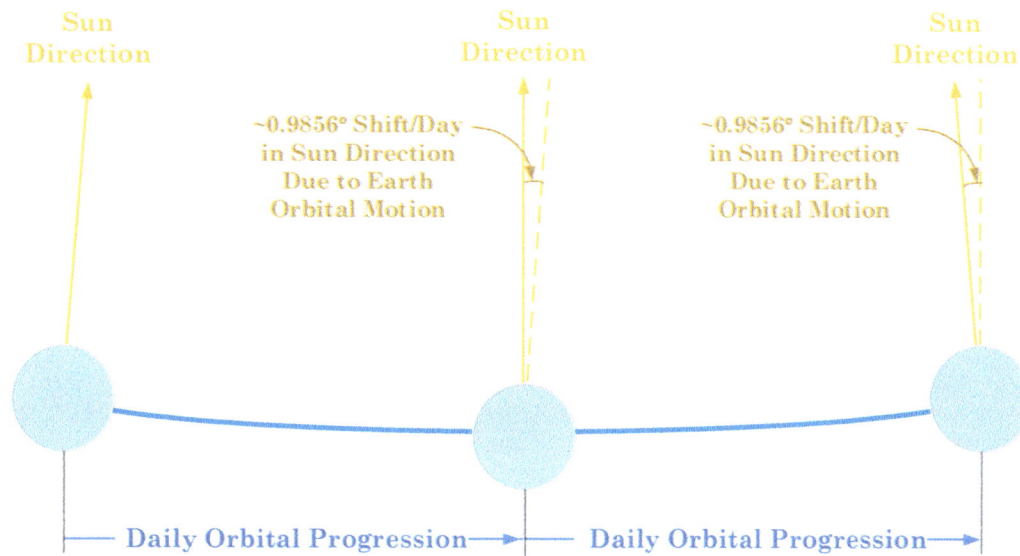

Figure 7-7 Earth Rotational and Orbital Motion Combined

7.3.2   Sun-Synchronous Orbit

A *sun-synchronous orbit* maintains a constant angle between the Sun and the orbital plane (a.k.a. the *beta angle*) as illustrated in figure 7-8. The angle between the Sun and angular momentum (i.e., the complement to the beta angle) must then also remain constant. When considering that the Earth orbits the Sun at approximately 0.9856 degrees per day, such an orbital property would violate the law of conservation of angular momentum for a two-body orbit. Such an orbital property is therefore only possible by leveraging the perturbative nodal drift rate induced by Earth oblateness as seen in section 7.1.1.1.

Equation 7.1 indicates the orbital inclination must be retrograde (i.e., greater than 90 degrees) for a positive nodal drift rate. Figure 7-1 likewise shows that (depending on altitude) an orbital inclination between 95 and 105 degrees will provide the 0.9856 degree per day nodal drift rate needed to maintain a constant beta angle throughout the year. The specific orbital inclination (i) for an orbit or semi-major axis a and eccentricity e is determined by solving equation 7.1 in terms of inclination. The nodal drift rate ($\dot{\Omega}$) is set to the Earth's orbital rate of 2π radians per year (365.25 days or 31557600 seconds). This equates to a nodal drift rate of $\dot{\Omega} \approx 1.991 \times 10^{-7}$ rad/s.

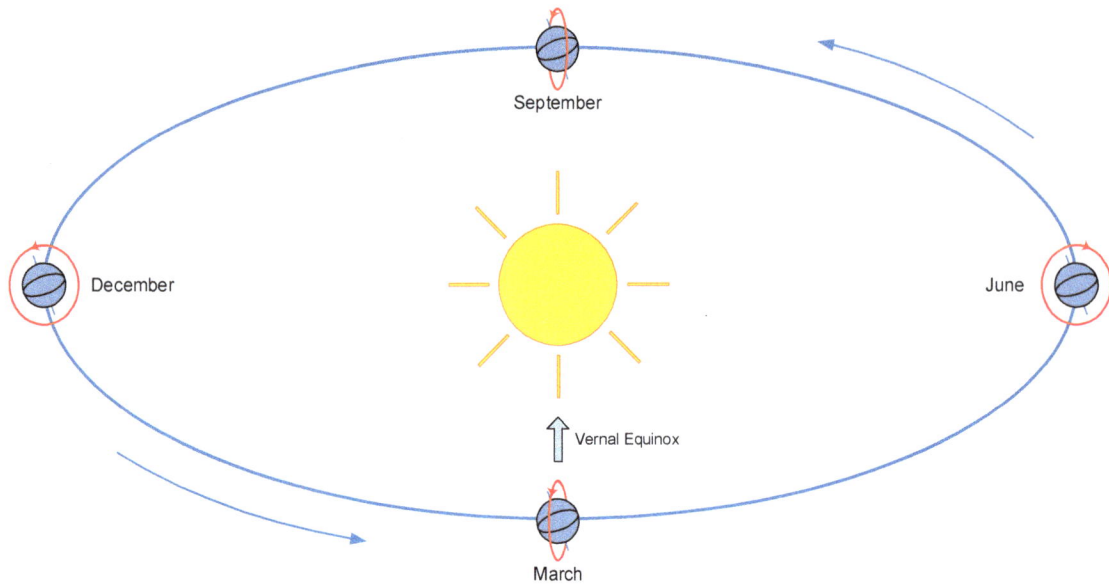

Figure 7-8 Sun-Synchronous Orbit

7.3.3 Molniya Orbit

Molniya (Молния) is the Russian word for lightning. The typically high latitude of the Russian mainland makes it advantageous to use several highly elliptical orbits with a 12-hour orbital period to have a continuous communications capability. The high eccentricity ($e > 0.7$) causes the satellite to spend most of the time near apoapsis with coverage to entire Russian mainland. Figure 7-9 illustrates a Molniya orbit.

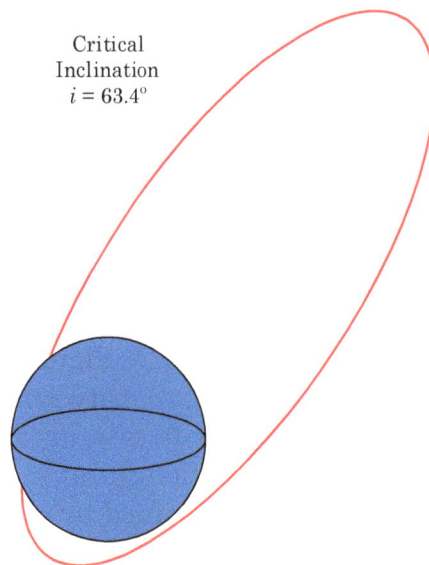

Figure 7-9 Molniya Orbit

The argument of periapsis is set at 270 degrees to place apoapsis at the orbit's highest latitude. The Molniya orbit uses a 63.4-degree inclination to avoid the apsidal drift perturbation that would otherwise occur due to Earth oblateness.

Key Terms:

Geostationary Earth Orbit: is a circular equatorial orbit at an altitude in which the orbital rate matches the Earth's rotational rate. Thus, it appears stationary to an Earth-bound observer.

Sun-Synchronous Orbit: is a retrograde polar orbit with perturbational nodal drift that matches the Earth's orbital rate around the Sun. Thus, its orbital plane maintains a constant Sun angle.

Molniya Orbit: is a highly elliptical 12-hour orbit at the 63.4° critical inclination that has no perturbational apsidal drift. Thus, apogee remains fixed to the northern hemisphere, facilitating communications through high latitude locations.

REFERENCES

1. Escobal, Pedro R., *Methods of Orbit Determination*, © 1965 John Wiley & Sons, Inc., ISBN 0-88275-319-3.
2. Escobal, Pedro R., *Methods of Astrodynamics*, © 1968 John Wiley & Sons, Inc., SBN 471-24528-3.
3. Curtis, Howard D, *Orbital Mechanics for Engineering Students*, Fourth Edition, © 2020 Elsevier, Ltd., ISBN 978-0-08-102133-0.
4. Herrick, Samuel, *Astrodynamics*, © 1971 by author, Van Nostrand Reinhold Company, Library of Congress Card Catalog Number 78-125199.
5. Hoots, Felix R. and Roehrich, Ronald L., *Models for Propagation of NORAD Element Sets*, Spacetrack Report No. 3, December 1980.
6. Vallado, David A., et al., *Revisiting Spacetrack Report #3*, AIAA 2006-6753, (https://celestrak.org/publications/AIAA/2006-6753/).
7. Kaplan, Marshall H., *Modern Spacecraft Dynamics & Control*, © 1976 John Wiley & Sons, Inc., ISBN 0-471-45703-5.
8. Vallado, David A., *Fundamentals of Astrodynamics and Applications*, Second Edition, © 2001 by author, Microcosm Press and Kluwer Academic Publishers, ISBN 1-881883-12-4.
9. Montenbruck, Oliver and Gill, Eberhard, *Satellite Orbits, Models, Methods, and Applications*, © 2000 Springer-Verlag, ISBN 3-540-67280-X.
10. Bate, Roger R. et al., *Fundamentals of Astrodynamics*, © 1971 Dover Publications, Inc., ISBN 0-486-60061-0.
11. Cunningham, Leland E., *On the Computing of the Spherical Harmonic Terms Needed During the Numerical Integration of the Orbital Motion of an Artificial Satellite*, 1970, Celestial Mechanics 2 pp. 207-216.

12. Gurfil, Pini and Seidelmann, P. Kenneth, *Celestial Mechanics and Astrodynamics: Theory and Practice*, © 2016 Springer-Verlag, ISBN 978-3-662-50368-3.
13. Kaula, William M., *Theory of Satellite Geodesy, Applications of Satellite Geodesy*, © 2000 by author, Dover Publications, Inc., ISBN 0-486-41465-5.
14. Farlow, Stanley J., *Partial Differential Equations for Scientists and Engineers,* © 1982 by author, Dover Publications, Inc., ISBN 0-486-67620-3.
15. Harper, Charlie, *Introduction to Mathematical Physics*, ©1976 Prentice-Hall, Inc., ISBN 0-89464-020-8.
16. Myint-U, Tyn, *Partial Differential Equations of Mathematical Physics,*_© 1973 Elsevier Scientific Publishing Company, ISBN 0-444-00132-8.
17. Beyer, William H., *CRC Standard Mathematical Tables*, 27th Edition, ©1973, CRC Press, ISBN 0-8493-0627-2.
18. Kreyszig, Erwin, *Advanced Engineering Mathematics*, ©1962 John Wiley & Sons, Inc., ISBN 0-471-15496-2.
19. Lorrain, Paul and Corson, Dale, *Electromagnetic Fields and Waves*, ©1970 W.H. Freeman Company, ISBN 0-7167-0331-9.

8 Trajectory Events Prediction

Trajectory events prediction facilitates the ability to determine when events such as time periods when ground antennas or optical trackers have a direct line-of-sight path with a spacecraft. While "brute force" methods can be used to determine the time when various events occur, such an approach is costly in terms of time and computing resources. Instead, elegant methods are devised using differential calculus of maxima and minima to quickly predict when various events occur.

8.1 Ground Site Prediction and Tracking

Events of interest for ground sites include acquisition prediction and radio frequency interference (RFI) screening. These typically provide predictions of when a ground antenna can communicate or perform other interactions with a spacecraft. The RFI screening determines time periods in which another spacecraft (or a radio source such as the Sun) would cross the antenna-to-spacecraft line-of-sight, causing potential communications interference.

8.1.1 Acquisition Prediction [Level I – Descriptive]

Predicting acquisition intervals and associated geometry are fundamental activities for space flight operations. This creates a schedule of when spacecraft are accessible to various ground sites. An acquisition event (sometimes referred to as a "pass") typically has information including *Acquisition of Signal* (AOS) information, the information pertinent to maximum elevation above the horizon, and the *Loss of Signal* (LOS) information.

Pertinent AOS information includes the date/time and azimuth (*Az*) of the AOS event. If there is a minimum elevation other than zero (i.e., AOS occurs either above or below the local horizon), the AOS elevation angle (*El*) should also be provided. Range (*Rng* or ρ) and range rate (*RngRate* or $\dot{\rho}$) may also be provided to facilitate signal strength estimates and Doppler correction to the communication radio frequency.

The spacecraft's maximum elevation during an acquisition event is sometimes referred to as the *culmination*. For circular orbits, the culmination will also correspond to spacecraft's closest range to the ground site during the pass. Pertinent culmination information includes the date/time, azimuth angle (*Az*), elevation angle (*El*), and the slant range (*Rng* or ρ).

Pertinent LOS information is like AOS information in that date/time and azimuth angle (*Az*) are the minimal parameters reported. When applicable, the LOS elevation angle (*El*), range (*Rng* or ρ) and range rate (*RngRate* or $\dot{\rho}$) are also reported. Figure 8-1 depicts an acquisition event on an *Az/El* radar plot.

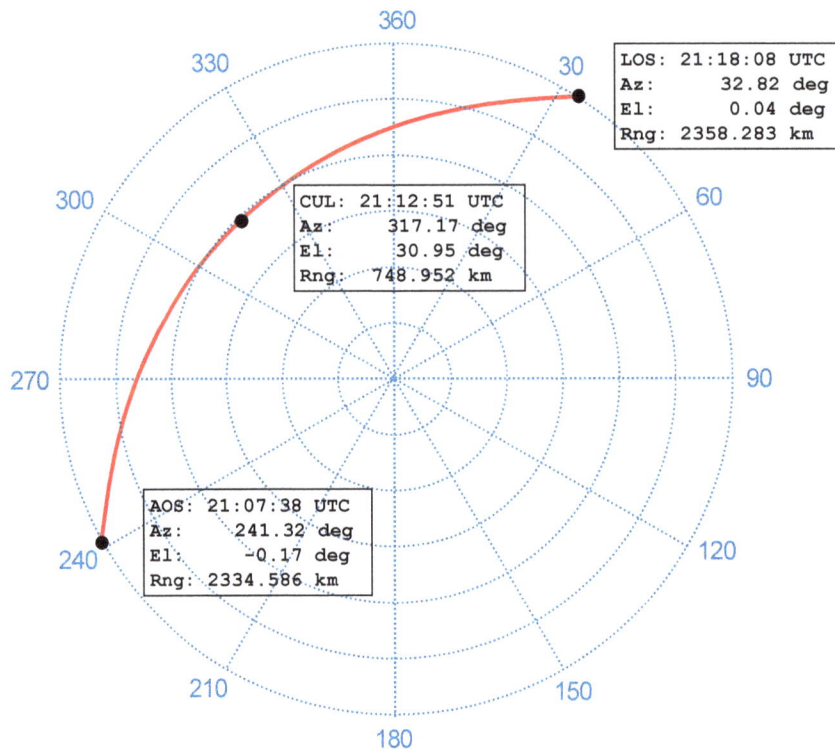

Figure 8-1 Acquisition Az/El Radar Plot

It should be noted that while an altitude/azimuth pedestal gimbaled to the Az/El angular geometry is common, there are alternative gimbal arrangements on tracking pedestals. Optical telescopes commonly have equatorial mounts gimbaled to right ascension (*Ra* or α) and declination (*Dec* or δ) angles. Some antennas are mounted on an alternative X/Y pedestal that moves the *keyhole* (i.e., gimbal lock region) from the overhead position of an altitude/azimuth pedestal to the horizon to facilitate communications continuity on high elevation passes. Regardless of the pedestal mount being used, the algorithm in the equations section can be used to make pass predictions.

Key Terms:

Acquisition of Signal (AOS): is the event where a ground antenna acquires the signal of a satellite rising above its horizon.
Culmination: is the maximum elevation a satellite reaches relative to a ground tracking site.
Keyhole: is the region where the azimuth rate exceeds ability of the tracking system to maintain line-of-sight pointing.
Loss of Signal (LOS): is the event where a ground antenna loses the signal of a satellite setting below its horizon.

> ➤ **Transition**: *You may continue this section with Acquisition Prediction at Level II, or you may skip to the beginning of the next section called Computing Look Angles.*

The acquisition prediction algorithm searches for relative maxima instances for the elevation angle. When a maximum elevation is determined to be above the local horizon, it performs a backward and forward search to determine the associated AOS and LOS times. The date/times for AOS, culmination, and LOS plus their corresponding geometric information is saved as an acquisition event.

The algorithm transforms the spacecraft position and velocity vectors to the Earth fixed coordinates (i.e., ECEF/GCRF) to allow the ground site position vector ($\bar{G}$) to remain constant. This transformation is available in the provided in the IAU SOFA software libraries. Figure 8-1 shows the relationship between the spacecraft position ($\bar{r}$), the ground site position, and the slant range vector ($\bar{\rho}$).

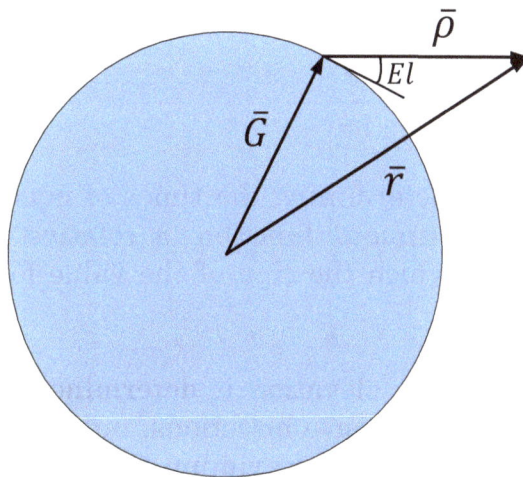

Figure 8-2 Range from Site and Spacecraft Positions

Equation 8.1 computes the slant range as a vector subtraction. Equation 8.2 computes the sine of the elevation angle using the dot product, recognizing that elevation is complementary to the angle between the ground site zenith and the slant range.

$$\bar{\rho} = \bar{r} - \bar{G} \qquad\qquad 8.1$$

$$\sin El = \hat{\rho} \cdot \hat{G} \qquad\qquad 8.2$$

The maximum elevation occurs when the time rate of change of the $\sin El$ is zero as computed in equation 8.5, with equations 8.3 and 8.4 providing preliminary quantities. Note that equation 8.5 presumes body-fixed vectors, causing zero derivatives of $\hat{G}$.

$$\dot{\rho} = \frac{\bar{\rho} \cdot \bar{v}}{\rho} \qquad\qquad 8.3$$

$$\dot{\hat{\rho}} = \frac{\rho\bar{v} - \dot{\rho}\bar{\rho}}{\rho^2} \qquad\qquad 8.4$$

$$\frac{d}{dt}(\sin El) = \dot{\hat{\rho}} \cdot \hat{G} \qquad\qquad 8.5$$

Minimum and maximum elevations occur when the time rate of change of the sine of the elevation angle is zero. The process involves sampling equation 8.5 every tenth of an orbital revolution. Equation 8.6 computes time between samples (Δt), which should be rounded to the nearest whole second. Equation 2.10 provides the time rate of change of true anomaly ($\dot{\theta}$).

$$\Delta t = \frac{\pi}{5\dot{\theta}} \qquad\qquad 8.6$$

The immediate interest is for determining the times of maximum elevation. Since equation 8.5 represents a continuous function, a relative maximum elevation is isolated in a time interval in which the sign of the value for equation 8.5 changes from positive to negative.

The actual time of the maximum elevation is determined by an iterative method such as a golden search or by successive bisections, until the value for equation 8.5 is sufficiently close to zero. When the maximum (i.e., culmination time) is found, the corresponding elevation angle is computed. If the elevation angle is above the horizon (or prescribed minimum valid elevation) a valid acquisition event was found.

When a valid acquisition event is identified, a reverse and forward search about the culmination time then commences to determine the AOS and LOS times. The applicable geometric data is computed for the AOS, culmination, and LOS times. If, however, the elevation at the culmination time is below the horizon (or minimum valid elevation), the equation 8.5 sampling resumes forward of the interval when the maximum was isolated until a valid acquisition event is found.

8.1.3 Computing Look Angles

Look angles are the pedestal gimbal angles to direct an antenna or optical tracking sensor to complement the range parameter, which is the magnitude of the slant range vector. The altitude/azimuth system is a common antenna system in which the gimbals are set to azimuth (Az) and elevation (El). The equatorial system is a common optical system in which the gimbals are set to right ascension (α) and declination (δ).

It is common to tabulate the look angles at regular intervals (such as every 10 seconds for example) during an acquisition pass as an alternative or complement to the radar plot. A tabulation would have a single line for each date/time, followed by the corresponding look angles and range. The first line is for the AOS, with each subsequent line occurring at a later user-specified time step until reaching LOS.

8.1.3.1 Altitude/Azimuth Look Angles [Level I - Descriptive]

The attitude/azimuth system has Az and El look angles related to the Cartesian ENU coordinate system. The ENU coordinates system is referenced to the geoid, aligning the "Up" axis with geodetic zenith, thus placing the "East" and "North" axes in the local horizontal plane.

The azimuth is measured from the true North direction clockwise about local zenith (i.e., the "up" axis of the ENU system). Azimuth has a range from 0° to 360°. Elevation is the angle above or below the local horizontal plane. Elevation has a range from −90° to +90°, with negative values looking below the horizon and positive values looking above the horizon.

➤ **Transition**: *You may continue this section with Altitude/Azimuth Look Angles at Level II, or you may skip to the beginning of the next section called Equatorial Look Angles.*

8.1.3.2 Altitude/Azimuth Look Angles [Level II – Equations]

Equation 8.7 transforms the slant range vector from ECEF to ENU coordinates. Equation 8.8 defines the transformation in terms of the ground site geodetic latitude (L) and longitude (λ) coordinates.

$$\bar{\rho}_{ENU} = C^{ENU}_{ECEF}\bar{\rho}_{ECEF} \qquad\qquad 8.7$$

$$C_{ECEF}^{ENU} = \begin{bmatrix} -\sin\lambda & \cos\lambda & 0 \\ -\cos\lambda\sin\lambda & -\sin\lambda\sin L & \cos L \\ \cos\lambda\cos L & \sin\lambda\cos L & \sin L \end{bmatrix} \qquad 8.8$$

Equation 8.9 relates the elevation to the ratio of the slant range vertical component to its magnitude. Equation 8.10 relates the azimuth to the slant range east and north components.

$$\sin El = \frac{\rho_U}{\rho} \qquad\qquad 8.9$$

$$\tan Az = \frac{\rho_E}{\rho_N} \qquad\qquad 8.10$$

Note that a two-argument inverse tangent function is needed to preserve the signs on the east and north slant range components to produce the correct azimuth quadrant.

8.1.3.3   Equatorial Look Angles [Level I – Descriptive]

An equatorial pedestal is inclined to the local site declination, aligning what would be the azimuthal axis with the polar rotation axis. This allows a single axis correction for rotation at the planetary rotation rate (ω). Corrections by the planetary rate about this axis allow the pedestal to dwell while directed to a specific location on the celestial sphere.

Rotation about this axis other than to counter planetary rotation changes the pointing in right ascension (α). Rotation about the perpendicular axis changes the pointing in declination (δ).

➤ **Transition**: *You may continue this section with Equatorial Look Angles at Level II, or you may skip to the beginning of the next section called Radio Frequency Interference Screening.*

8.1.3.4   Equatorial Look Angles [Level II – Equations]

The equatorial coordinates are inertial in nature. The first step is therefore to convert the slant range vector from ECEF to ECI coordinates. The IAU SOFA software library transformation from ITRF to GCRS is a rigorous method to convert the slant range to an ECI frame that is parallel to the ICRF. Given the slant range in GCRS coordinates, equation 8.11 relates declination to the $\hat{k}$ component and magnitude of the GCRS slant range. Likewise, equation 8.12 relates right ascension to the GCRS slant range $\hat{\imath}$ and $\hat{\jmath}$ components.

$$\sin \delta = \frac{\rho_k}{\rho} \qquad\qquad 8.11$$

$$\tan \alpha = \frac{\rho_j}{\rho_i} \qquad\qquad 8.12$$

Note that a two-argument inverse tangent function is needed to preserve the signs on the $\hat{\imath}$ and $\hat{\jmath}$ slant range components to produce the correct right ascension quadrant.

8.1.4 Radio Frequency Interference Screening [Level I – Descriptive]

The potential for Radio Frequency Interference (RFI) exists while communicating with a spacecraft from a ground antenna when a different spacecraft operating on a similar frequency crosses the line-of-sight path as illustrated in figure 8-3. The prediction considers a small user-defined cone angle around the signal path. The conflict interval begins when the secondary spacecraft enters the cone and end ends when it exits the cone.

RFI conflicts can only occur when the ground antenna has a clear signal path with the intended spacecraft. Therefore, an efficient technique when predicting RFI conflicts constrains the search within pre-determined acquisition events.

The search proceeds by determining the relative minima of line-of-sight angles between the primary and secondary spacecraft, relative to the ground antenna. Similarly, RFI conflicts for the Sun as the secondary object may be computed if its radio emissions could interfere with signal reception. The minimum reportable data for an RFI conflict event are the times of entry, closest signal path alignment, and exit as well as the event's minimum angle between the signal paths.

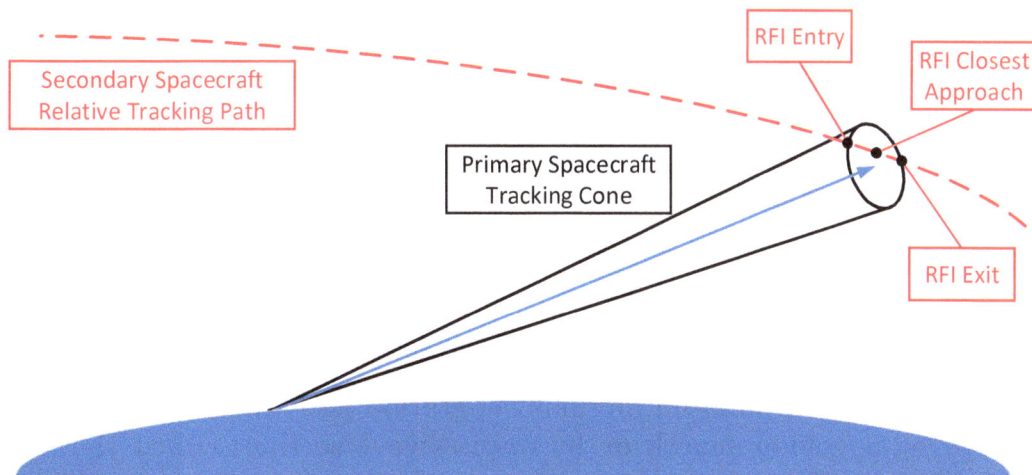

Figure 8-3 RFI Conjunction

Key Term:

> **Radio Frequency Interference (RFI)**: can occur when the lines-of-sight of two satellites using similar frequencies cross relative to a ground antenna.

> ➤ **Transition**: *You may continue this section with Radio Frequency Interference Screening at Level II, or you may skip to the beginning of the next section called Atmospheric Refraction.*

8.1.5   Radio Frequency Interference Screening [Level II – Equations]

An RFI event requires the minimum angular line-of-sight separation ($\Delta\alpha$) between the primary and secondary objects to be less than the user-defined cone angle. Equation 8.13 relates the angular separation to the primary ($\bar{\rho}_1$) and secondary ($\bar{\rho}_2$) line-of-sight unit vectors.

$$\cos\Delta\alpha = \hat{\rho}_1 \cdot \hat{\rho}_2 \qquad\qquad 8.13$$

Evaluations are sampled from the beginning to the end of the acquisition event time interval at every 1/30 of a revolution for the orbit with the faster true anomaly rate. Equation 8.14 computes time between samples *(Δt)*, which should be rounded to the nearest whole second.   Equation 2.10 provides the time rate of change of true anomaly ($\dot\theta$).   The true anomaly separation ($\Delta\theta$) AOS and LOS must also be evaluated to determine whether it is less than the user cone angle.

$$\Delta t_1 = \frac{\pi}{15\dot\theta_1} \qquad \Delta t_2 = \frac{\pi}{15\dot\theta_2} \qquad \Delta t = min(\Delta t_1, \Delta t_2) \qquad\qquad 8.14$$

The relative minimum $\Delta\alpha$ occurs at the maximum $\cos\Delta\alpha$. The minimum angular separation occurs at the time of the zero value of the cosine derivative in equation 8.15.   Equations 8.3 and 8.4 can be used to compute the time derivatives of the line-of-sight unit vectors.

$$\frac{d}{dt}(\cos\alpha) = \hat{\rho}_1 \cdot \dot{\hat{\rho}}_2 + \dot{\hat{\rho}}_1 \cdot \hat{\rho}_2 \qquad\qquad 8.15$$

Since equation 8.15 is a continuous function, a relative minimum line-of-sight angle is isolated in a time interval in which the value for equation 8.15 changes from positive to negative.

The actual time of the minimum angular separation is determined by an iterative method such as a golden search or by successive bisections, until the value for equation 8.15 is sufficiently close to zero.   When the minimum is found, the corresponding separation angle (α) is computed. If the angle is within the user cone angle, an RFI conflict event was found.

When an RFI conflict is identified, a reverse and forward search about the minimum time then commences to determine the entry and exit times. If the conflict straddles the AOS and/or the LOS, these events can define the entry and/or exit times.

If there is no minimum detected, the entry and exit times should be inspected to see if an RFI conflict persisted through the pass, with the minimum outside pass interval.

8.1.6 Atmospheric Refraction [Level I – Descriptive]

Tracking a spacecraft from a ground site located on a celestial body that has a significant atmosphere (such as Earth) has refraction as an added dynamic. The effects of refraction have a delaying property that effectively increases both the elevation angle and slant range. The azimuth angle is not affected by refraction.

For elevation, the effect causes the AOS to occur earlier and the LOS to occur later. The effects have the greatest magnitude near the horizon where the signal has the longest path through the atmosphere. At zenith, there is no effect on elevation and only a slight change to range. But since the effects persist throughout the pass, the predictions should modify the slant range direction when predicting acquisition passes and RFI events for Earth-based ground sites.

The range lengthening will be apparent in radio signals causing a slight delay in communications propagation and an increase to any ranging measurements. The corrective models tend to be empirical, requiring a unique model to be developed for each celestial body with an atmosphere.

➢ **Transition**: *You may continue this section with Atmospheric Refraction at Level II, or you may skip to the beginning of the next section called Space Platform Prediction and Tracking.*

8.1.7 Atmospheric Refraction [Level II – Equations]

The Hopfield tropospheric refraction model as modified by Goad and Saastamoinen (as presented by Montenbruck and Gill, 2000) provides accurate Earth-based elevation and range corrections. The model requires input environmental parameters for the tracking station: Temperature (K), barometric pressure (hPa), relative humidity (as a fraction between 0.0 and 1.0), and the local geoidal radius (m). The normal of the station position vector can be used as the radius. The model also requires inputs of the tracking line-of-sight elevation and slant range (m) and the signal wavelength (m) if it is in the optical range.

Since the model requires site meteorological data, the most accurate results are achieved by installing a weather station at the site to provide real-time corrections to measured tracking data. But if the site does not have a weather station or there is not easy access to the site's weather information, approximations are better than not modeling refraction effects. In such instances, effective predictions have been made by using the monthly historic average values for the site.

Table 8-1 provides an example refraction profile for a 1000-km circular Earth orbit for a span of elevation angles (like that provided by Montenbruck and Gill, 2000). The slant range corresponds to the site to spacecraft distance for the various elevation angles. The meteorological data represents example mid-latitude conditions.

Table 8-1 Refraction Profile for an Example Earth Orbit												
El (deg)	0	1	3	5	7	10	15	20	30	40	50	70	90
ΔEl (deg)	0.606	0.423	0.250	0.172	0.130	0.094	0.063	0.047	0.030	0.020	0.014	0.006	0
ΔRng (m)	90	63	47	37	19	14	9	7	5	4	3	3	2
Note: Conditions are for ranges corresponding to a 1000-km circular Earth orbit. T = 288 K P = 1013.25 hPa Rel. humid. = 0.75												

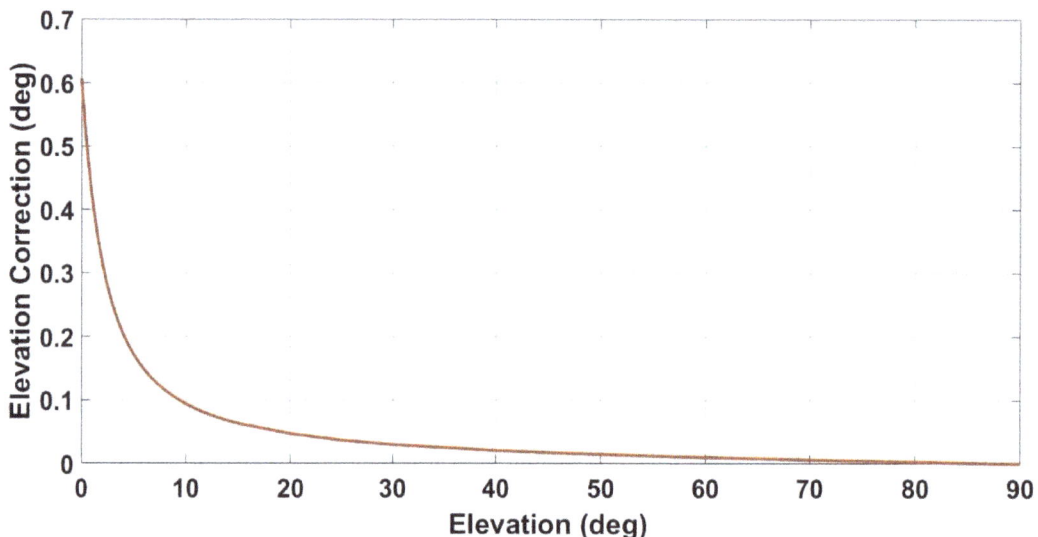

Figure 8-4 Elevation Refraction for Example Earth Orbit

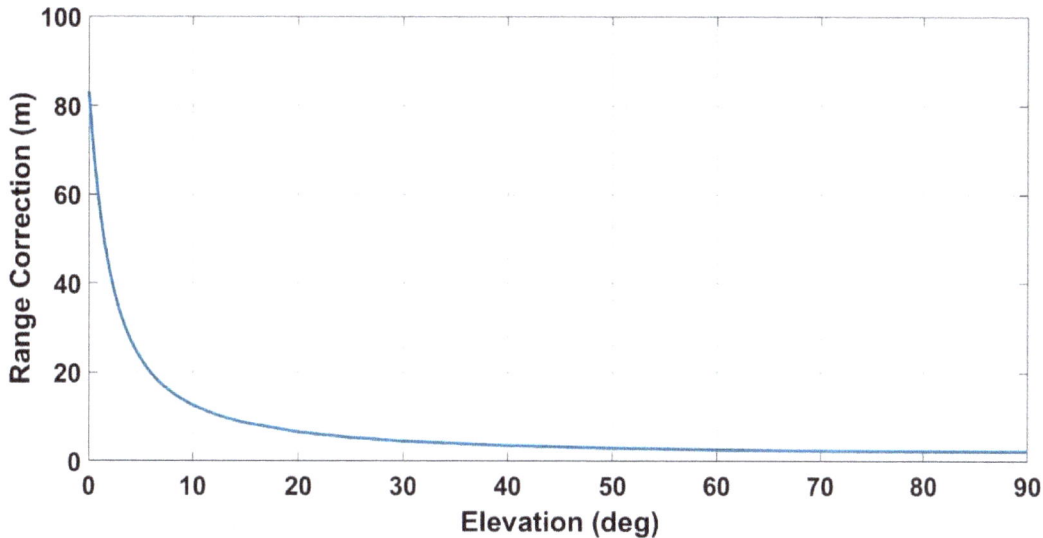

Figure 8-5 Range Refraction for Example Earth Orbit

Equation 8.16 converts the temperature (T) from Kelvins to degrees Celsius (T_C). Equation 8.17 computes the partial pressure of water vapor (e) using the Celsius temperature and fraction of humidity saturation (f_h).

$$T_C = T - 273.15 \qquad\qquad 8.16$$

$$e = 6.10 \cdot f_h \cdot exp\left(\frac{17.15 T_C}{234.7 + T_C}\right) \qquad\qquad 8.17$$

Equations 8.18 and 8.19 compute the dry (N_1) and wet (N_2) atmospheric refractive indexes using temperature (T) and pressure (P, e) parameters. Note that the wet refractive index is applicable only to radio frequencies. When optical frequencies the wet refractive index is set to zero ($N_2 = 0$).

$$N_1 = \frac{77.264 \cdot P}{T} \qquad\qquad 8.18$$

$$N_2 = \left(\frac{371900}{T^2} - \frac{12.92}{T}\right) \cdot e \qquad\qquad 8.19$$

Equations 8.20 and 8.21 compute the dry (h_1) and wet (h_2) tropospheric heights (in meters). Note that h_2 *is only computed for radio frequencies.*

$$h_1 = \frac{5 \cdot 0.002277 \cdot P}{N_1 \times 10^{-6}} \qquad\qquad 8.20$$

$$h_2 = \frac{5 \cdot 0.002277 \cdot P}{N_2 \times 10^{-6}} \left(\frac{1255}{T} + 0.05 \right) \cdot e \qquad 8.21$$

Equation 8.22 computes the range correction ($\Delta \rho$), using a ninth order polynomial.

$$\Delta \rho = C_\rho \sum_{j=1}^{2} \frac{N_j}{10^6} \sum_{i=1}^{9} \frac{\alpha_{ij} r_j^i}{i} \qquad 8.22$$

Equation 8.23 is the constant in front of the range correction. The λ' is a scaled signal wavelength with $\lambda' = \lambda/(1\mu m)$. Thus, $(1/\lambda')^2$ can be set to zero for radio frequencies.

$$C_\rho = \left[\frac{170.2649}{173.3 - (1/\lambda')^2} \right] \left[\frac{78.8828}{77.624} \right] \left[\frac{173.3 + (1/\lambda')^2}{173.3 - (1/\lambda')^2} \right] \qquad 8.23$$

Equation 8.24 provides the distances to the top of the dry ($j = 1$) and wet ($j = 2$) troposphere using the elevation angle (El), local geoid radius (R_E) and the applicable tropospheric height.

$$r_j = \sqrt{\left(R_E + h_j \right)^2 - (R_E \cos El)^2} - R_E \sin El \qquad 8.24$$

Equations 8.25 and 8.26 provide the computation of the range correction α_{ij} coefficients.

$$a_j = -\frac{\sin El}{h_j} \qquad b_j = -\frac{\cos^2 El}{2h_j R_E} \qquad 8.25$$

$$
\begin{aligned}
\alpha_{1j} &= 1 \\
\alpha_{2j} &= 4a_j \\
\alpha_{3j} &= 6a_j^2 + 4b_j \\
\alpha_{4j} &= 4a_j \left(a_j^2 + 3b_j \right) \\
\alpha_{5j} &= a_j^4 + 12a_j^2 b_j + 6b_j^2 \\
\alpha_{6j} &= 4a_j b_j \left(a_j^2 + 3b_j \right) \\
\alpha_{7j} &= b_j^2 \left(6a_j^2 + 4b_j \right) \\
\alpha_{8j} &= 4a_j b_j^3 \\
\alpha_{9j} &= b_j^4
\end{aligned}
\qquad 8.26
$$

Equation 8.27 computes the elevation correction.

$$\Delta El = C_{El}\frac{4\cos El}{\rho}\left[\sum_{j=1}^{2}\frac{N_j}{10^6 h_j}\left(\sum_{i=1}^{7}\left[\frac{\beta_{ij}r_j^{i+1}}{i(i+1)} + \frac{\beta_{ij}r_j^i}{i}(\rho - r_j)\right]\right)\right] \qquad 8.27$$

Equation 8.28 is the constant in front of the elevation correction.

$$C_{El} = \left[\frac{170.2649}{173.3 - (1/\lambda')^2}\right]\left[\frac{78.8828}{77.624}\right] \qquad 8.28$$

Equation 8.29 provides the β_{ij} coefficients for the elevation correction.

$$\begin{aligned}
\beta_{1j} &= 1 \\
\beta_{2j} &= 3a_j \\
\beta_{3j} &= 3\left(a_j^2 + b_j\right) \\
\beta_{4j} &= a_j\left(a_j^2 + 6b_j\right) \\
\beta_{5j} &= 3b_j\left(a_j^2 + b_j\right) \\
\beta_{6j} &= 3a_j b_j^2 \\
\beta_{7j} &= b_j^3
\end{aligned} \qquad 8.29$$

8.2   Space Platform Prediction and Tracking

Events of interest involving only spacecraft include spacecraft-to-spacecraft spatial close approach conjunctions and spacecraft eclipse events.

8.2.1   Spacecraft-to-Spacecraft Conjunctions [Level I – Descriptive]

Spacecraft-to-spacecraft spatial close approach conjunctions determine the relative minimum physical separations between the two spacecraft that is within a user-defined threshold of interest. The most common use for these predictions is collision avoidance. Figure 8-6 illustrates trajectory progression through a relative minimum conjunction.

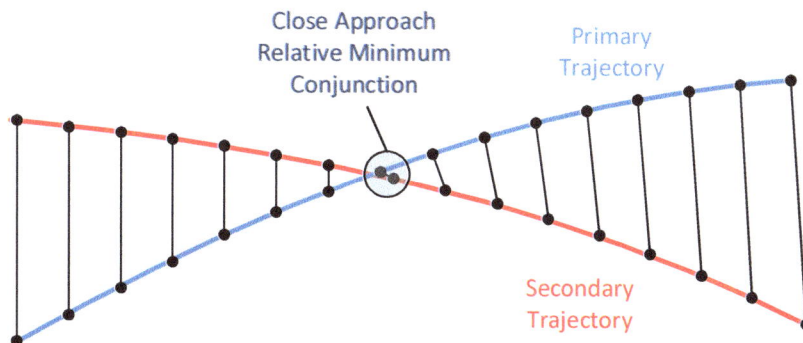

Figure 8-6 Spatial Close Approach Relative Minimum

225

> ➢ **Transition**: *You may continue this section with Atmospheric Refraction at Level II, or you may skip to the beginning of the next section called Eclipse Prediction.*

8.2.2   Spacecraft-to-Spacecraft Conjunctions [Level II – Equations]

Equation 8.30 computes square of the spatial separation vector ($\Delta \bar{r}$) between the primary spacecraft position ($\bar{r}_1$) and the secondary spacecraft position ($\bar{r}_2$). Equation 8.31 is half the time rate of change of the Δr^2. The spacecraft approach each other when the value of equation 8.31 is negative and recede from each other when the value is positive. Therefore, the relative minimum occurs for a zero value of equation 8.31 that is bracketed by a change in sign from negative to positive.

$$\Delta r^2 = (\bar{r}_1 - \bar{r}_2) \cdot (\bar{r}_1 - \bar{r}_2) = r_1^2 + r_2^2 - 2r_1 r_2 \qquad 8.30$$

$$\frac{d}{dt}\left(\frac{\Delta r^2}{2}\right) = \bar{r}_1 \cdot \dot{\bar{r}}_1 + \bar{r}_2 \cdot \dot{\bar{r}}_1 - \bar{r}_1 \cdot \dot{\bar{r}}_2 - \bar{r}_2 \cdot \dot{\bar{r}}_1 \qquad 8.31$$

Equation 8.32 is sampled in time steps approximately one fifth the smaller period of the two orbits which should be rounded to the nearest whole second. Equation 2.10 provides the time rate of change of true anomaly ($\dot{\theta}$) used in equation 8.32.

$$\Delta t_1 = \frac{2\pi}{5\dot{\theta}_1} \qquad \Delta t_2 = \frac{2\pi}{5\dot{\theta}_2} \qquad \Delta t = min(\Delta t_1, \Delta t_2) \qquad 8.32$$

A relative minimum is bracketed by negative to positive sign change in equation 8.31. The time of the relative minimum (i.e., the closest approach distance) is determined by an iterative method such as a golden search or by successive bisections, until the value for equation 8.5 is sufficiently close to zero. When the minimum is found the close approach distance is computed from the square root of equation 8.30. Relative minima that are less than the user-specified threshold are reported.

8.2.3   Eclipse Prediction [Level I – Descriptive]

Eclipses occur when a celestial body blocks the sunlight (an event called *occultation*). Predicting eclipses supports spacecraft thermal and power collection management as well as more accurately modeling solar radiation pressure for high fidelity orbit prediction. Spacecraft in LEO usually experience Earth occulting eclipses each orbital revolution, with durations that may approach half the orbital period. Spacecraft in *cislunar space* generally experience eclipses with both the Earth and Moon as occulting bodies.

226

Predictions of interest are the eclipse type and the transition times between full sunlight and the eclipse types. There are four eclipse regions, illustrated in figure 8-7, which include the three eclipse types and full sunlight. The three eclipse types are umbral, penumbral, and annular.

An umbral eclipse is one in which the occulting body fully blocks the Sun. While in the umbra, a spacecraft is in complete darkness. An umbral eclipse is always bracketed by penumbral eclipse periods in which the occulting body partially blocks the Sun. While in the penumbra, a spacecraft receives partial sunlight. Spacecraft at low altitudes typically experience umbral eclipses. Spacecraft at medium and higher altitudes may experience eclipses that never reach the umbral stage.

Annular eclipses only occur at higher altitudes. An annular eclipse is one in which the angular subtense of the occulting body (from the observer's point of view) is less than that of the Sun. It is therefore not possible for the occulting body to completely block the Sun. At maximum extent, the Sun's profile is a ring (or annulus) of light with the occulting body blocking the Sun's center portion. As with an umbral eclipse, an annular eclipse is always bracketed by penumbral eclipse periods.

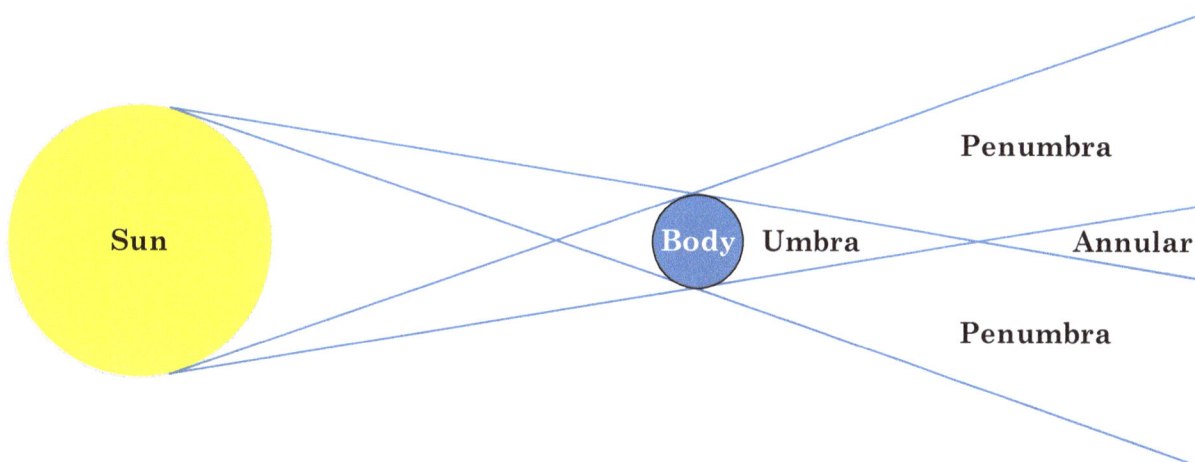

Figure 8-7 Eclipse Zones

Key Terms:

Annular Eclipse: is an eclipse type in which the apparent size of the occulting is too small to completely block the Sun. When the Sun and the occulting body are in line with the observer, the sunlight forms a ring (or annulus).
Occultation: occurs when one celestial body blocks the viewing path of another celestial body.
Penumbral Eclipse: is an eclipse type in which the occulting body partially blocks the sunlight.
Umbral Eclipse: is an eclipse type in which the occulting body fully blocks the sunlight.

➤ **Transition**: *You may continue this section with Eclipse Prediction at Level II, or you may skip to the beginning of the next chapter called Orbit Determination.*

8.2.4   Eclipse Prediction [Level II – Equations]

Eclipse predictions involve determining the degree of anti-alignment between the Sun, the occulting body, and the spacecraft. The maximum anti-alignment occurs when all three are in-line, with the Sun and spacecraft on opposite sides of the occulting body. Since the computations focus around the occulting body, the positions of the Sun ($\bar{r}_{Sun,B}$) and the spacecraft ($\bar{r}_{S/C,B}$) are represented relative to that body. If the occulting body is different from the trajectory's centrally attracting body, the spacecraft trajectory position ($\bar{r}$) must be related to the occulting body's position ($\bar{r}_B$) relative to the central body by vector subtraction as provided in equation 8.33. Likewise, vector subtraction will also be needed to relate the Sun's position ($\bar{r}_{Sun}$) to the occulting body as is done in equation 8.34.

$$\bar{r}_{S/C,B} = \bar{r}_B - \bar{r} \qquad \qquad 8.33$$

$$\bar{r}_{Sun,B} = \bar{r}_B - \bar{r}_{Sun} \qquad \qquad 8.34$$

Similarly, equations 8.35 and 8.36 are the spacecraft and Sun velocities relative to the occulting body.

$$\dot{\bar{r}}_{S/C,B} = \dot{\bar{r}}_B - \dot{\bar{r}} \qquad \qquad 8.35$$

$$\dot{\bar{r}}_{Sun,B} = \dot{\bar{r}}_B - \dot{\bar{r}}_{Sun} \qquad \qquad 8.36$$

Equation 8.37 computes the cosine of the separation angle (θ) which has its maximum negative extent at anti-alignment. The range of values is $-1 \leq \theta \leq 1$. Equation 8.38 is the time rate of change of $\cos\theta$.

$$\cos\theta = \hat{r}_{Sun,B} \cdot \hat{r}_{S/C,B} \qquad \qquad 8.37$$

$$\frac{d}{dt}(\cos\theta) = \hat{r}_{Sun,B} \cdot \dot{\hat{r}}_{S/C,B} + \dot{\hat{r}}_{Sun,B} \cdot \hat{r}_{S/C,B} \qquad \qquad 8.38$$

Equations 8.39 and 8.40 provide the time rates of change of the spacecraft and Sun relative unit vectors.

$$\dot{\hat{r}}_{S/C,B} = \frac{\dot{\bar{r}}_{S/C,B} - \dot{r}_{S/C,B}\hat{r}_{S/C,B}}{r_{S/C,B}} \qquad \qquad 8.39$$

$$\hat{\dot{r}}_{Sun,B} = \frac{\dot{\bar{r}}_{Sun,B} - \dot{r}_{Sun,B}\hat{r}_{Sun,B}}{r_{Sun,B}}$$ 8.40

Equations 8.41 and 8.42 provide the corresponding relative range rates.

$$\dot{r}_{S/C,B} = \hat{r}_{S/C,B} \cdot \dot{\bar{r}}_{S/C,B}$$ 8.41

$$\dot{r}_{Sun,B} = \hat{r}_{Sun,B} \cdot \dot{\bar{r}}_{Sun,B}$$ 8.42

Equation 8.38 is sampled to isolate the maximum $\cos\theta$ value, which is bracketed between a negative to positive sign transition. Equation 8.43 provides the sampling time step of every twentieth of an orbital revolution which should be rounded to the nearest whole second.

$$\Delta t = \frac{2\pi}{5\dot{\theta}}$$ 8.43

The time of maximum anti-alignment (is determined by an iterative method such as a golden search or by successive bisections, until the value for equation 8.38 is sufficiently close to zero. The eclipse zone (sunlight, penumbra, umbra, annular) for the maximum anti-alignment needs to be determined next.

8.2.5 Eclipse Zone Determination [Level II – Equations]

The eclipse conical shadow model is defined by Montenbruck and Gill (2000). It determines the observer's eclipse zone, by the geometry in figure 8-8. The eclipse plane is normal to the line between the Sun and the occulting body at a distance containing the spacecraft.

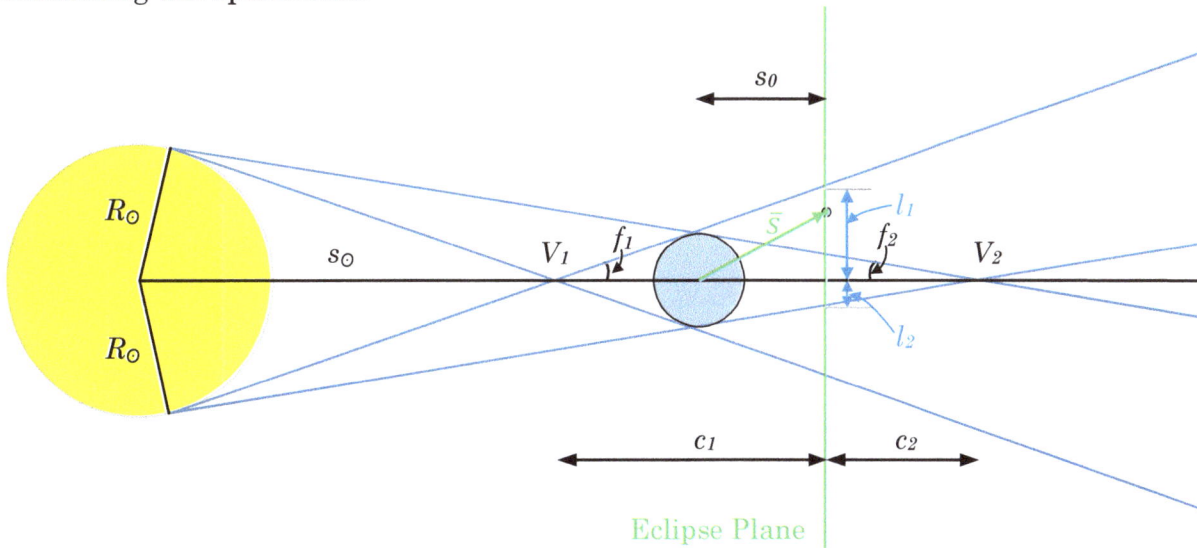

Figure 8-8 Conical Shadow

Equation 8.44 is the Sun's position relative to the occulting body. Equation 8.45 is the spacecraft position relative to the occulting body.

$$\bar{s}_\odot = \bar{r}_\odot - \bar{r}_B \qquad\qquad 8.44$$

$$\bar{s} = \bar{r} - \bar{r}_B \qquad\qquad 8.45$$

Equation 8.46 is the eclipse plane distance from the occulting body center.

$$s_0 = -\frac{\bar{s} \cdot \bar{s}_\odot}{s_\odot} \qquad\qquad 8.46$$

Equations 8.47 and 8.48 relate the apex angles of the penumbral and umbral shadow cones to the radii of the Sun ($R_\odot$) and occulting body (R_B) and the Sun's distance to the occulting body ($s_\odot$).

$$\sin f_1 = \frac{R_\odot + R_B}{s_\odot} \qquad\qquad 8.47$$

$$\sin f_2 = \frac{R_\odot - R_B}{s_\odot} \qquad\qquad 8.48$$

Equations 8.49 and 8.50 are the distances of the penumbral and umbral apex vertices V_1 and V_2 to the eclipse plane.

$$c_1 = \frac{R_B}{\sin f_1} + s_0 \qquad\qquad 8.49$$

$$c_2 = \frac{R_B}{\sin f_2} - s_0 \qquad\qquad 8.50$$

Equations 8.51 and 8.52 are the distances of the penumbral and umbral cones from the shadow axis at the eclipse plane distance.

$$l_1 = c_1 \tan f_1 \qquad\qquad 8.51$$

$$l_2 = c_2 \tan f_2 \qquad\qquad 8.52$$

The eclipse zone is determined from the geometry using the computed parameters. If the eclipse plane distance is less than or equal to zero ($s_0 \le 0$), the spacecraft is in full sunlight since it is between the Sun and the occulting body. Otherwise, more sophisticated checks are needed to determine eclipse zone.

Equation 8.53 computes the spacecraft distance (l) from the shadow axis.

$$l = \sqrt{s^2 - s_0^2}$$

$$8.53$$

The various eclipse zones are determined by the following rules:

- If the spacecraft's distance from the shadow axis is greater than the penumbral cone threshold ($l > l_1$), the spacecraft is in full sunlight.
- If the spacecraft's distance from the shadow axis is between the umbral and penumbral cone thresholds ($l_2 \leq l \leq l_1$), the spacecraft is in the penumbral zone.
- If the spacecraft's distance from the shadow axis is less than the umbral cone threshold, the spacecraft is inside one of the umbral cones. One more test is needed to determine whether the spacecraft is in the umbral or annular eclipse zone:
  - If the umbral cone threshold is positive ($l_2 \geq 0$), the spacecraft is in the umbral zone.
  - If the umbral cone threshold is negative ($l_2 < 0$), the spacecraft is in the annular zone.

The spacecraft illumination fraction (v) is 1.0 in full sunlight and 0.0 in the umbral zone. Partial illumination ($0.0 \leq v \leq 1.0$) occurs in the penumbral and annular zone and the illumination fraction is computed by the shadow function.

8.2.6   Eclipse Shadow Function [Level II – Equations]

The eclipse shadow function, as presented by Montenbruck and Gill (2000), determines the illumination fraction (v) when the spacecraft is in the penumbral eclipse zone. This is a geometric solution of the overlapping area of two circular disks as illustrated in figure 8-9.

Equation 8-53 computes the Sun's angular radius as perceived at the spacecraft's relative position. Equation 8-54 computes the occulting body's angular radius as perceived at the spacecraft's relative position.

$$a = \sin^{-1}\frac{R_\odot}{|\bar{r}_\odot - \bar{r}|}$$

$$8\text{-}53$$

$$b = \sin^{-1}\frac{R_B}{s}$$

$$8\text{-}54$$

Equation 8-55 computes the angular separation between the Sun and the occulting body as perceived at the spacecraft's relative position.

$$c = \cos^{-1}\frac{-\bar{s}\cdot(\bar{r}_\odot - \bar{r})}{s\left|\bar{r}_\odot - \bar{r}\right|}$$

8-55

The Sun's area occulted by the occluding body is the sum of overlapping areas enclosed by CDC' and CFC'. There is overlap if the $|a - b| < c < a + b$ inequality is satisfied.

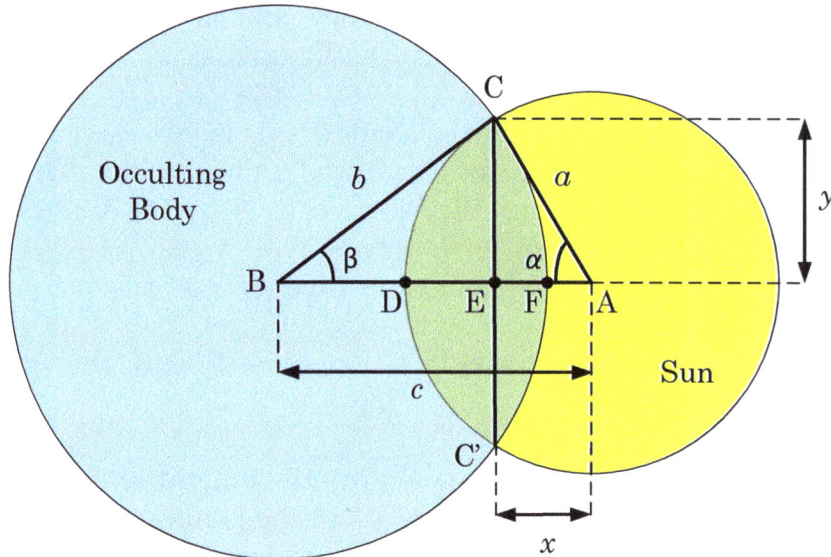

Figure 8-9 Sun Occultation by a Celestial Body

Equation 8-56 defines the occulted area, which considers the areas of the circular segments bounded by BCF and ACD versus the areas of triangles BCE and ACE.

$$A = 2(A_{\text{BCF}} - A_{\text{BCE}}) + 2(A_{\text{ACD}} - A_{\text{ACE}})$$

8-56

Equation 8-57 computes the distance AE which is the distance from the Sun's center to the $\overline{CC'}$ chord. Equation 8-58 computes the distance $\overline{CE}$ which is half the $\overline{CC'}$ chord length.

$$x = \frac{c^2 + a^2 - b^2}{2c}$$

8-57

$$y = \sqrt{a^2 - x^2}$$

8-58

Equations 8-59 and 8-60 provide the formulas for the sector and triangular areas for the right side of equation 8-56. The left side of expressions are computed in the same manner.

$$A_{\text{ACD}} = \frac{1}{2}\alpha a^2$$

8-59

$$A_{\text{ACE}} = \frac{1}{2}xy \qquad\qquad 8\text{-}60$$

Equation 8-61 computes the overlap area (A) between the Sun and the occulting body, consistent with equation 8-56.

$$A = a^2 \cos^{-1}\left(\frac{x}{a}\right) + b^2 \cos^{-1}\left(\frac{c-x}{b}\right) - cy \qquad\qquad 8\text{-}61$$

Equation 8-62 computes the spacecraft illumination fraction.

$$v = 1 - \frac{A}{\pi a^2} \qquad\qquad 8\text{-}62$$

Equation 8-63 computes the spacecraft illumination fraction in the annular eclipse zone.

$$v = 1 - \frac{b^2}{a^2} \qquad\qquad 8\text{-}63$$

REFERENCES

1. Seidelmann, P. Kenneth (editor), *Explanatory Supplement to the Astronomical Almanac*, © 1992 University Science Books, ISBN 0-935702-68-7.
2. Montenbruck, Oliver and Gill, Eberhard, *Satellite Orbits, Models, Methods, and Applications*, © 2000 Springer-Verlag, ISBN 3-540-67280-X.
3. Beyer, William H., *CRC Standard Mathematical Tables*, 27th Edition, © 1981 Chemical Rubber Company, ISBN 0-8493-0627-2.
4. Hoots, Felix R., Crawford, Linda L., and Roehrich, Ronald L., *An Analytic Method to Determine Future Close Approaches Between Satellites*, Celestial Mechanics 33 (1984), 143-158.

9 Orbit Determination

Orbit Determination (OD) is the process by which a trajectory's parameters are characterized mathematically from measurement data. Measurement data traditionally consisted of ground-based optical or radio tracking data. More recently, measurements are made from radio-based tracking spacecraft, such as Global Navigation Satellite System (GNSS) constellations.

All measurements taken from instruments have associated errors. Thus, preliminary estimates made from minimal data will need to be refined. OD refines the trajectory parameters using statistical processes and considers the presence of errors not only in the measurement process, but also in the prediction process (due to approximations in the perturbation modeling). Thus, orbital parameter maintenance by statistical OD is an ongoing process throughout the mission.

9.1 Initial Orbit Determination

Initial Orbit Determination (IOD) is a process by which a preliminary mathematical characterization of trajectory parameters is accomplished from minimal information. While a wide variety of methods exist, the discussion here is restricted to practical methods that use three position vectors as inputs, namely the Gibbs and Herrick-Gibbs methods. Readers interested in other techniques and derivations are referred to sources such as Escobal, Vallado, or Bate, et. al. Figure 9-1 illustrates the three-position vector IOD problem.

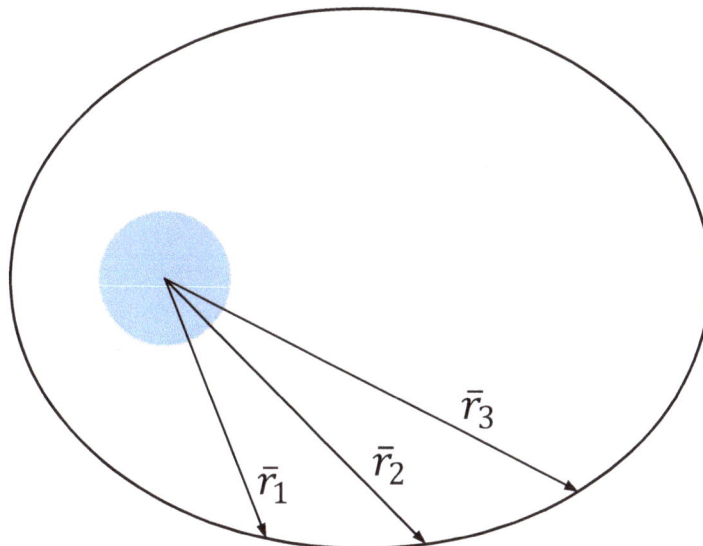

Figure 9-1 Three Position Initial Orbit Determination

9.1.1 Gibbs Method [Level I – Descriptive]

The Gibbs IOD method requires three non-zero coplanar position vectors within a single orbital revolution as the input. This method is geometric in nature since it does not consider the relative time associated with the positions. This method is the most accurate of the three-position methods when the angular spacing between vectors is wide (i.e., greater than 5 degrees).

> ➢ **Transition**: *You may continue this section with Gibbs Method at Level II, or you may skip to the beginning of the next section called Herrick Gibbs Method.*

9.1.2 Gibbs Method [Level II – Equations]

The Gibbs method relies only on geometry and the central attracting body's gravitational parameter (μ). The method also presumes the three position vectors ($\bar{r}_1$, $\bar{r}_2$, $\bar{r}_3$) are coplanar and in sequence of orbital progression. Equations 9.1 through 9.3 compute cross products of the position vector pairs, which are colinear with the angular momentum direction.

$$\bar{c}_{12} = \bar{r}_1 \times \bar{r}_2 \qquad\qquad 9.1$$

$$\bar{c}_{31} = \bar{r}_3 \times \bar{r}_1 \qquad\qquad 9.2$$

$$\bar{c}_{23} = \bar{r}_2 \times \bar{r}_3 \qquad\qquad 9.3$$

Equations 9.4 through 9.6 compute three auxiliary vectors $\bar{N}$, $\bar{D}$, and $\bar{S}$.

$$\bar{N} = r_1 \bar{c}_{23} + r_2 \bar{c}_{31} + r_3 \bar{c}_{12} \qquad\qquad 9.4$$

$$\bar{D} = \bar{c}_{12} + \bar{c}_{31} + \bar{c}_{23} \qquad\qquad 9.5$$

$$\bar{S} = (r_2 - r_3)\bar{r}_1 + (r_3 - r_1)\bar{r}_2 + (r_1 - r_2)\bar{r}_3 \qquad\qquad 9.6$$

Equation 9.7 computes a derived in-plane auxiliary vector $\bar{B}$ that represents the intrack direction relative to $\bar{r}_2$. Equation 9.8 computes an auxiliary scalar quantity (L).

$$\bar{B} = \bar{D} \times \bar{r}_2 \qquad\qquad 9.7$$

$$L = \sqrt{\frac{\mu}{DN}} \qquad\qquad 9.8$$

Equation 9.9 computes the velocity vector corresponding to $\bar{r}_2$.

$$\bar{v}_2 = L\left(\frac{\bar{B}}{r_2} + \bar{S}\right) \qquad 9.9$$

The combination of $\bar{r}_2$ and $\bar{v}_2$ is a state vector that defines the orbit.

9.1.3 Herrick-Gibbs Method [Level I – Descriptive]

The Herrick-Gibbs IOD method is a variant of the Gibbs method that requires three non-zero coplanar position vectors and their associated times. This method also requires knowledge of the central body's gravitational parameter. The Herrick-Gibbs method is more accurate when there is narrow angular spacing between the vectors (i.e., less than 5 degrees), as would typically occur in from single pass of a ground tracking station.

> ➤ **Transition**: *You may continue this section with Herrick Gibbs Method at Level II, or you may skip to the beginning of the next section called Statistical Orbit Determination.*

9.1.4 Herrick Gibbs Method [Level II – Equations]

The Herrick-Gibbs method relies on the times (t_1, t_2, t_3) associated with the three position vectors $(\bar{r}_1, \bar{r}_2, \bar{r}_3)$ as well as the central attracting body's gravitational parameter (μ). Knowledge of the relative times between the vectors, computed in equations 9.10 through 9.12, provides better numerical conditioning for closely spaced positions.

$$\Delta t_{31} = t_3 - t_1 \qquad 9.10$$

$$\Delta t_{32} = t_3 - t_2 \qquad 9.11$$

$$\Delta t_{21} = t_2 - t_1 \qquad 9.12$$

Equation 9.13 computes the velocity vector corresponding to $\bar{r}_2$.

$$\begin{aligned}
\bar{v}_2 = -\Delta t_{32}\left(\frac{1}{\Delta t_{21}\Delta t_{31}} + \frac{\mu}{12r_1^3}\right)\bar{r}_1 + \Delta t_{21}\left(\frac{1}{\Delta t_{32}\Delta t_{31}} + \frac{\mu}{12r_3^3}\right)\bar{r}_3 \\
+ (\Delta t_{32} - \Delta t_{21})\left(\frac{1}{\Delta t_{21}\Delta t_{32}} + \frac{\mu}{12r_2^3}\right)\bar{r}_2
\end{aligned} \qquad 9.13$$

The combination of $\bar{r}_2$ and $\bar{v}_2$ is a state vector that defines the orbit.

Key Term:

> **Initial Orbit Determination**: is a fit of positions or directions to compute a preliminary orbital estimate.

9.2 Statistical Orbit Determination

Statistical Orbit Determination (OD) is a process that recognizes inherent noise (random error) in all measurements. The processes rely on the noise approximating a Gaussian distribution. By over-determining the problem with a large sample size of measurements, the OD process filters the noise and computes the statistically most likely orbit to represent the measurements.

OD is inherently a non-linear problem. Nevertheless, the OD has the linear least-squares process as its underlying method. The implementations employ linearized approximations to allow solution of the non-linear problem. To do this effectively, modifications are made to the processes that are detailed in the specific methods.

Key Term:

> **Statistical Orbit Determination**: is a is the filtering process by which the statistically most likely orbital parameters are computed from an over-determined set of noisy measurement data.

9.2.1 OD Data Considerations

Since the OD processes are formulated to filter out zero-mean Gaussian noise, larger quantities of measurement data tend to produce more accurate estimates. But there is a point of diminishing returns in which higher quantities of data, the processing of which slows the computations, provides trivial improvement to the estimation. The focus is thus on using enough data, while optimizing the data quality.

The most obvious improvement to data quality is using lower noise measurements (i.e., measurements made with higher accuracy instrumentation). While this does improve the estimation, it can lead to the unfortunate short-sighted approach of completely avoiding using lower accuracy measurements. Since the OD filters are formulated to weigh higher accuracy measurements commensurately more favorably than lower accuracy measurements, the inclusion of some well-placed lower quality measurements results in an improved estimation quality.

It should be noted that while the use of lower accuracy measurements is often driven by economics and budget, it might alternatively be driven by factors such as availability. Since a correctly weighted OD filter suffers no estimation accuracy issues by including lower accuracy measurements, it is a good practice to supplement high accuracy measurements with some of a lower accuracy, when there the data set is lacking.

A measurement data set that is "lacking" could have issues other than data quantity. Equally important are the temporal and geometric data diversity. Diversity in the time domain is critical. Too short of a time span biases the trajectory estimates to artificially fit toward the measurement noise. Longer time spans force the estimate to ignore local measurement noise in favor of getting an end-to-end fit. In general, a minimum time span for a reasonable OD fit is half an orbital revolution, with a data set spanning several revolutions providing an excellent fit. The half revolution equivalent for hyperbolic trajectories is twice the time of flight from periapsis to the semi-latus rectum.

Geometric diversity is similarly important because it promotes a solution that fits the whole trajectory. Conversely, measurements clustered into a narrow angular arc of the trajectory can tend to bias the estimate to a narrow geometric feature. With greater the geometric diversity in the measurements, the OD filter will tend to better fit the estimate to the whole trajectory. Geometric diversity is best characterized by measurements taken with the greatest distribution in argument of latitude.

9.2.2 OD System Architecture

An OD system architecture has at least one variety of statistical filters. The two major categories are batch and sequential filters. The type of filter(s) used depends on the mission requirements. It is common for a ground-based command and control system to implement both filter types as presented in the figure 9-2 flow diagram.

9.2.2.1 Batch versus Sequential Filters

A batch filter fits the trajectory state to a measurement data arc of finite duration. Since the batch filter's perspective of the trajectory is limited to a discrete data interval, it is sometimes referred to as a Finite Impulse Response (FIR) filter. Batch filter strengths include the ability to detect and eliminate measurement outliers and that they can start with only a moderately accurate initial orbital estimate . An important batch filter feature is the production of a variance-covariance matrix as part of the process. The variance-covariance matrix is, among other things, useful for initializing a sequential filter.

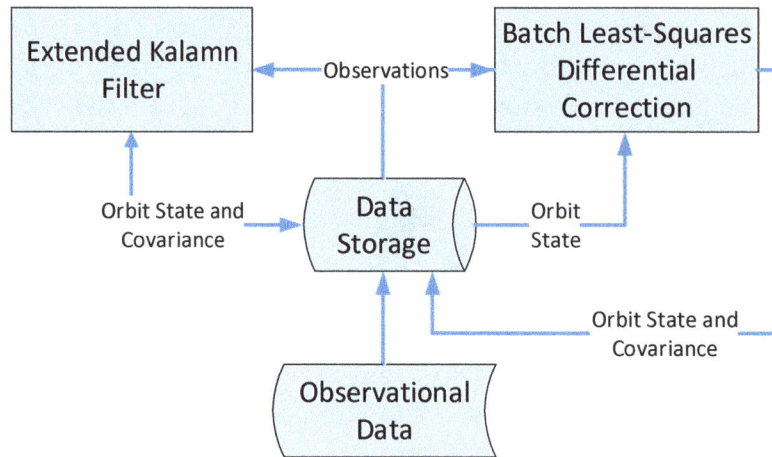

Figure 9-2 General OD Processing Flow

A sequential OD process such as an Extended Kalman Filter (EKF) maintains the trajectory estimate by performing an update to both the trajectory state and the variance-covariance matrix when a new measurement (or set of measurements) is provided. A sequential filter is influenced by all measurements since its initialization. As such it is sometimes referred to as an Infinite Impulse Response (IIR) filter. A well-implemented sequential filter tends to better forecast future trajectory states than its batch counterpart. However, sequential filters also tend to be unforgiving when an essential detail is not rigorously implemented. In a practical sense, such details may be difficult to implement with the desired level of rigor, resulting in a filter that may diverge or otherwise require frequent re-initialization.

Key Terms:

Batch OD Filter: is a statistical OD that refines a trajectory estimate using a discrete measurement data set in a simultaneous fitting process.
Sequential OD Filter: is a statistical OD that refines a trajectory estimate using only measurements from a single discrete time, processing continuously as the measurements become available.

9.2.2.2 Trajectory Solution States

The choice of trajectory solution state representation is sometimes ignored in OD filter design. Since the Cartesian vector tends to be the most convenient trajectory representation, they are sometimes used as the default. However, Cartesian vectors suffer from a fundamental disadvantage as the OD solution state since all six parameters are constantly changing over time. This dynamic situation requires a fairly accurate input orbital state to ensure stable convergence.

Orbit representations with few fast-changing parameters are more tolerant to inaccuracy in the input orbital state. For example, Keplerian elements have only a single fast-changing state (i.e., true, or mean anomaly), with the remainder changing slowly due to perturbational effects. However, as previously mentioned in section 3.6, Keplerian elements are an unstable representation for nearly equatorial and/or nearly circular orbits. Instead, an equinoctial element representation is recommended, but modified to use the semi-major axis reciprocal (α) to accommodate all trajectory types.

9.2.3   Least-Squares Batch Differential Correction

A least-squares differential correction (LSDC) is a batch filtering process. It uses the same least-squares mathematics used by linear regression to fit the best estimate of the slope and intercept of a straight line to a set of x, y measurement points. A fundamental difference, however, is that unlike linear regression an OD process is a non-linear problem. The LSDC adapts the least-squares process as a linearized update to the non-linear problem essentially as a multi-dimensional Newton-Raphson like process. While linear regression can estimate a solution in a single step, the LSDC can only approximate a linear correction to a pre-existing orbital state. Since the correction is only approximate, the process must iterate (as does Newton-Raphson) until the correction produces diminishing returns.

9.2.3.1   LSDC Process [Level I – Descriptive]

The LSDC requires a reasonable approximation to the trajectory (i.e., an a priori trajectory estimate). It is capable of refining that estimate by statistically fitting a large sample of measurements. The process computes *residuals* which are the differences between the measurements (observations) and what the current trajectory state and the measurement model would predict the measurement to be at the observed time.

The process seeks to update the trajectory estimate to a state that minimizes the sum of the squares of all the residuals. The fitting process computes the sensitivity of each measurement to slight changes in each of the trajectory state parameters. These partial derivatives with respect to each state parameter are arranged into the *Jacobian* matrix. Since the Jacobian matrix has one row of partial derivatives for each measurement, there are more measurements than there are state trajectory parameters, making the fit an overdetermined problem. Because of this and that the Jacobian matrix is not square, the correction to trajectory state is computed using a *Moore-Penrose Pseudo Inverse*.

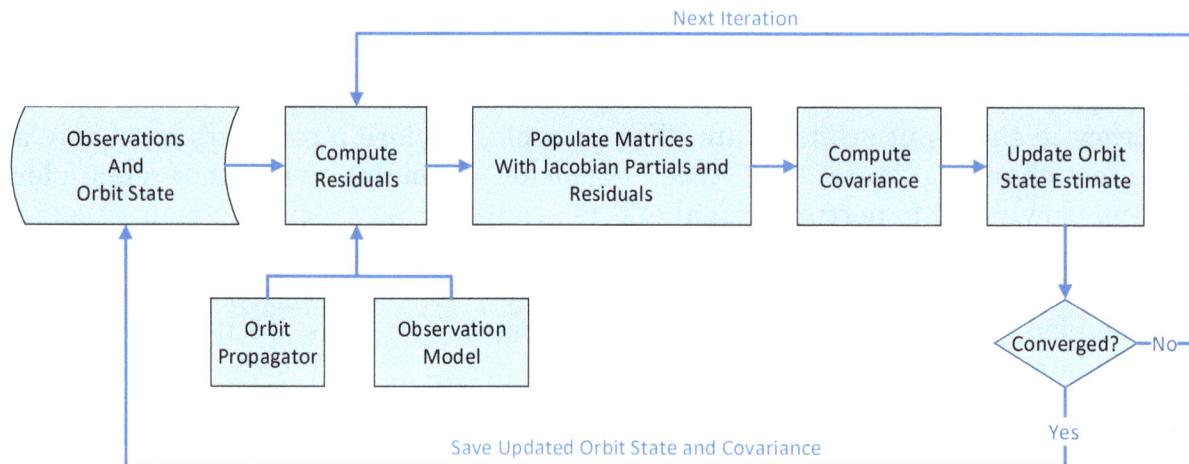

Figure 9-3 Least Squares Batch OD Processing Flow

A portion of the pseudo inverse produces the *covariance matrix*, consisting of the estimated statistical *variances* of each state parameter on the main diagonal with covariance element off the diagonal. A variance value is the square is a statistical standard deviation. A covariance element is the product of the standard deviations of the corresponding state elements and the degree of cross correlation (i.e., cross-dependency) between the elements. The state update is computed as the product of the covariance matrix and the residuals.

The correction process is iterative, typically continuing until the state correction reaches *diminishing returns*. Diminishing returns are conditions in which no meaningful update is being made. This often consists of achieving at least one of multiple criteria, such the state update becoming negligibly small or the difference between consecutive updates becoming negligibly small.

Key Terms:

> **Covariance Matrix**: is a square matrix resulting from a least-squares estimation that has variance terms on the main diagonal and covariance terms off the diagonal.
> **Jacobian Matrix**: is a matrix with elements representing the sensitivities of measurement values to variations in the trajectory state.
> **Moore-Penrose Pseudo-Inverse**: is mathematical process used to compute the pseudo-inverse of a non-square matrix.

➤ **Transition**: *You may continue this section with LSDC Process at Level II, or you may skip to the beginning of the next section called Kalman Filtering.*

Equation 9.14 is the LSDC state update equation. The $(A^TwA)^{-1}$ is the weighted covariance matrix produced as a portion of the Moore-Penrose Pseudo Inverse. The covariance matrix has dimensions $n \times n$, with n being the number of state elements the LSDC is correcting.

$$\Delta \bar{x} = (A^TwA)^{-1}A^Tw\bar{b}$$
$$\text{9.14}$$

Equation 9.15 is the Jacobian matrix (A) of partial derivatives of the measurement at the observed time, derived from the trajectory state estimate at the state epoch time. Thus, the partial derivatives represent the sensitivity of each measurement to slight changes in each of the [epoch] state parameters. Each row of the Jacobian matrix represents the partial derivative of a single measurement, with the corresponding columns being the partials for that measurement with respect to one state element. Given that, the Jacobian matrix has dimensions $m \times n$, the number of rows equals the number of measurements (m) and the number of columns equals the number of state elements (n) being corrected.

One might notice that the Jacobian has a well-defined number of columns, but an open-ended number of rows. While dynamic array dimensioning is available in most programming languages, knowledge of the matrix multiplication process allows the A^TwA matrix product to optionally be constructed in a single $n \times n$ array space as each partial derivative is computed. Likewise, the $A^Tw\bar{b}$ array can be coincidentally constructed in a single $n \times 1$ array space. The method for constructing these arrays will be covered in a subsequent section.

$$A = \begin{bmatrix} \dfrac{\partial M_1}{\partial x_1} & \dfrac{\partial M_1}{\partial x_2} & \cdots & \dfrac{\partial M_1}{\partial x_n} \\ \dfrac{\partial M_2}{\partial x_1} & \dfrac{\partial M_2}{\partial x_2} & \cdots & \dfrac{\partial M_2}{\partial x_n} \\ \vdots & \vdots & \ddots & \vdots \\ \dfrac{\partial M_m}{\partial x_1} & \dfrac{\partial M_m}{\partial x_2} & \cdots & \dfrac{\partial M_m}{\partial x_n} \end{bmatrix}$$
$$\text{9.15}$$

Equation 9.16 is the measurement weighting matrix consisting of the reciprocals of each measurement's statistical variance, which is the square of the measurement standard deviation (σ). The measurement standard deviation represents the statistical accuracy of the measuring instrument, though this can include any additional factors affecting the measurement accuracy under the observational conditions.

$$\mathbf{w} = \begin{bmatrix} \dfrac{1}{\sigma_1^2} & 0 & \cdots & 0 \\ 0 & \dfrac{1}{\sigma_2^2} & \cdots & 0 \\ \vdots & \vdots & \ddots & \vdots \\ 0 & 0 & \cdots & \dfrac{1}{\sigma_m^2} \end{bmatrix} \qquad 9.16$$

Equation 9.17 is the residuals matrix which has $m \times 1$ dimensions. Each residual (z_i) is the difference between a measurement as observed (M_i) and the expected measurement value (C_i), based on the current trajectory state estimate and the measurement model. The expected measurement value requires the trajectory state to be propagated to the measurement time and the measurement model applied.

$$\bar{b} = \begin{bmatrix} M_1 - C_1 \\ M_2 - C_2 \\ \vdots \\ M_m - C_m \end{bmatrix} = \begin{bmatrix} z_1 \\ z_2 \\ \vdots \\ z_m \end{bmatrix} \qquad 9.17$$

9.2.3.2.1 Numerical Partial Derivatives

Analytic measurement partial derivatives may be computed efficiently but require an effort in determining the correct formulation. Numerical partials provide an alternative that leverages the process used to compute the expected measurement value for the residuals computation.

Numerical partial derivatives consider the effects a slight change to a single parameter of the current state estimate has to the expected measurement value. This provides the measurements' sensitivity to changes in that element.

A perturbed trajectory state estimate applies a small perturbation (δx_j) to the jth state parameter. Two perturbed states are created: one with the perturbation added to the jth state parameter ($\bar{x}^+$) and one with the small perturbation subtracted from the jth state parameter ($\bar{x}^-$). Both perturbed trajectory states are propagated to the measurement time where the measurement model is applied. The expected measurement corresponding to the $\bar{x}^+$ perturbed state is C_i^+ and the expected measurement from the $\bar{x}^-$ perturbed state is C_i^-.

Equation 9.18 provides the method for computing the numerical partial derivative as the difference between the expected measurements (C_i^+ and C_i^-) divided by twice the perturbation value ($2\delta x_j$).

$$\frac{\partial M_i}{\partial x_j} = \frac{C_i^+ - C_i^-}{2\delta x_j} \qquad\qquad 9.18$$

Each measurement has n numerical partial derivatives computed; each computing the measurement's sensitivity to one parameter of the solution state. The measurement's n partials are the *ith* row of the Jacobian Matrix.

9.2.3.2.2 Efficient Array Construction

The $A^T wA$ and $A^T w\bar{b}$ arrays may be constructed incrementally as the partial derivatives and residuals are computed for each measurement. Equation 9.19 populates the $A^T wA$ diagonal elements while equation 9.20 populates its off-diagonal elements. Equation 9.21 populates the $A^T w\bar{b}$ vector components.

$$[A^T wA]_{i,i} = \sum_{k=1}^{m} w_k \left(\frac{\partial M_k}{\partial x_i}\right)^2 \qquad\qquad 9.19$$

$$[A^T wA]_{i,j} = [A^T wA]_{j,i} = \sum_{k=1}^{m} w_k \frac{\partial M_k}{\partial x_i} \frac{\partial M_k}{\partial x_j} \qquad\qquad 9.20$$

$$\left[A^T w\bar{b}\right]_i = \sum_{k=1}^{m} w_k \frac{\partial M_k}{\partial x_i} b_k \qquad\qquad 9.21$$

9.2.3.2.3 Measurement Outliers

The least-squares process is tolerant to measurement outliers provided their respective frequencies versus magnitudes are consistent with a Gaussian distribution. One technique to help maintain a reasonable distribution is for the LSDC to automatically eliminate measurements that are considered statistical outliers using an *n-sigma editing process*.

Equation 9.22 computes the weighted Root Mean Square (RMS) of the residuals for the measurement type ($r_{m,RMS}$). The RMS is used because, different from a standard deviation, it includes non-zero mean residual values (i.e., residuals biases) caused by errors in the trajectory state. Including the residuals biases prevents the filter from falsely determining measurements to be outliers before the filter's initial iterations remove most of these biases.

$$r_{m,RMS} = \sqrt{\frac{1}{m} \sum_{i=1}^{m} w_{m,i}^2 r_{m,i}^2} \qquad\qquad 9.22$$

The filtering process removes the influence of any observations that are outside the filtering threshold. The editing threshold (ε_{thresh}) applies a user-specified n-multiplier to the residuals RMS as provided in equation 9.23.

$$\varepsilon_{m,thresh} = n \cdot r_{m,RMS} \qquad\qquad 9.23$$

Any measurement with a residual value greater than the editing threshold (i.e., $r_{m,j} > \varepsilon_{m,thresh}$) is edited out.

9.2.3.2.4 Matrix Inversion

Lower Upper (LU) decomposition is a simple straight forward matrix inversion method. This will usually work well in the LSDC, provided a stable set of solution state elements is chosen. It is good practice to have a backup matrix inversion method, such as Singular Value Decomposition (SVD), in the event the primary method is unable to invert the matrix.

9.2.3.2.5 LSDC Convergence

Since the non-linear LSDC is an iterative process, it requires a method to indicate when to stop. The most common stopping criterion is when subsequent iterations produce no better results than their predecessors – i.e., when diminishing returns are produced. Since the LSDC seeks to minimize the sum of the squares of the residuals, the weighted RMS of the residuals of *all* measurements (r_{RMS}) should be smaller on each iteration to justify continuing the process. It is also recommended that a maximum number of iterations be specified ($iter_{max}$) be specified to preclude the possibility of an infinite loop.

Thus, a suggested LSDC stopping criteria is established when at least one of the following conditions is met:
1. The absolute value of the difference between the residual RMS (r_{RMS}) of the previous ($i-1$) and current (i) iterations is sufficiently small. This indicates a convergence condition.

$$\left| \frac{r_{RMS}(i-1) - r_{RMS}(i)}{r_{RMS}(i-1)} \right| \leq \epsilon$$

2. The residual RMS of the current iteration is virtually zero, which is also a convergence condition.

$$|r_{RMS}(i)| < \varepsilon$$

3. The maximum number of iterations is reached. This is a non-convergence stopping point in which the implementation might allow the operator to modify the execution parameters and resume for a finite number of additional iterations.

9.2.4   Kalman Filtering [Level I – Descriptive]

Kalman filtering is a sequential filtering process. Since OD is a non-linear process, it uses the non-linear EKF variant of the process to compute the best statistical estimate of the trajectory parameters and the associated covariance matrix.

9.2.4.1   Kalman Filtering [Level I – Descriptive]

A Kalman filter uses the same mathematics as the LSDC, but with a somewhat different approach. As an analogy, arithmetic mean (average) values may be computed by two different approaches. An approach most familiar is summing the values and dividing the sum by the number of values. That would correspond to the LSDC batch method. The alternative approach, corresponding to a sequential filter, is the moving average. With the moving average, the first value is the current sum and there is only one value. As each new value is introduced, it is added to the sum with the current count of values incremented. The current estimate for the mean value is the current sum, divided by the current count. While this analogy is not perfect, it provides a reasonable illustration of how sequential filters differ from batch filters.

Since the estimates in Kalman filters are constantly evolving, the current state estimate and covariance matrix are updated with each new measurement introduced. The filter performs a balancing act between two noise (i.e., error) sources: process (or prediction) noise and measurement noise. Process noise indicates the deterioration of prediction accuracy in the "plant model" as a function of time. In the OD case, the trajectory propagation method is the plant model.

The Kalman gain is the arbiter, determining the relative credibility of the estimated state versus that of the measurements. A high Kalman gain indicates low confidence in the state estimate, allowing the measurement residuals to have a greater influence in updating the state estimate. In contrast, a low Kalman gain indicates a high confidence in the state estimate, reducing the influence the measurement residuals have in updating the state estimate. The Kalman gain is computed by a rigorous process that juxtaposes the confidence in the state estimate (covariance and process noise at the new measurement time) with the measurement noise.

Well-designed Kalman filters often have high gain / low state prediction accuracy when first initialized. However, as measurements come in, the covariance converges down to a steady-state accuracy level and is held at that level by the process noise. Convergence can be accelerated by seeding the Kalman Filter with a state estimate and covariance matrix produced by a batch LSDC, as is widespread practice. The LSDC as part of an OD system also provides a re-initialization capability should the Kalman filter begin diverging due to some anomaly.

➢ **Transition**: *You may continue this section with Kalman Filtering at Level II, or you may skip to the beginning of the next section called Measurement Models.*

9.2.4.2 Kalman Filtering [Level II – Equations]

Table 9-1 identifies and explains the symbols used in the equations and diagrams. The symbols chosen are compact and are the same as or like those commonly used in other references.

Table 9-1 Kalman Filter Symbol Nomenclature and Explanations			
Name	**Symbol**	**Size**	**Comments**
State Vector at Time k	$\bar{x}_k$	n	The vector size [n] represents the number of state parameters to be solved.
Predicted State Estimate at Time k	$\hat{x}_k^-$	n	The predicted state is an open-loop update to a previously estimated state, propagated to the current time.
Current [corrected] State Estimate at Time k	$\hat{x}_k^+$	n	The corrected state is a closed-loop update to the predicted state, using the measurements (or observations) as feedback.
State Transition Matrix at Time k	Φ_k	$n \times n$	For the linear filter, the state transition extrapolates (propagates) the current state estimate to a past or future time. For a non-linear filter, the state transition represents the sensitivity in the plant model's ability to propagate the state, to errors in the estimated state.
Covariance Matrix at Time k	P_k	$n \times n$	Diagonal elements are the estimated state elements variances (σ_i^2). Off-diagonal elements are the covariance terms between the ith and j^{th} state vector parameters. The covariance terms are the product of corresponding state element standard

Name	Symbol	Size	Comments
			deviations (σ_i and σ_j) and the correlation coefficient (μ_{ij}) between the two state elements.
Predicted Covariance at Time k	P_k^-	$n \times n$	The predicted covariance is an open-loop estimate of the growth in the variance and covariance elements due to propagation of errors in the estimated state elements as well as inaccuracy of the prediction (plant) model.
Corrected Covariance at Time k	P_k^+	$n \times n$	The corrected covariance is a closed-loop update to the predicted variance and covariance elements.
Process Noise at Time k	Q_k	$n \times n$	Matrix with variances and covariances representing the statistical accuracy of the plant model's (state transition) ability to predict the state a specified time.
Observation (measurement) Noise at time k	R_k	$m \times m$	Diagonal matrix with the on-diagonal variances representing the noise in the measurements (observations).
Vector of Measurements (Observations) at Time k	$\bar{z}_k$	m	The vector size (m) represents the number of measurements processed at time k.
Observation Model (Jacobian) matrix for Time k	H_k	$m \times n$	The Jacobian matrix representing the sensitivity of the observation model to small variations in the state elements.
Kalman Gain Matrix at Time k	K_k	$n \times m$	Transforms the measurements (or observations) into a correction to the state vector that is appropriately weighted based on confidence in the state versus measurement accuracy.

Figure 9-4 illustrates the Kalman filter general processing flow. Introducing new observations (measurements) triggers the update cycle.

The filter's initial phase is predictive: the current state estimate and covariance are propagated to the new observation time. The plant model dictates the state update, and for a linear system model may be fully defined by the state transition matrix. However, for non-linear systems such as OD, the plant model is instead used to compute the state transition matrix. The state transition propagates covariance growth (i.e., accuracy deterioration) over the prediction interval due to the state uncertainties. The process noise adds additional covariance growth, reflecting the inability of the propagator to exactly predict even a perfect trajectory state.

The observation model predicts the noise-free observation values from the predicted state. The observation residuals are the difference between the observed and the predicted measurements (i.e., observed minus computed).

The Jacobian matrix provides the observation model's sensitivity to variations in the state elements. It describes variations in the predicted observations with respect to small variations in the state elements.

The Kalman gain matrix is computed from the predicted covariance matrix, the Jacobian matrix, and the observation noise variances. It is used to both transform the measurement residuals to a state correction vector and to update the covariance matrix.

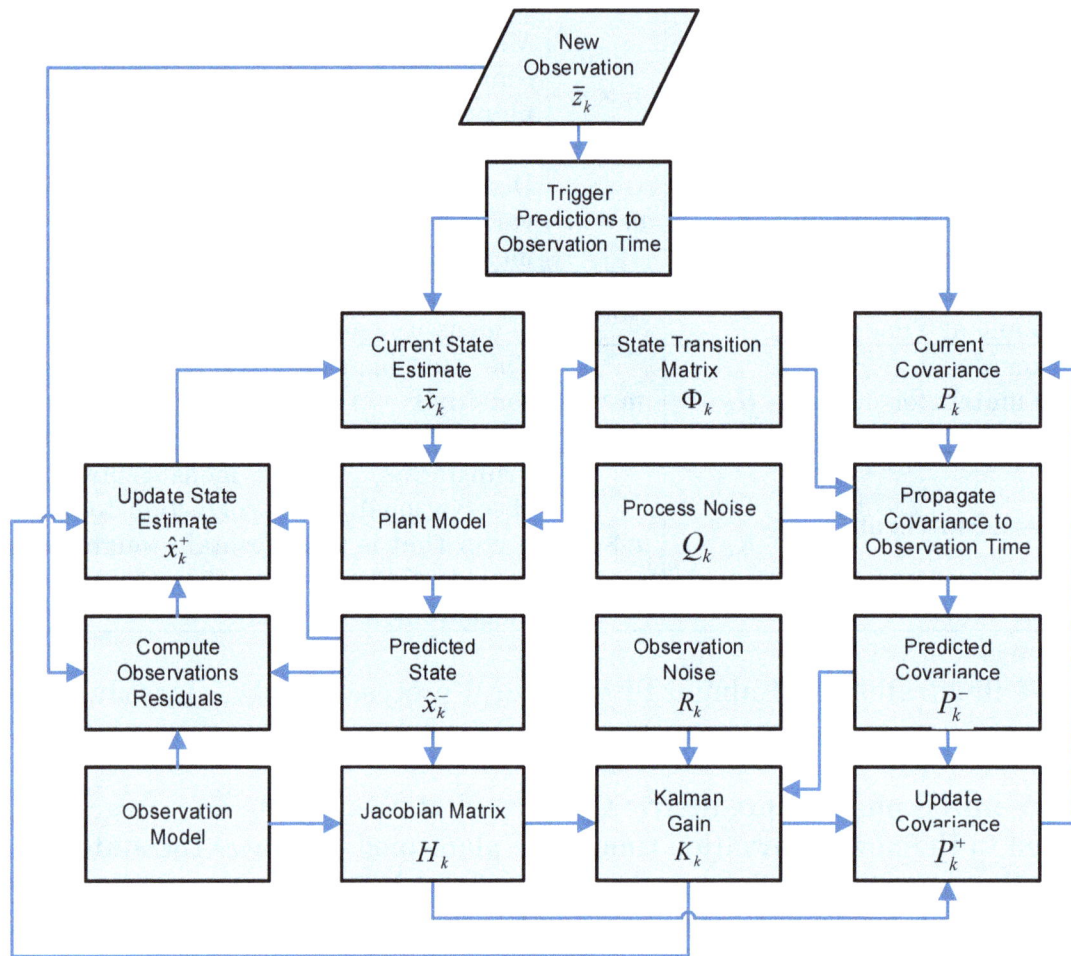

Figure 9-4 Extended Kalman Filter OD Processing Flow

Equation 9.24 computes the state prediction to the current (i.e., observation) time for a linear Kalman filter using the state transition matrix ($\mathbf{\Phi_k}$) operating on the previously corrected state estimate.

$$\hat{x}_{k+1}^- = \Phi_k \hat{x}_k^+ \qquad\qquad 9.24$$

However, since the OD is a non-linear process, the state prediction is made using the trajectory propagation as the plant model. The non-linear state transition (i.e., partial derivatives of state elements at the observation time, with respect to the state elements prior to the propagation) are typically computed numerically using the plant model.

Equation 9.25 computes the covariance growth over the prediction interval from the covariance matrix from the previous estimate (P_k^+), using the state transition matrix (Φ_{k+1}) and the process noise (Q_{k+1}).

$$P_{k+1}^- = \Phi_{k+1} P_k^+ \Phi_{k+1}^T + Q_{k+1} \qquad\qquad 9.25$$

The Kalman gain matrix (K_{k+1}) computed in equation 9.26 provides the transformation through the measurement model matrix (H_{k+1}) of the current measurement residuals into a state correction.

$$K_{k+1} = P_{k+1}^- H_{k+1}^T \left(H_{k+1} P_{k+1}^- H_{k+1}^T + R_{k+1} \right)^{-1} \qquad\qquad 9.26$$

The Kalman gain is weighted by several factors including the predicted covariance matrix (P_{k+1}^-) and the measurement noise variances (R_{k+1}). Accurate measurements (having low noise variances) will tend to be weighted greater in the correction since they contribute to a higher Kalman gain. Larger covariance matrices likewise indicate lower confidence in the state estimate, thus contributing to higher Kalman gains.

Equation 9.27 is the corrected state estimate $(\hat{x}_{k+1}^+)$ using the Kalman gain and the measurement residuals (i.e., the parenthetical quantity) to update the estimated trajectory state.

$$\hat{x}_{k+1}^+ = \hat{x}_{k+1}^- + K_{k+1}(\bar{z}_{k+1} - H_{k+1}\hat{x}_{k+1}^-) \qquad\qquad 9.27$$

Equation 9.28 updates the predicted covariance using the identity matrix, the Kalman gain, and the measurement model matrix.

$$P_{k+1}^+ = (I - K_{k+1}H_{k+1})P_{k+1}^- \qquad\qquad 9.28$$

Equation 9.29 is a more numerically stable (but equivalent) form of equation 9.28 provided by Bucy and Joseph. Equation 9.29 is the preferred formulation, with equation 9.28 presented since it has a more intuitive form.

$$P_{k+1}^+ = P_{k+1}^- - P_{k+1}^- H_{k+1}^T \widetilde{R}_{k+1}^{-1} H_{k+1} P_{k+1}^- \qquad\qquad 9.29$$

Equation 9.30 defines the $\tilde{R}_{k+1}^{-1}$ term in equation 9.29.

$$\tilde{R}_{k+1}^{-1} = H_{k+1}P_{k+1}^{-}H_{k+1}^{T} + R_{k+1} \qquad 9.30$$

Explanations of and formulations for the various vectors and matrices used as inputs to equations 9.24 through 9.30 are provided in the following sections.

9.2.4.2.1 Estimated State

The state vector in equation 9.31 is a one-dimensional array containing the parameters solved in the filter.

$$\bar{x} = \begin{bmatrix} x_1 \\ x_2 \\ \vdots \\ x_n \end{bmatrix} \qquad 9.31$$

For OD, the state vector would be the orbital parameters.

9.2.4.2.2 State Transition Matrix

Equation 9.32 is the state transition matrix, which provides the plant model (i.e., trajectory propagation) model's prediction of the state elements at the measurement time, with respect to small variations in the epoch (pre-propagated) state elements.

$$\Phi_{k+1} = \begin{bmatrix} \dfrac{\partial x_1(t_{k+1})}{\partial x_1(t_k)} & \dfrac{\partial x_1(t_{k+1})}{\partial x_2(t_k)} & \cdots & \dfrac{\partial x_1(t_{k+1})}{\partial x_n(t_k)} \\ \dfrac{\partial x_2(t_{k+1})}{\partial x_1(t_k)} & \dfrac{\partial x_2(t_{k+1})}{\partial x_2(t_k)} & \cdots & \dfrac{\partial x_2(t_{k+1})}{\partial x_n(t_k)} \\ \vdots & \vdots & \ddots & \vdots \\ \dfrac{\partial x_n(t_{k+1})}{\partial x_1(t_k)} & \dfrac{\partial x_n(t_{k+1})}{\partial x_2(t_k)} & \cdots & \dfrac{\partial x_n(t_{k+1})}{\partial x_n(t_k)} \end{bmatrix} \qquad 9.32$$

9.2.4.2.3 Covariance Matrix

Covariance matrix is an abbreviation for what is a variance-covariance matrix. Diagonal elements are the estimated variances (or squares of the standard deviations) of the estimated state parameters. The off-diagonal elements are the covariance terms between the ith and j^{th} state elements. The covariance terms are the products of the corresponding state parameter standard deviations (σ_i and σ_j) and the correlation coefficient (μ_{ij}) between the two state parameters.

$$\mathbf{P} = \begin{bmatrix} \sigma_1^2 & \mu_{12}\sigma_1\sigma_2 & \cdots & \mu_{1n}\sigma_1\sigma_n \\ \mu_{12}\sigma_1\sigma_2 & \sigma_2^2 & \cdots & \mu_{2n}\sigma_2\sigma_n \\ \vdots & \vdots & \ddots & \vdots \\ \mu_{1n}\sigma_1\sigma_n & \mu_{2n}\sigma_2\sigma_n & \cdots & \sigma_n^2 \end{bmatrix} \qquad 9.33$$

9.2.4.2.4 Observation Vector

The observation vector is an ordered one-dimensional array containing measurements from the observing instruments.

$$\bar{z} = \begin{bmatrix} z_1 \\ z_2 \\ \vdots \\ z_n \end{bmatrix} \qquad 9.34$$

9.2.4.2.5 Noise Variance Matrices

Noise is represented in square matrices. Process noise is the trajectory propagation error due to inexactness in the prediction method. Measurement noise represents the typical inaccuracy in the measurements.

9.2.4.2.5.1 Process Noise

Process noise represents the deterioration in the trajectory propagator's ability to accurately predict as a function of time. Some implementations represent the diagonal elements as quadratic polynomials in time ($\sigma_{\Phi i}^2 = c_{0,i} + c_{1,i}\Delta t + c_{2,i}\Delta t^2$). Process noise for completely independent state elements will be a diagonal matrix.

$$\mathbf{Q} = \begin{bmatrix} \sigma_{\Phi 2}^2 & 0 & \cdots & 0 \\ 0 & \sigma_{\Phi 2}^2 & \cdots & 0 \\ \vdots & \vdots & \ddots & 0 \\ 0 & 0 & \cdots & \sigma_{\Phi n}^2 \end{bmatrix} \qquad 9.35$$

9.2.4.2.5.2 Measurement Noise

Measurement noise is the typical errors in the measurements due to the inaccuracy of the observing instrument. If the measurements are independent, there should be no correlation between measurement types and the noise will be represented by a diagonal matrix.

$$R = \begin{bmatrix} \sigma_{1m}^2 & 0 & \cdots & 0 \\ 0 & \sigma_{2m}^2 & \cdots & 0 \\ \vdots & \vdots & \ddots & \vdots \\ 0 & 0 & 0 & \sigma_{zm}^2 \end{bmatrix} \qquad 9.36$$

9.2.4.2.6 Jacobian Matrix

The Jacobian matrix provides the sensitivity of the measurement model's prediction of the measurement values. They are the partial derivatives of the measurements at their predicted times with respect to respect to small variations in the epoch (pre-propagated) state elements.

$$H_{k+1} = \begin{bmatrix} \dfrac{\partial z_1(t_{k+1})}{\partial x_1(t_k)} & \dfrac{\partial z_1(t_{k+1})}{\partial x_2(t_k)} & \cdots & \dfrac{\partial z_1(t_{k+1})}{\partial x_n(t_k)} \\ \dfrac{\partial z_2(t_{k+1})}{\partial x_1(t_k)} & \dfrac{\partial z_2(t_{k+1})}{\partial x_2(t_k)} & \cdots & \dfrac{\partial z_2(t_{k+1})}{\partial x_n(t_k)} \\ \vdots & \vdots & \ddots & \vdots \\ \dfrac{\partial z_m(t_{k+1})}{\partial x_1(t_k)} & \dfrac{\partial z_m(t_{k+1})}{\partial x_2(t_k)} & \cdots & \dfrac{\partial z_m(t_{k+1})}{\partial x_n(t_k)} \end{bmatrix} \qquad 9.37$$

9.2.5 Covariance Analysis

The standard deviations of the variance (diagonal) covariance matrix terms are general indicators of the state parameters' estimation quality. The correlation matrix is computed from the covariance matrix by dividing each element by the corresponding standard deviations of the diagonal elements of its row and column.

$$M = \begin{bmatrix} 1 & \mu_{12} & \cdots & \mu_{1n} \\ \mu_{12} & 1 & \cdots & \mu_{2n} \\ \vdots & \vdots & \ddots & \vdots \\ \mu_{1n} & \mu_{2n} & \cdots & 1 \end{bmatrix} \qquad 9.38$$

Correlation coefficients always have a value in the range from negative to positive one ($-1.0 \leq \mu_{ij} \leq +1.0$). Correlation indicates both the strength of relationship and the relative direction of change between two state parameters. A positive correlation indicates one state parameter (x_i) tends to change in the same direction as the other state parameter (x_j). A negative correlation indicates one state parameter (x_i) tends to change in the opposite direction as the other state parameter (x_j). Zero correlation indicates the two state parameters change independently of each other.

9.2.6 Measurement Models

An OD filter is only as good as its measurement models. A measurement model is a mathematical algorithm that represents factors affecting the measurement. A good measurement model considers and calibrates out systematic biases that would tend to skew the OD process and mask themselves as unmodeled noise.

There are numerous types of trajectory measurement methods and instruments, each with their own unique features. It is beyond the scope of this work to provide a set of models. Instead, the remainder of this section provides examples of factors that should be considered when modeling tracking measurements.

9.2.6.1 Common Measurement Types

This section gives examples of measurement types used to track and characterize spacecraft trajectories. Understanding these forms is the basis for creating mathematical measurement models.

9.2.6.2 Surface-Based Tracking System Types

Surface-based tracking is typically performed by telescopes or antennas fixed to a celestial body surface. The ground site antenna coordinates should be converted to body-fixed Cartesian coordinates as done for Earth coordinates in section 3.14. Note that each celestial body has its own unique equatorial radius flattening factor.

It is straightforward to transform the trajectory position and velocity to the celestial body coordinate system such that a body fixed slant range vector and look angles can be computed as is done in section 8.1.3. It is most convenient when the antenna gimbals are aligned to the standard *Az/El* look angles. When this is not the case, the gimbal angles need to be reformulated based on the actual gimbal arrangement.

Earth-based tracking antennas are subject to refraction as was modeled in section 8.1.6 being anticipated for elevation and range measurements. Tracking antennas on celestial bodies with significant atmospheres should anticipate the need to correct for atmospheric refraction.

At the time of this writing, the most common space-based tracking consists of radio ranging from Navigation Satellite Timing and Ranging (NAVSTAR) Global Positioning System (GPS) and related Global Navigation Satellite System (GNSS) payloads.

The most accurate use of GPS and other GNSS satellite signals is direct ranging measurements. However, this requires decoding the navigation signals and reading and propagating the ephemerides for the spacecraft originating the signals. Because of this inconvenience, many users choose to use self-contained GPS or GNSS receivers and receive processed position Cartesian vectors as the measurements.

Regardless of the method used, it should be noted that space-based measurements indicate the range or position of the receiving antenna phase center. Trajectory propagation predicts the location and motion of a spacecraft center of mass. Thus, an offset vector should be determined from the antenna phase center to the vehicle center-of mass in spacecraft body coordinates. This vector should be transformed using the spacecraft attitude knowledge to the measurement coordinate system. From there, the range or position information should be corrected to represent the equivalent quantities relative to the spacecraft center of mass.

REFERENCES

1. Escobal, Pedro R., *Methods of Orbit Determination*, © 1965 John Wiley & Sons, Inc., ISBN 0-88275-319-3.
2. Vallado, David A., *Fundamentals of Astrodynamics and Applications*, Second Edition, © 2001 by author, Microcosm Press and Kluwer Academic Publishers, ISBN 1-881883-12-4.
3. Bate, Roger R. et al., *Fundamentals of Astrodynamics*, © 1971 Dover Publications, Inc., ISBN 0-486-60061-0.
4. Herrick, Samuel, *Astrodynamics*, © 1971 by author, Van Nostrand Reinhold Company, Library of Congress Card Catalog Number 78-125199.
5. Tapley, Byron, et al., *Statistical Orbit Determination*, © 2004 Elsevier, Inc., ISBN 0-12-683630-2.
6. Gelb, Arthur (editor), *Applied Optimal Estimation*, © 1974 The Analytical Sciences Corporation, ISBN 0-262-57048-3.
7. Zarchan, Paul and Musoff, Howard, *Fundamentals of Kalman Filtering, A Practical Approach*, Volume 190 Progress in Astronautics and Aeronautics, © 2000 American Institute of Aeronautics and Astronautics (AIAA), ISBN 1-56347-455-7.
8. Wiesel, William E., *Modern Orbit Determination*, © 2003 by author, Aphelion Press, ISBN 978-4536119-8-2.

9. Gibbs, Bruce P., *Advanced Kalman Filtering, Least-Squares, and Modeling*, © 2011 John Wiley & Sons, 978-0-470-52970-6.
10. Strang, Gilbert and Borre, Kai, *Linear Algebra, Geodesy, and GPS*, © 1977 by authors, Wellesley-Cambridge Press, ISBN 0-9614088-6-3.
11. Wertz, James R. (editor), *Spacecraft Attitude Determination and Control*, © 1978 Kluwer Academic Publishers, ISBN 90-277-1204-2.
12. Gurfil, Pini and Seidelmann, P. Kenneth, *Celestial Mechanics and Astrodynamics: Theory and Practice*, © 2016 Springer-Verlag, ISBN 978-3-662-50368-3.
13. Broucke, R.A. and Cefola, P.J., *On the Equinoctial Elements*, Celestial Mechanics 5 (1972) 303-310, © 1972 D. Reidel Publishing Company.
14. Danielson, D.A. et al., *Semianalytic Satellite Theory*, Mathematics Department, Naval Postgraduate School, Monterey, CA 93943.
15. Bierman, Gerard J., *Factorization Methods for Discrete Sequential Estimators*, © 1977 by author, Dover Publications, Inc., ISBN 0-486-44981-5.

10 Interplanetary Trajectories

Interplanetary trajectories provide a transition from a trajectory relative to one planet to a trajectory relative to another planet. The trajectories relative to the planets are localized, leaving most of the flight as a trajectory relative to the Sun. Thus, the primary planning starts from a solar trajectory from the first planet's position at the departure date and time to the second planet's position at the arrival date and time.

10.1 Interplanetary Travel Characteristics

Interplanetary trajectories typically travel from one solar system celestial body to another (usually planets), with the Sun being the central attracting celestial body. These are closed (elliptical) trajectories for most cases.

As we depart from or approach a planet, the spacecraft encounters a gravitational transition. The transition occurs when the planet's and the Sun's gravitational field strength are equal. While actual gravitational handover is gradual, the point at which the two have equal gravitational strength is the demarcation in which the central attracting body changes from planet to Sun (on a planetary departure) or from Sun to planet (on a planetary arrival). The radius where the planet's gravity equals that of the Sun defines the celestial planet's *sphere of influence*. The spacecraft crosses the sphere of influence virtually on a hyperbolic asymptote with respect to the planet. For planning purposes, it is presumed to cross on the asymptote at the hyperbolic excess speed (v_∞).

Since the trajectory is always modeled relative to the central attracting body, there are three distinct trajectories to consider with interplanetary travel: the trajectory of the spacecraft departing the initial planet, the heliocentric interplanetary transfer, and the trajectory when the spacecraft arrives at the target planet.

10.1.1 Interplanetary Transfers [Level I – Descriptive]

Interplanetary trajectories initially focus on heliocentric transfers since most of the interplanetary travel is under the Sun's gravitational influence. Most of the solar system planets are in nearly circular orbits around the Sun and most are orbiting in nearly the same plane, as evidenced by the celestial body orbital parameters in Appendix A. This allows a common interplanetary trajectory to be approximated by a heliocentric Hohmann transfer.

Preliminary planning involves determining an opportunity for a heliocentric Hohmann transfer between the two approximately circular planetary orbits as illustrated in figure 10-1. This requires evaluation of the synodic period between the planets for favorable planetary phasing as was done with the rendezvous planning in section 6.2. Notice for this case that the target planet is in a higher heliocentric orbit than the departure planet, giving it a slower orbital rate than the departure planet. Thus, the heliocentric departure timing is planned, giving the target planet a head start so it meets up with the spacecraft at the transfer orbit aphelion.

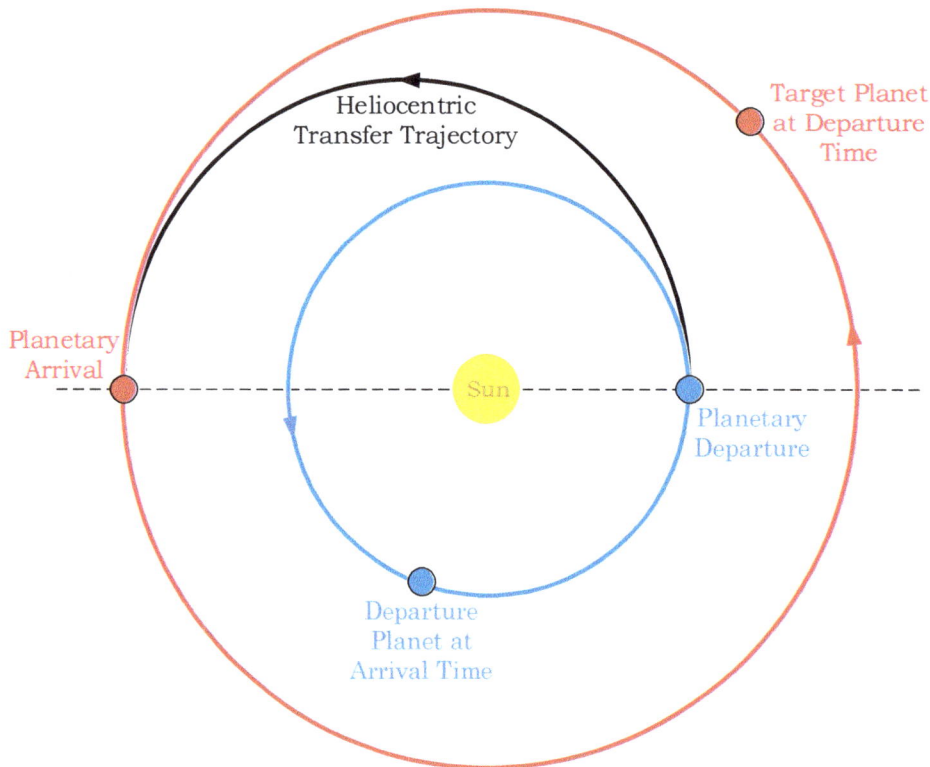

Figure 10-1 Heliocentric Hohmann Transfer

After the heliocentric Hohmann transfer is planned, the next task is determining the departure trajectory from the departure planet and the arrival trajectory for the destination (or target) planet. Departing the first planet's gravity field requires an escape (i.e., hyperbolic trajectory) which is aimed to blend into the beginning of the Hohmann transfer trajectory. Similarly, the Hohmann transfer's completion point is blended into the target planet arrival trajectory which will also be hyperbolic.

The initial estimates of these trajectories consider only the gravity of the central attracting body using two-body analysis. The process that blends these trajectories at the gravitational crossovers is called the *method of patched conics*. Using this method, the blending of the two-body conics is the first approximation, which is iteratively refined with high fidelity perturbational models using numerical integration to produce an executable mission profile.

A heliocentric Hohmann transfer from an outer to an inner planet is the reverse of the inner-to-outer transfer as illustrated in figure 10-2. The outer planet requires the same lead angle as required when departing from the inner planet. As with all

Hohmann sequences transferring from a higher to a lower orbital radius, the spacecraft must reduce its heliocentric speed to transition from a high circular orbit at the outer planet orbital radius, achieving a perihelion at the inner planet's orbital radius. A second heliocentric speed reduction is required at perihelion to achieve a circular orbit at the inner planet's orbital radius.

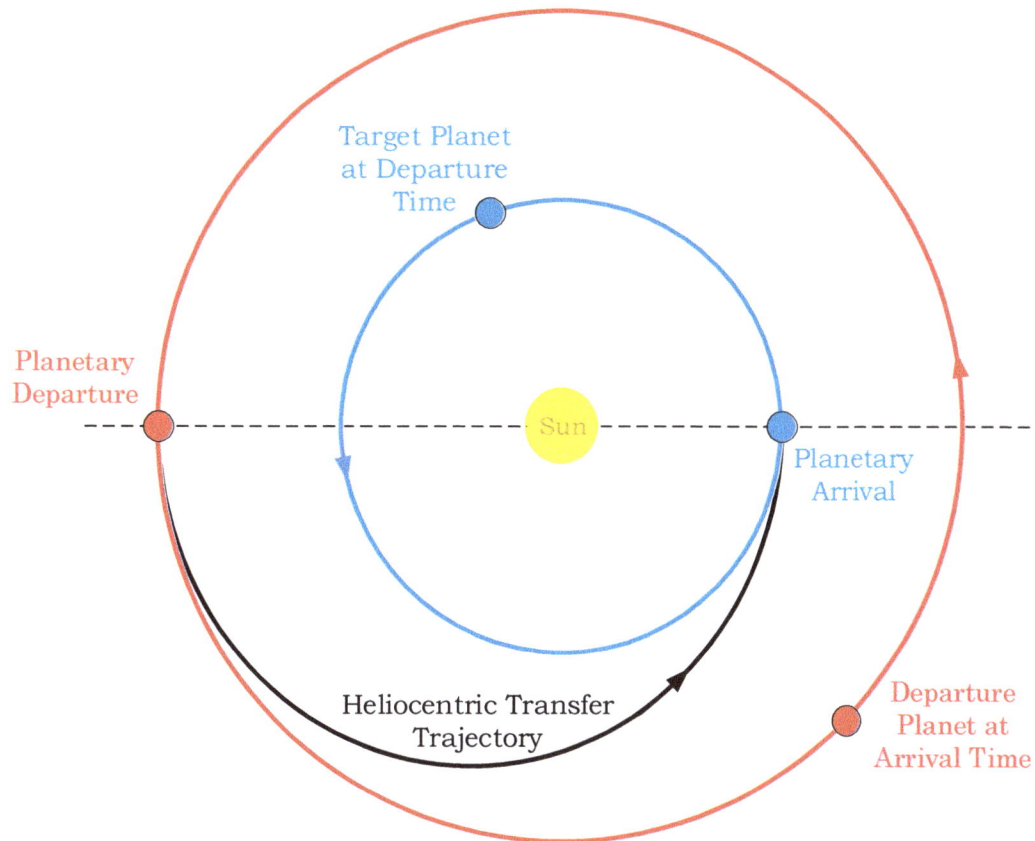

Figure 10-2 Heliocentric Hohmann Transfer from Outer Planet

Key Terms:

Method of Patched Conics: is a piece-wise fit of conic section trajectories that are patched at the transition between celestial body spheres of influence.
Sphere of Influence: is the zone surrounding a celestial body within which its gravity is stronger than that of a more massive celestial body.

➤ **Transition**: *You may continue this section with Interplanetary Transfers at Level II, or you may skip to the beginning of the next section called Launch Window.*

10.1.2 Interplanetary Transfers [Level II – Equations]

Interplanetary transfers begin by planning the transfer between the departure and target planets. The process herein considers the Hohmann transfer since it is the most efficient in terms of having a minimum required speed change (Δv) and thus need minimal propellant. Because of the efficiency advantage, the scope herein focuses on the Hohmann transfer.

There are non-Hohmann transfers as well, most notably type I and type II transfers. Type I transfers travel less than 180° and type II transfer travel between 180° and 360°. The potential advantage of a type I transfer for a human occupied spacecraft is the reduction of life support consumables needed with a shorter transit time. However, for this to be advantageous, the additional propellant mass needed for an increased departure speed and arrival deceleration must offset the mass savings in the life support consumables.

Current trends focus strongly on closed life support systems that recycle air, water, and food. Such advances are beneficial regardless of whether a type I transfer can be made feasible. The transition to type I transfers will occur in the future when advanced propulsion capabilities create a clear advantage. At that point, the transfers will be planned using solutions to the Lambert problem.

It should be emphasized that the algorithms presented in this section are analytic first approximations. These are the starting estimates for more sophisticated, high fidelity numerical solutions.

10.1.2.1 Interplanetary Hohmann Transfer

An interplanetary Hohman transfer is a heliocentric trajectory between circular planetary orbits. This makes it an elliptical transfer orbit that is cotangential to the two circular orbital radii. Since the Sun is the central attracting body, $\mu_\odot = 132{,}712{,}440\ km^3/s^2$ is the gravitational parameter. The aphelion (r_a) and perihelion (r_p) radii are the applicable planetary orbital radii in kilometers. Equation 10.1 is the Hohmann transfer semimajor axis (a_T), equation 10.2 is the Hohmann transfer eccentricity (e_T), and equation 10.3 is half the orbital period which is the interplanetary transit (τ_T) time.

$$a_T = \frac{r_a + r_p}{2} \qquad\qquad 10.1$$

$$e_T = \frac{r_a - r_p}{r_a + r_p} \qquad\qquad 10.2$$

$$\tau_T = \pi \sqrt{\frac{a_T^3}{\mu_\odot}}$$

$$10.3$$

Equation 10.3 computes the interplanetary transit time (τ_T) in seconds units. This may be converted to days by dividing τ_T by 86,400, which is the number of seconds in a day.

10.1.2.2 Hohmann Transfer Apse Speeds

The size and shape of the Hohmann transfer ellipse determines the speeds needed at periapsis ($v_{p,T}$) and apoapsis ($v_{a,T}$). Equation 10.4 computes the speeds at the transfer orbit apse points.

$$v_{p,T} = \sqrt{\frac{\mu_\odot}{a_T}\left[\frac{1+e_T}{1-e_T}\right]} \qquad v_{a,T} = \sqrt{\frac{\mu_\odot}{a_T}\left[\frac{1-e_T}{1+e_T}\right]}$$

$$10.4$$

10.1.2.3 Departure Planet Phasing Angle

The departure phase angle ($\Delta\theta_D$) is the true anomaly difference that the planet with the larger orbital radius needs as a head start on the planet with the smaller orbital radius. This angular head start sets up the timing, so the spacecraft meets up with the target planet when it reaches the arrival apse. The lead angle and its timing are the based on the transfer orbit period ($T_T = 2\tau_T$) and the target orbit period (T_2). Equation 10.5 computes the mean motions for the departure and target planet orbits based on the departure (T_1) and target (T_2) planetary orbital periods. Equation 10.6 computes the departure phase angle at the beginning of the Hohmann transfer.

$$n_1 = \frac{2\pi}{T_1} \qquad n_2 = \frac{2\pi}{T_2}$$

$$10.5$$

$$\Delta\theta_T = \pi - n_2 T_T$$

$$10.6$$

Equation 10.7 is the synodic period, which is the time between favorable alignments ($\Delta\theta_T$) for entry into a Hohmann transfer between the departure and target planet.

$$T_{syn} = \frac{T_1 T_2}{|T_1 - T_2|} = \frac{2\pi}{|n_1 - n_2|}$$

$$10.7$$

The synodic period is the same for each planetary pair, regardless of which is the departure, and which is the target planet.

Key Term:

> **Synodic Period**: is the time between two orbits in which the phasing between their true anomalies repeats.

10.1.2.4 Launch Window

Launch is usually restricted to the times around which the Hohmann transfer plane passes through the launch site. Most launch sites have launch azimuth restrictions due to range safety and the corridors in which boosters can be dropped. Thus, launch windows occur when the launch site is near the orbital plane and an available launch azimuth permits insertion into the transfer orbit plane.

There may be some additional flexibility on the choice of launch azimuth, provided the launch booster has sufficient margin to include an adjustment to the insertion orbital plane.

10.1.3 Planetary Departure Trajectory [Level I – Descriptive]

When departing from one planet to another on a Hohmann transfer, the departure direction depends on whether the heliocentric radius is to be increased or decreased. Figure 10-1 illustrated a transfer to an outer planet, corresponding to an increase to the heliocentric radius. An increase establishing the aphelion radius requires an increase to the heliocentric speed. Figure 10-3 illustrates the geometry needed, in which the hyperbolic departure asymptote is parallel to and in the same direction as the planet's orbital velocity. In this instance, the heliocentric speed is the sum of the planet speed and the hyperbolic excess speed (v_∞).

Note also that the Sun's direction is perpendicular to the planet's orbital path and the departure asymptote. This establishes a zero value for the flight path angle producing the heliocentric transfer's periapsis.

A Hohmann transfer to an inner planet has a corresponding decrease to the heliocentric radius. Figure 10-4 illustrates the geometry needed. in which the hyperbolic departure asymptote is parallel to and in the opposite direction of the planet's orbital velocity. In this instance, the heliocentric speed is the difference between the planet speed and the hyperbolic excess speed (v_∞).

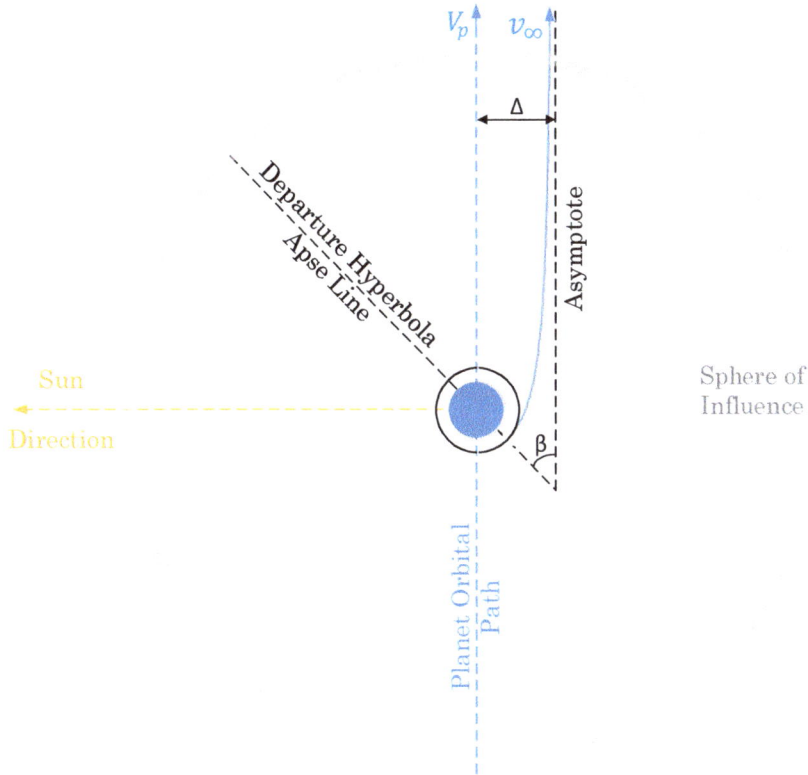

Figure 10-3 Departure Geometry to Outer Planet

When transferring from an outer-to-inner planet as was illustrated in figure 10-2, the heliocentric speed must decrease to establish a perihelion radius equal to that of the inner planet's orbit.

➢ **Transition**: *You may continue this section with Planetary Departure Trajectory at Level II, or you may skip to the beginning of the next section called Planetary Arrival Trajectory.*

10.1.4 Planetary Departure Trajectory [Level II – Equations]

The heliocentric departure trajectory typically begins with a circular parking orbit around the departure planet. A circular parking orbit radius defines the periapsis radius for heliocentric trajectory. The first step is computing the hyperbolic speed at the departure point, which defines the spacecraft speed, relative to the planet, required at exit of the sphere of influence.

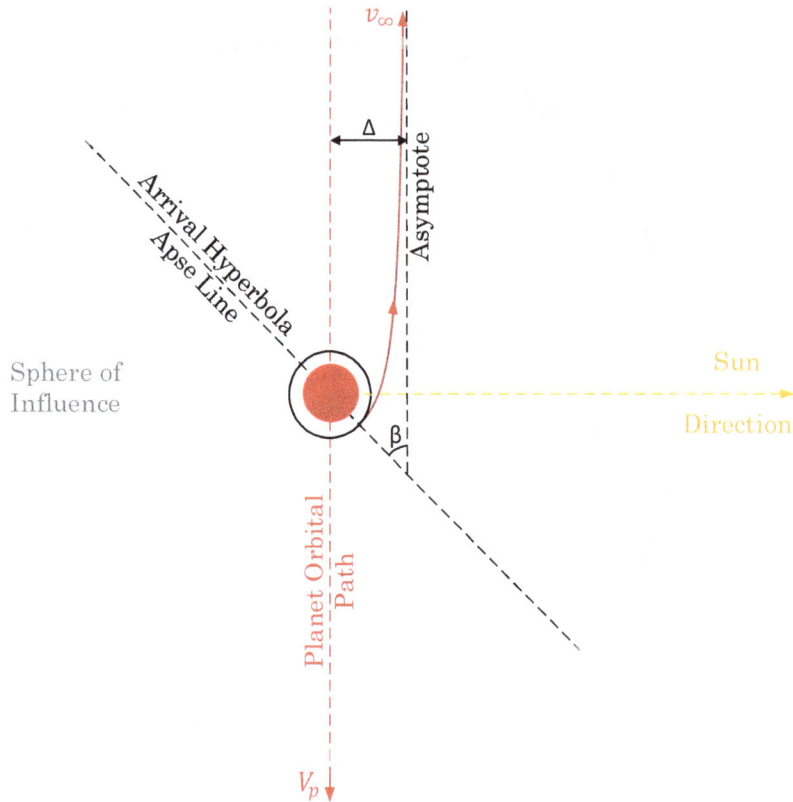

Figure 10-4 Departure Geometry to Inner Planet

Equation 10.8 is the hyperbolic excess speed ($v_{\infty,p}$) for a departure at the heliocentric transfer orbit perihelion, with V_p being the planet's heliocentric speed. This corresponds to the v_∞ needed for an inner to outer planet departure as was illustrated in figure 10-3. Equation 10.9 is the hyperbolic excess speed ($v_{\infty,a}$) for a departure at the heliocentric transfer orbit aphelion. This corresponds to the v_∞ needed for an outer to inner planet departure as was illustrated in figure 10-4.

$$v_{\infty,p} = v_{p,T} - V_p \qquad\qquad 10.8$$

$$v_{\infty,a} = v_{a,T} + V_p \qquad\qquad 10.9$$

These speeds may be computed as scalars since a Hohmann transfer is used and thus the events occur at the apse points where the flight path angles are zero.

Equation 10.10 solves for the hyperbolic semimajor axis (a_h) using equation 2.125 using parameters at infinity. Known inputs are $v = v_\infty$ and $r = \infty$, and the departure planet's gravitational parameter (μ).

$$a_h = -\frac{\mu}{v_\infty^2} \qquad\qquad 10.10$$

Equation 10.11 is the hyperbolic eccentricity (e_h), determined from the periapsis radius (r_p) and the semimajor axis. The periapsis radius is that of the circular parking orbit.

$$e_h = 1 - \frac{r_p}{a_h} \qquad\qquad 10.11$$

Equation 10.12 relates the hyperbolic trajectory's periapsis by the β angle between the hyperbolic line of apsides and the departure asymptote.

$$\cos\beta = \frac{1}{e_h} \qquad\qquad 10.12$$

Equation 10.13 relates the true anomaly (θ_∞) of the outbound hyperbolic trajectory asymptote to the eccentricity. The inverse cosine has two solutions; the outbound asymptote has a true anomaly less than 180°.

$$\cos\theta_\infty = -\frac{1}{e_h} \qquad\qquad 10.13$$

Equation 10.14 determines the hyperbolic trajectory's aiming radius (Δ), which is the distance between the departure planet and asymptote.

$$\Delta = |a|\sqrt{e_h^2 - 1} \qquad\qquad 10.14$$

The aiming radius value is less critical for the departure than it is for the arrival hyperbolic trajectory.

10.1.5 Planetary Arrival Trajectory [Level I – Descriptive]

Spacecraft arrival geometry for a Hohmann transfer is less intuitive than the departure geometry. The spacecraft speed relative to that of the planet determines whether the approach will be ahead of or behind the planet. If the spacecraft's speed is slower than the planet, it must approach ahead of the planet, allowing the planet to catch up and establish the encounter. If the spacecraft speed is faster than the planet, it must approach behind the planet, overtaking it to establish the encounter. Visualizing in either case references the encounter from the planet's point of view using the spacecraft relative velocity within the sphere of influence.

Spacecraft arrival to an outer planet needs to be slightly ahead of the planet. The transfer orbit aphelion speed will be slower than that of the planet, causing the relative spacecraft velocity to be opposite that of the planet. The spacecraft has a hyperbolic trajectory relative to the planet, observed once it enters the planet's sphere of influence.

The hyperbolic excess speed (v_∞) is the sum of the planet's speed and the spacecraft's aphelion speed, which is also the relative velocity. Since the planet's speed is faster than the aphelion speed, the planet catches up with the spacecraft at the sphere of influence leading edge as illustrated in figure 10-5. The spacecraft's relative velocity is in the opposite direction of the planet's orbital velocity since the planet is overtaking. What may seem counterintuitive is that while both the spacecraft and the planet are moving in the same direction in the heliocentric coordinate system, the relative velocity is opposite in the planet coordinate system.

Spacecraft arrival to an inner planet needs to be slightly behind the planet. The transfer orbit perihelion speed will be faster than that of the planet, causing the relative spacecraft velocity to be in the same direction as the planet. The spacecraft has a hyperbolic trajectory relative to the planet, observed once it enters the planet's sphere of influence.

The hyperbolic excess speed (v_∞) is the difference between the spacecraft's perihelion speed and the planet's speed, which is also the relative velocity. Since the planet's speed is slower than the spacecraft perihelion speed, the spacecraft overtakes the planet at the sphere of influence trailing edge as illustrated in figure 10-6. The spacecraft's relative velocity is in the direction of the planet's orbital velocity since the spacecraft is overtaking.

➤ **Transition**: *You may continue this section with Planetary Arrival Trajectory at Level II, or you may skip to the beginning of the next section called B-Plane Targeting.*

10.1.6 Planetary Arrival Trajectory [Level II – Equations]

The heliocentric departure trajectory terminates upon entry of the target planet sphere of influence. The heliocentric arrival speed affects the periapsis radius for the subsequent heliocentric trajectory. Equation 10.15 is the hyperbolic excess speed ($v_{\infty,a}$) for an arrival at the heliocentric transfer orbit aphelion, with V_p being the planet's heliocentric speed. Equation 10.16 is the hyperbolic excess speed ($v_{\infty,p}$) for an arrival at the heliocentric transfer orbit perihelion.

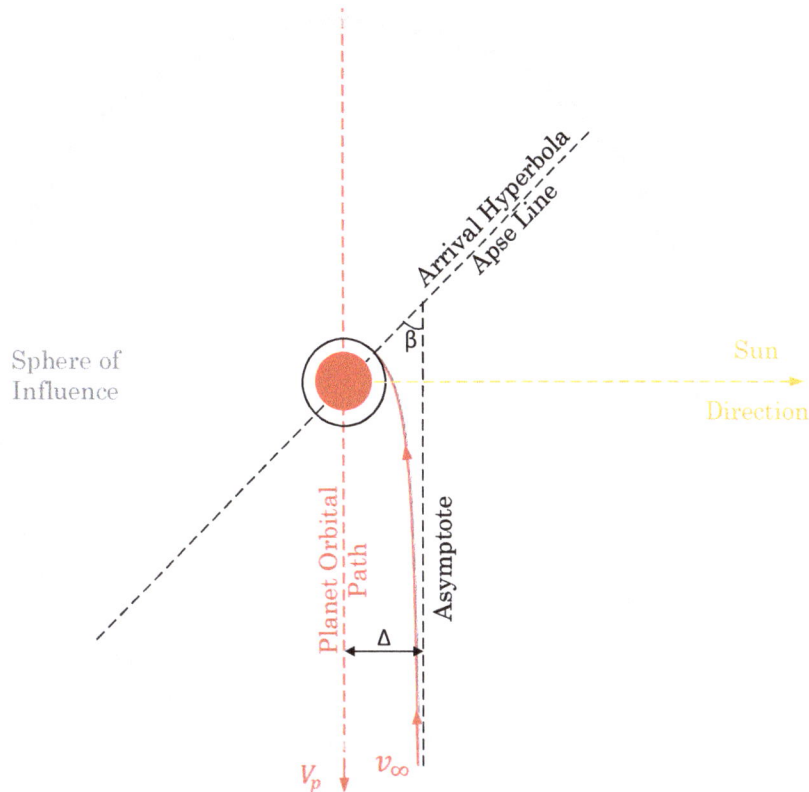

Figure 10-5 Arrival Geometry for Outer Planet

$$v_{\infty,p} = v_{a,T} + V_p \qquad\qquad 10.15$$

$$v_{\infty,p} = v_{p,T} - V_p \qquad\qquad 10.16$$

These speeds may be computed as scalars since a Hohmann transfer is used and thus the events occur at the apse points where the flight path angles are zero.

Equation 10.17 solves for the hyperbolic semimajor axis (a_h) using equation 2.125 using parameters at infinity. Known inputs are $v = v_\infty$ and $r = \infty$, and the arrival planet's gravitational parameter (μ).

$$a_h = -\frac{\mu}{v_\infty^2} \qquad\qquad 10.17$$

The hyperbolic eccentricity (e_h) is set so the hyperbolic trajectory will have the targeted periapsis radius in equation 10.18. This radius depends on the mission requirements such as whether an orbit will be established, or landing will be made.

$$e_h = 1 - \frac{r_p}{a_h} \qquad\qquad 10.18$$

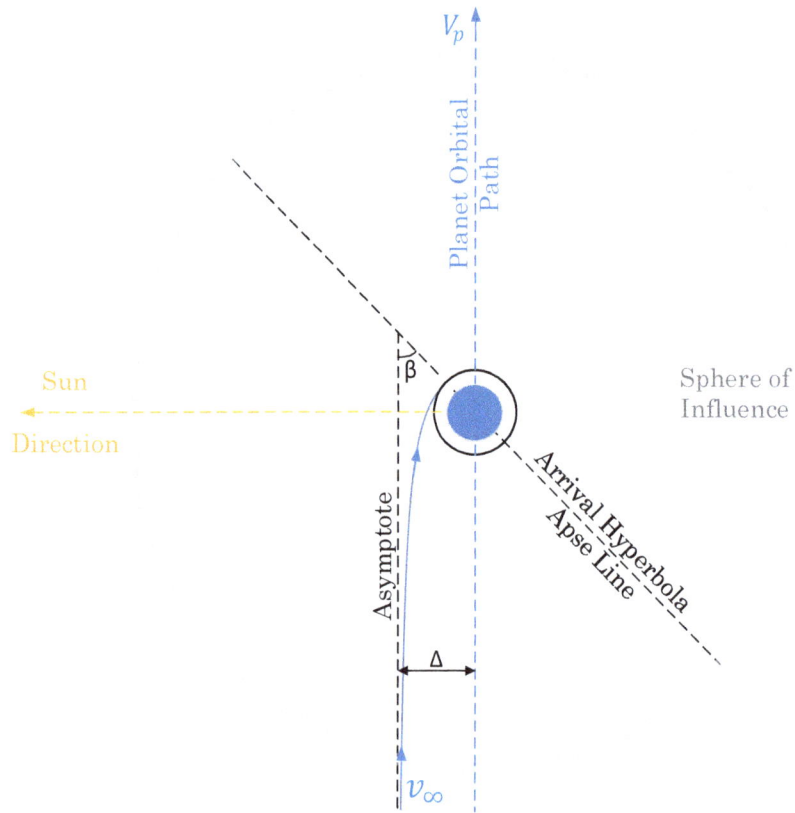

Figure 10-6 Arrival Geometry for Inner Planet

Equation 10.19 computes the hyperbolic aiming radius (Δ) which is used to establish the periapsis radius. The aiming radius is the magnitude of a $\bar{B}$ targeting vector, which will be covered more in the B-plane Targeting section.

$$\Delta = |a|\sqrt{e_h^2 - 1} \qquad \qquad 10.19$$

Equations 10.20 and 10.21 compute the true anomaly of the inbound asymptote (θ_∞) and the β angle between the hyperbolic line of apsides and the inbound asymptote. The inbound asymptote true anomaly is the inverse cosine solution that is greater than 180°. The two true anomaly solutions ($\theta_\infty, \theta'_\infty$) are symmetric about the apse line such that $\theta'_\infty = 2\pi - \theta_\infty$.

$$\cos\theta_\infty = -\frac{1}{e_h} \qquad \qquad 10.20$$

$$\cos\beta = \frac{1}{e_h} \qquad \qquad 10.21$$

Equation 10.22 is the hyperbolic trajectory angular momentum (h_h) magnitude.

$$h_h = \frac{\mu}{v_\infty} \sqrt{e_h^2 - 1}$$ 10.22

10.1.7 B-Plane Targeting [Level I – Descriptive]

Planetary hyperbolic trajectories are targeted using the *B-plane*. The B-plane passes through the arrival planet center of mass as its origin and is perpendicular to the hyperbolic trajectory plane as illustrated in figure 10-7.

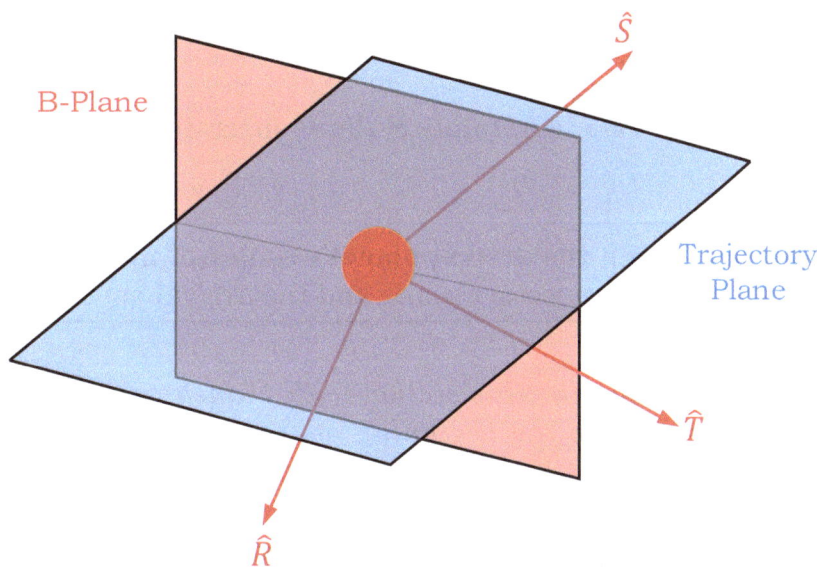

Figure 10-7 B-Plane Coordinate System

The B-Plane trajectory axes are defined by the $\hat{R}$, $\hat{S}$, $\hat{T}$ unit vectors, with $\hat{R}$ and $\hat{T}$ in the B-Plane and $\hat{S}$ parallel to the arrival asymptote, normal to the B-Plane. The B-vector provides the two-dimensional position where the inbound asymptote intersects the B-Plane as illustrated in figure 10-8. The $\hat{T}$ axis is in the planet's x, y plane and the $\hat{R}$ is perpendicular to the x, y plane.

The B-vector ($\bar{B}$) has components in the $\hat{R}$ and $\hat{T}$ directions (B_R and B_T respectively). Since it locates where the inbound asymptote crosses the B-plane, its magnitude equals the inbound hyperbolic trajectory's aiming radius (Δ). The magnitude of $\bar{B}$ controls the hyperbolic trajectory's periapis; the B_R and B_T components control the hyperbolic trajectory's inclination relative to the planet. Thus, how the inbound asymptote is aimed fashions the inbound trajectory around the target planet.

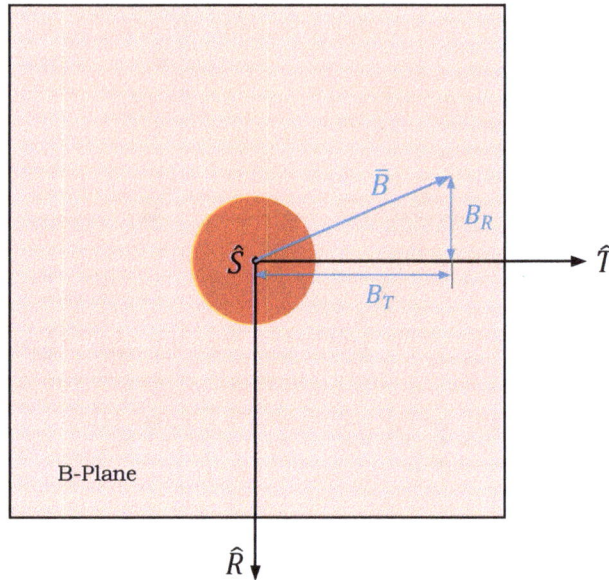

Figure 10-8 Asymptote B-Plane Intersection

Key Term:

> **B-Plane**: is a plane through the arrival planet's center of mass that is mutually perpendicular to the arrival trajectory plane and the arrival asymptote.

➤ **Transition**: *You may continue this section with Planetary Arrival Trajectory at Level II, or you may skip to the beginning of the next section called Sphere of Influence.*

10.1.8 B-Plane Targeting [Level II – Equations]

Placing the spacecraft in the desired hyperbolic trajectory relative to the target planet is an iterative process. Equation 10.23 computes the hyperbolic angular momentum ($\bar{h}_h$) from a candidate position ($\bar{r}_h$) and velocity ($\bar{v}_h$) state vector set. Equation 10.24 computes the eccentricity vector ($\bar{e}_h$) from the position and velocity.

$$\bar{h}_h = \bar{r}_h \times \bar{v}_h \qquad\qquad 10.23$$

$$\bar{e}_h == \frac{\bar{r}_h \times (\bar{r}_h \times \bar{v}_h)}{\mu} - \frac{\bar{r}_h}{r_h} \qquad\qquad 10.24$$

The eccentricity vector ($\bar{e}_h$) is directed toward the hyperbolic periapsis and thus its direction is offset from the asymptote by the β angle (equation 10.21). Equation 10.25 thus orients the B-plane $\hat{S}$ vector.

$$\hat{S} = \hat{e}_h \cos\beta + \left(\hat{h}_h \times \hat{e}_h\right) \sin\beta \qquad\qquad 10.25$$

Equation 10.26 computes the $\hat{T}$ direction, recognizing the planet's z-axis has components $\hat{z} = [0 \quad 0 \quad 1]^T$. The $\hat{R}$ vector completes the orthonormal set.

$$\hat{T} = \frac{\hat{S} \times \hat{z}}{|\hat{S} \times \hat{z}|} = \frac{1}{\sqrt{S_x^2 + S_y^2}} \begin{bmatrix} S_y \\ -S_x \\ 0 \end{bmatrix}$$

10.26

$$\hat{R} = \hat{S} \times \hat{T}$$

10.27

Equation 10.28 characterizes the angle (φ) the B-plane is tilted relative to the planet's z-axis. Angle φ is the minimum hyperbolic orbital inclination ($\cos i_h$) relative to the planet.

$$\sin \varphi = \hat{S} \cdot \hat{z}$$

10.28

The aiming radius (Δ) was computed from equation 10.19 during the design of the in-bound trajectory. Equation 10.29 relates Δ and B_T to the desired inclination (i_h) of the inbound hyperbolic trajectory with the applicable constraint. Equation 10.30 relates the B_R component to Δ and B_T.

$$\cos i_h = \frac{B_T}{\Delta} \cos \varphi \qquad (i_h \geq \varphi)$$

10.29

$$B_R = \pm \sqrt{\Delta^2 - B_T^2}$$

10.30

10.1.9 Sphere of Influence [Level I – Descriptive]

The Sun is the dominant gravity source in most of the solar system. Thus, the Sun will be the central gravitational body for solar system transit except for small spheres of influence surrounding the planets and other celestial bodies. A spacecraft in heliocentric orbit that is approaching another celestial body will experience a progressively increasing gravitational influence by that body. If it gets close enough, it will cross a threshold after which the celestial body's gravitational influence is stronger as the Sun's gravitational influence becomes progressively weaker. The opposite is true for a spacecraft departing to interplanetary space, where the planet's gravity weakens as the Sun's influence increases until the Sun becomes the dominant gravitation source. The sphere of influence (SOI) is the transition radius about the less massive body in which the gravitational influence from both celestial bodies is equal.

Determining the SOI transition radius is a three-body problem analysis that applies not only to the Sun and planets, but other celestial body pairs. One celestial body is less massive than the more massive body about which it orbits. The three-body problem considers that each of the two celestial bodies exerts a gravitational acceleration on the spacecraft proportional to their mass and the square of their distance as was seen in Chapter 1's first equation. But the two celestial bodies are simultaneously exerting a gravitational force on each other. The simultaneous gravitational influence of all three bodies approximates the sphere of influence radius.

> ➤ **Transition**: *You may continue this section with Planetary Arrival Trajectory at Level II, or you may skip to the beginning of the next section called Planetary Encounters.*

10.1.10   Sphere of Influence [Level II – Equations]

Equation 10.31 computes the SOI radius (r_{SOI}) by the Laplace method. The radius is computed in terms of the mass ratio of the more massive primary (m_p) celestial body to the less massive secondary (m_s) celestial body.

$$r_{SOI} = r_s \left(\frac{m_s}{m_p} \right)^{2/5}$$

10.31

The r_s is the radius at which the less massive celestial body orbits about the more massive celestial body.

10.2 Planetary Encounters

Planetary encounters occur within the destination planet's SOI. The events planned during a planetary encounter are dictated by the mission requirements. There can be one of three goals: planetary impact, planetary capture (i.e., orbit), or planetary flyby.

Planetary impact is the simplest in that the hyperbolic trajectory periapsis is less than the planet's physical radius (or atmospheric radius if applicable). This is planned by setting up an appropriate aiming radius (Δ) or by a thrust event on the inbound trajectory that reduces radius. Planetary capture and flybys are covered in the ensuing sections.

10.2.1 Planetary Capture [Level I – Descriptive]

Planetary capture is an event in which the inbound hyperbolic trajectory has energy removed, transitioning it to an orbit around the planet. Curtis (2020) shows periapsis to be the most efficient location in which to perform the retrograde velocity change (Δv) to establish orbital capture. Figure 10-7 illustrates the associated geometry.

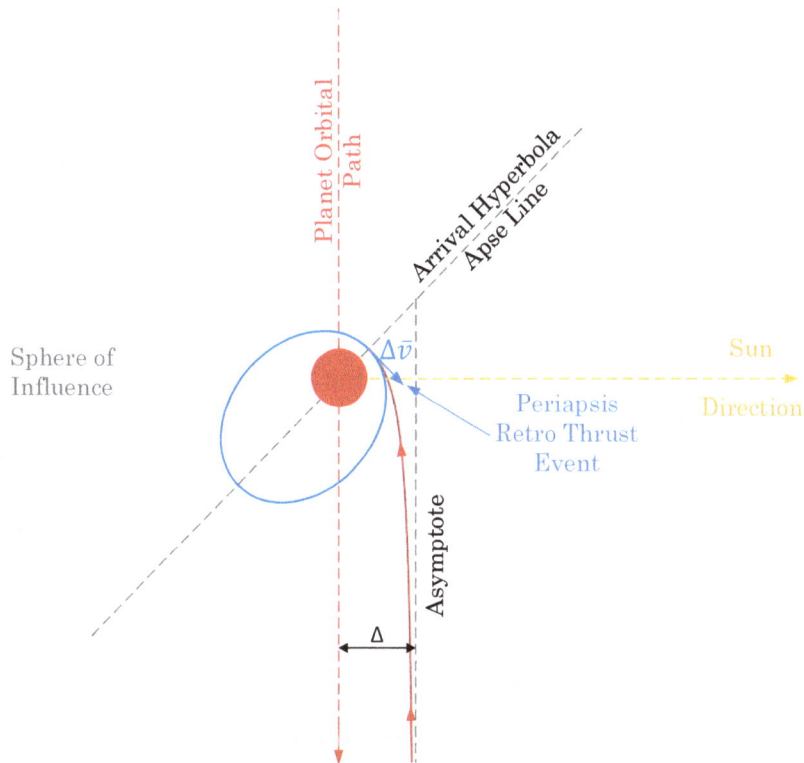

Figure 10-9 Capture to Planetary Orbit

> ➢ **Transition**: *You may continue this section with Planetary Capture at Level II, or you may skip to the beginning of the next section called Planetary Flyby.*

10.2.2 Planetary Capture [Level II – Equations]

The velocity change to transition from the hyperbolic trajectory to an elliptical orbit is the difference between the two periapsis speeds since the flight path angle is zero for both at periapsis. Equation 10.32 is the hyperbolic periapsis speed ($v_{p,h}$) as a function of angular momentum magnitude (h_h) and periapsis radius (r_p). Equation 10.33 is the corresponding elliptical periapsis speed ($v_{p,e}$) which also requires a

275

user-specified semi-major axis (a_e). Equation 10.34 is the speed change (Δv) needed to transition.

$$v_{p,h} = \frac{h_h}{r_p} \qquad\qquad 10.32$$

$$v_{p,e} = \sqrt{\mu \left[\frac{2}{r_p} - \frac{1}{a_e} \right]} \qquad\qquad 10.33$$

$$\Delta v = \sqrt{\mu \left[\frac{2}{r_p} - \frac{1}{a_e} \right]} - \frac{h_h}{r_p} \qquad\qquad 10.34$$

10.2.3 Planetary Flyby [Level I – Descriptive]

A planetary flyby is an encounter that has both inbound and outbound hyperbolic segments. A flyby can be as simple as swinging by the planet and exiting the sphere of influence on the outbound leg. More complex flybys can include a thrust event to enhance the flyby effects. Flybys may perform science experiments during the planetary encounter and/or may be used to alter the spacecraft's heliocentric trajectory. Design of the flyby geometry directly affects characteristics of the post flyby heliocentric trajectory.

The general presumption is the inbound asymptote is not necessarily aligned (or anti-aligned) with the planet's orbital path around the Sun. Thus, the orientations of the inbound and outbound asymptotes, relative to the planet, as well as the hyperbolic apsidal line need to be determined using vector or trigonometric analysis.

General conclusions may be drawn from the geometry, considering that the hyperbolic excess speed (v_∞) relative to the planet is the same at the entry and exit of the planet's SOI. The entry velocity is clearly redirected by the turn angle (δ) when determining the outbound velocity direction relative to the planet. While the speed relative to the planet remains unchanged, the heliocentric speed almost always changes due to a planetary encounter.

When one performs the vector addition or subtraction from heliocentric to the planetary coordinate system and reverses the process for the turned outbound velocity back to the heliocentric frame, the speed (velocity magnitude) change becomes apparent. This may also be viewed geometrically by examining the flyby geometries.

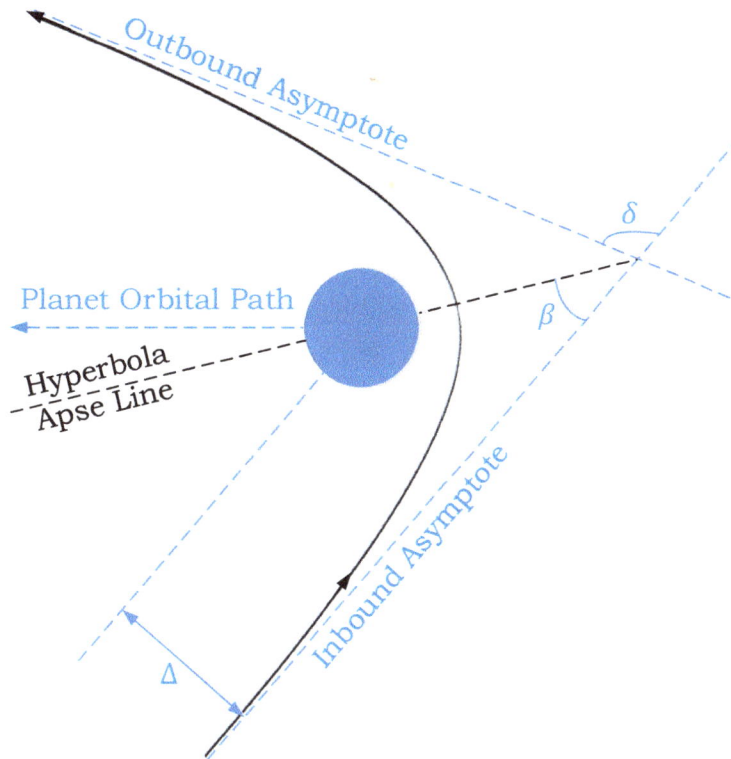

Figure 10-10 Trailing-Side Flyby of a Planet

Figure 10-10 is a trailing-side flyby, meaning periapsis is behind the planet's orbital path. For this encounter, the faster-moving planet was positioned for the hyperbolic trajectory to swing behind (or opposite) the planet's orbital path. The velocity redirection from a component opposing the planet's path to having a component along the planet's orbit path results in a faster heliocentric velocity than prior to the planetary encounter. This represents what is commonly known as a slingshot velocity increase.

Figure 10-11 is a leading-side flyby, meaning periapsis is ahead of the planet's orbital path. For this encounter, the faster-moving planet was positioned for the hyperbolic trajectory to swing in front of the planet's orbital path. The velocity redirection from a component in the same direction as the planet's path to having a component opposite the planet's orbit path results in a slower heliocentric velocity than prior to the planetary encounter.

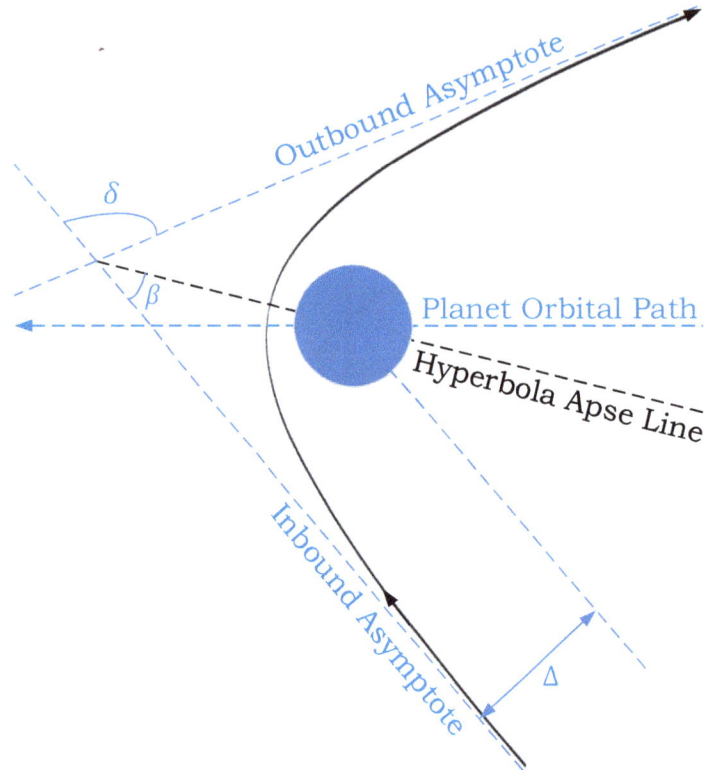

Figure 10-11 Leading-Side Flyby of a Planet

> ➢ **Transition**: *You may continue this section with Planetary Capture at Level II, or you may skip to the beginning of the next section called Post Flyby Trajectories.*

10.2.1 Planetary Flyby [Level II – Equations]

Planetary flyby encounters are evaluated using vector mathematics. For this analysis, basic geometry including the heliocentric arrival velocity ($\bar{v}$) and aiming radius (Δ) are known.

The inbound velocity orientation depends on the heliocentric flight path angle at arrival. Equation 10.35 relates the heliocentric velocity ($\bar{v}$) with the hyperbolic excess velocity ($\bar{v}_\infty$) and the planetary velocity ($\bar{v}_p$) at the SOI entry and exit.

$$\bar{v} = \bar{v}_p + \bar{v}_\infty \qquad\qquad 10.35$$

The spacecraft enters at the planet's SOI radius ($r = r_{SOI}$) on the arrival asymptote at the hyperbolic excess velocity ($\bar{v}_\infty$). Equation 10.36 is the hyperbolic trajectory's energy. Equation 10.37 is the hyperbolic trajectory's semi-major axis.

$$\varepsilon = \frac{v_\infty^2}{2} - \frac{\mu_p}{r_{SOI}}$$

10.36

$$a = \frac{\mu_p}{2\varepsilon}$$

10.37

Equation 10.38 computes the hyperbolic trajectory's eccentricity. Equation 10.39 computes the true anomaly of the inbound and outbound asymptotes.

$$e = \sqrt{1 + \frac{\Delta^2}{a^2}}$$

10.38

$$\theta_{\infty,IN} = 2\pi - \cos^{-1}\left(-\frac{1}{e}\right) \qquad \theta_{\infty,OUT} = \cos^{-1}\left(-\frac{1}{e}\right)$$

10.39

Equation 10.40 computes the angle between the asymptotes (β) and periapsis and equation 10.41 computes the turn angle (δ).

$$\beta = \cos^{-1}\left(\frac{1}{e}\right)$$

10.40

$$\delta = 2\sin^{-1}\left(\frac{1}{e}\right)$$

10.41

Equation 10.42 computes the hyperbolic trajectory's angular momentum magnitude and equation 10.43 computes the periapsis radius.

$$h = \sqrt{\mu_p a(e^2 - 1)}$$

10.42

$$r_p = \frac{h^2}{\mu_p(1 + e)}$$

10.43

10.3 Post Flyby Trajectories

A spacecraft returns to the heliocentric realm post flyby upon exiting the planet's sphere of influence. The planetary flyby results in an acceleration that alters the heliocentric trajectory. This acceleration changes trajectory's velocity, typically in terms of both speed (magnitude) and direction.

A post flyby trajectory results from the planet's position and the heliocentric velocity at the SOI exit. For a simplified flyby in the ecliptic plane, the hyperbolic trajectory turn angle provides the change to the velocity direction within the SOI. This must be evaluated by a triangular vector diagram as indicated in equation 10.35. The triangle solution determines the outbound heliocentric velocity.

REFERENCES

1. Curtis, Howard D, *Orbital Mechanics for Engineering Students*, Fourth Edition, © 2020 Elsevier, Ltd., ISBN 978-0-08-102133-0.
2. Chobotov, Vladimir A., *Orbital Mechanics*, © 1991 American Institute of Aeronautics and Astronautics (AIAA), ISBN 1-56347-007-1.
3. Brown, Charles D., *Spacecraft Mission Design*, Second Edition, © 1998 American Institute of Aeronautics and Astronautics (AIAA), ISBN 1-56374-262-7.
4. Prussing, John E. and Conway, Bruce A., *Orbital Mechanics*, © 1993 Oxford University Press, ISBN 0-19-507834-9.
5. Roy, Archie E., *Orbital Motion*, Third Edition Student Text, © 1988 by author, ISBN 0-85274-229-0.
6. Wie, Bong, *Space Vehicle Dynamics and Control*, © 1998 American Institute of Aeronautics and Astronautics (AIAA), ISBN 1-56347-261-9.
7. Bate, Roger R. et. al, *Fundamentals of Astrodynamics*, © 1971 Dover Publications, Inc., ISBN 0-486-60061-0.

11 Circular Restricted Three Body Problem

The circular restricted three body problem was first studied by mathematician and physicist Joseph-Louis Lagrange. The top-level assumptions are there are two celestial bodies with the primary significantly more massive than the secondary, the two bodies have circular orbits about their mutual center of mass (barycenter), and that the third body mass is insignificant in comparison to both celestial bodies. It is applicable to a first order approximation of an Earth-Moon-Spacecraft system or a Sun-Earth-Spacecraft system. Familiarity with the circular restricted three body problem is crucial for planning Earth-Moon mission trajectories since, unlike planetary trajectories, there is not an intermediate heliocentric transfer with which to plan.

11.1 Three Body Coordinate System [Level I – Descriptive]

The circular restricted three body problem has two celestial bodies of significant mass, with the third body being a spacecraft or other object with mass that is insignificant relative to the celestial bodies. The two celestial bodies orbit each other with circular paths about the system center of mass (or barycenter). Evaluation of the problem follows the derivation presented by Curtis (2020).

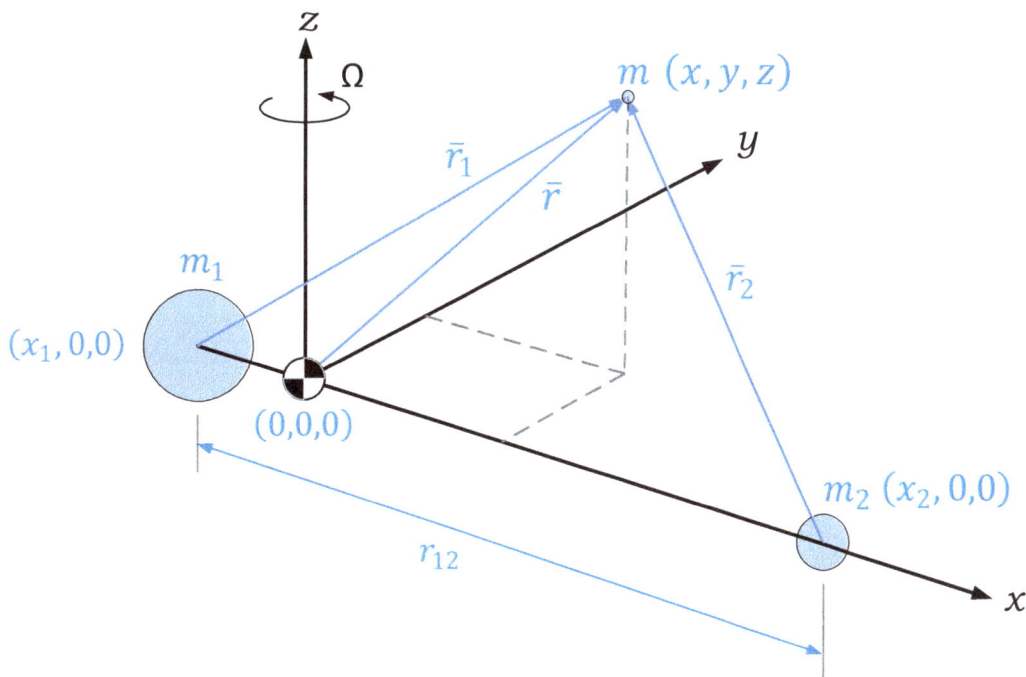

Figure 11-1 Circular Restricted Three Body Coordinates

The coordinate system origin is at the barycenter as illustrated in figure 11-1. The x-axis remains in line with the two celestial bodies, directed from the more massive

m_1 body, through the origin toward the m_2 smaller body. Thus, both celestial bodies are stationary relative to the rotating coordinate frame.

The z-axis is normal (i.e., perpendicular to) the orbital motion. The y-axis completes the orthogonal system. The spacecraft position is described by three vectors, with $\bar{r}$ being the position relative to the coordinate origin, $\bar{r}_1$ being the position relative to the more massive celestial body, and $\bar{r}_2$ being the position relative to the less massive celestial body. The distance between the celestial bodies is r_{12}, which is always oriented along the x-axis.

Key Term:

Barycenter: is the center of mass of a system of celestial bodies.

> ➢ **Transition**: *You may continue this section with Three Body System Motion at Level II, or you may skip to the beginning of the next section called Lagrange Points.*

11.2 Three Body System Motion [Level II – Equations]

Due to the circular motion, the coordinate system has a constant inertial angular velocity (Ω) directed through the z-axis as expressed in equation 11.1. Equation 11.2 is the system rotational period (T).

$$\bar{\Omega} = \Omega \hat{k} \qquad\qquad 11.1$$

$$T = \frac{2\pi}{\Omega} \qquad\qquad 11.2$$

The total system mass is the sum of the two celestial body masses ($M = m_1 + m_2$) and the system gravitational constant is $\mu = GM$. Equation 11.3 expresses the angular velocity in terms of the system gravitational constant (μ) and the distance between the celestial bodies, using the orbital period equation.

$$\Omega = \sqrt{\frac{\mu}{r_{12}^3}} \qquad\qquad 11.3$$

The celestial body masses (m_1, m_2) have coordinates along the x-axis relative to the origin that is also the system center of mass.

$$m_1 x_1 + m_2 x_2 = 0 \qquad\qquad 11.4$$

Since m_2 has a distance r_{12} from m_1 in the $+x$ direction:

$$x_1 = x_2 + r_{12}$$

11.5

Equations 11.6 through 11.9 define dimensionless mass ratios π_1 and π_2.

$$x_1 = -\pi_2 r_{12}$$

11.6

$$x_2 = \pi_1 r_{12}$$

11.7

$$\pi_1 = \frac{m_1}{m_1 + m_2}$$

11.8

$$\pi_2 = \frac{m_2}{m_1 + m_2}$$

11.9

Equations 11.10 through 11.12 are the three scalar equations of motion for the circular restricted three body problem.

$$\ddot{x} - 2\Omega\dot{y} - \Omega^2 x = -\frac{\mu_1}{r_1^3}(x + \pi_2 r_{12}) - \frac{\mu_2}{r_2^3}(x - \pi_1 r_{12})$$

11.10

$$\ddot{y} + 2\Omega\dot{x} - \Omega^2 y = -\frac{\mu_1}{r_1^3}y - \frac{\mu_2}{r_2^3}y$$

11.11

$$\ddot{z} = -\frac{\mu_1}{r_1^3}z - \frac{\mu_2}{r_2^3}z$$

11.12

No analytic solution has been found for these equations of motion. However, their evaluation provides insight into the system behavior.

➢ **Transition**: *You may continue this section with Three Body System Motion at Level III, or you may skip to the beginning of the next section called Lagrange Points.*

11.3 Three Body System Development [Level III – Derivation]

The motivation for evaluating the circular restricted three body problem is to characterize the motion of mass m relative to masses m_1 and m_2. Equations 11.13 and 11.14 provide the position of mass m relative to masses m_1 and m_2 respectively. Equation 11.15 is the position of mass m relative to the system center of mass.

$$\bar{r}_2 = (x - x_1)\hat{\imath} + y\hat{\jmath} + z\hat{k} = (x + \pi_2 r_{12})\hat{\imath} + y\hat{\jmath} + z\hat{k}$$

11.13

$$\bar{r}_2 = (x - \pi_1 r_{12})\hat{\imath} + y\hat{\jmath} + z\hat{k} \qquad\qquad 11.14$$

$$\bar{r} = x\hat{\imath} + y\hat{\jmath} + z\hat{k} \qquad\qquad 11.15$$

Equation 11.16 is the inertial velocity of m, expressed in terms of its velocity relative to the inertial frame defined by: $\bar{v}_{rel} = \dot{x}\hat{\imath} + \dot{y}\hat{\jmath} + \dot{z}\hat{k}$ and the inertial frame rotation rate ($\bar{\Omega}$). The vector $\bar{v}_{cm}$ is the center of mass inertial velocity.

$$\dot{\bar{r}} = \bar{v}_{cm} + \bar{\Omega} \times \bar{r} + \bar{v}_{rel} \qquad\qquad 11.16$$

Equation 11.17 expresses the inertial acceleration of mass m using the five-term acceleration equation.

$$\ddot{\bar{r}} = \bar{a}_{cm} + \dot{\bar{\Omega}} \times \bar{r} + \bar{\Omega} \times (\bar{\Omega} \times \bar{r}) + 2\bar{\Omega} \times \bar{v}_{rel} + \bar{a}_{rel} \qquad\qquad 11.17$$

The center of mass between two co-moving objects may be considered constant, making $\bar{a}_{cm} = 0$. (The constant center of mass velocity is due to an assumption in the problem of no external forces being present.) Likewise, since the angular rate ($\bar{\Omega}$) is constant, $\dot{\bar{\Omega}} = 0$. Equation 11.18 is the inertial acceleration with these simplifications.

$$\ddot{\bar{r}} = \bar{\Omega} \times (\bar{\Omega} \times \bar{r}) + 2\bar{\Omega} \times \bar{v}_{rel} + \bar{a}_{rel} \qquad\qquad 11.18$$

Equation 11.19 substitutes the three-body equations into 11.18. Equation 11.20 results from the cross-product evaluation, and equation 11.21 results from factoring.

$$\ddot{\bar{r}} = \Omega\hat{k} \times \left[\Omega\hat{k} \times (x\hat{\imath} + y\hat{\jmath} + z\hat{k})\right] + 2\Omega\hat{k} \times (\dot{x}\hat{\imath} + \dot{y}\hat{\jmath} + \dot{z}\hat{k}) + (\ddot{x}\hat{\imath} + \ddot{y}\hat{\jmath} + \ddot{z}\hat{k}) \qquad 11.19$$

$$\ddot{\bar{r}} = -\Omega^2(x\hat{\imath} + y\hat{\jmath}) + 2\Omega\dot{x}\hat{\jmath} - 2\Omega\dot{y}\hat{\imath} + \ddot{x}\hat{\imath} + \ddot{y}\hat{\jmath} + \ddot{z}\hat{k} \qquad\qquad 11.20$$

$$\ddot{\bar{r}} = (\ddot{x} - 2\Omega\dot{y} - \Omega^2 x)\hat{\imath} + (\ddot{y} + 2\Omega\dot{x} - \Omega^2 y)\hat{\jmath} + \ddot{z}\hat{k} \qquad\qquad 11.21$$

Equation 11.22 expresses Newton's Second Law for the spacecraft mass m.

$$m\ddot{\bar{r}} = \bar{F}_1 + \bar{F}_2 \qquad\qquad 11.22$$

Equations 11.23 and 11.24 express the two gravitational forces exerted by the celestial bodies.

$$\bar{F}_1 = -\frac{Gm_1 m}{r_1^3}\bar{r}_1 = -\frac{\mu_1 m}{r_1^3}\bar{r}_1 \qquad\qquad 11.23$$

$$\bar{F}_2 = -\frac{Gm_2m}{r_2^3}\bar{r}_2 = -\frac{\mu_2 m}{r_2^3}\bar{r}_2 \qquad 11.24$$

Equation 11.25 expresses Newton's Second Law in terms of the gravitational forces in equations 11.23 and 11.24.

$$\ddot{\bar{r}} = -\frac{\mu_1}{r_1^3}\bar{r}_1 - \frac{\mu_2}{r_2^3}\bar{r}_2 \qquad 11.25$$

Equation 11.26 is formed by substituting the right side of equation 11.25 into the left side of equation 11.21. Equations 11.13 and 11.14 are first substituted for $\bar{r}_1$ and $\bar{r}_2$ in equation 11.25.

$$-\frac{\mu_1}{r_1^3}(x + \pi_2 r_{12})\hat{\imath} + y\hat{\jmath} + z\hat{k} - \frac{\mu_2}{r_2^3}(x - \pi_1 r_{12})\hat{\imath} + y\hat{\jmath} + z\hat{k}$$
$$= (\ddot{x} - 2\Omega\dot{y} - \Omega^2 x)\hat{\imath} + (\ddot{y} + 2\Omega\dot{x} - \Omega^2 y)\hat{\jmath} + \ddot{z}\hat{k} \qquad 11.26$$

Equations 11.27 through 11.29 separate the components by the respective $(\hat{\imath}, \hat{\jmath}, \hat{k})$ coordinate axes into three scalar equations of motion.

$$\ddot{x} - 2\Omega\dot{y} - \Omega^2 x = -\frac{\mu_1}{r_1^3}(x + \pi_2 r_{12}) - \frac{\mu_2}{r_2^3}(x - \pi_1 r_{12}) \qquad 11.27$$

$$\ddot{y} + 2\Omega\dot{x} - \Omega^2 y = -\frac{\mu_1}{r_1^3}y - \frac{\mu_2}{r_2^3}y \qquad 11.28$$

$$\ddot{z} = -\frac{\mu_1}{r_1^3}z - \frac{\mu_2}{r_2^3}z \qquad 11.29$$

These results confirm equations 11.10 through 11.12.

11.4 Lagrange Points

Lagrange points, also known as Libration points, are interesting artifacts resulting from the circular restricted body equations of motion. Lagrange points are the positions in the three-body system in which have zero net acceleration (i.e., combining gravitational and centrifugal acceleration) in the rotating coordinate frame.

11.4.1 Lagrange Point Geometry [Level I – Descriptive]

Analysis of the circular restricted three body problem locates five distinct Lagrange points (denoted L_1, L_2, L_3, L_4, L_5) that are illustrated in figure 11-2. Lagrange points

L_1, L_2, and L_3 are colinear with the two celestial bodies of masses m_1 and m_2. Lagrange points L_4 and L_5 have locations that form equilateral triangles with the positions of the two celestial bodies.

Lagrange points L_1, L_2, and L_3 are *unstable*, whereas Lagrange points L_4 and L_5 are *stable*. A mass positioned at an unstable Lagrange point with zero velocity (relative to the rotating coordinate frame) will only remain at that position if it is placed perfectly with zero velocity error. This is analogous with attempting to balance a marble at the top of an inverted hemispherical bowl. Conversely, a mass positioned at a stable Lagrange point with zero velocity will tend to remain at that position. Slight imperfections in location or velocity will tend to cause motion about the Lagrange point with the mass eventually settling at the Lagrange point. This is analogous to a marble being placed near the bottom (valley) of a hemispherical bowl.

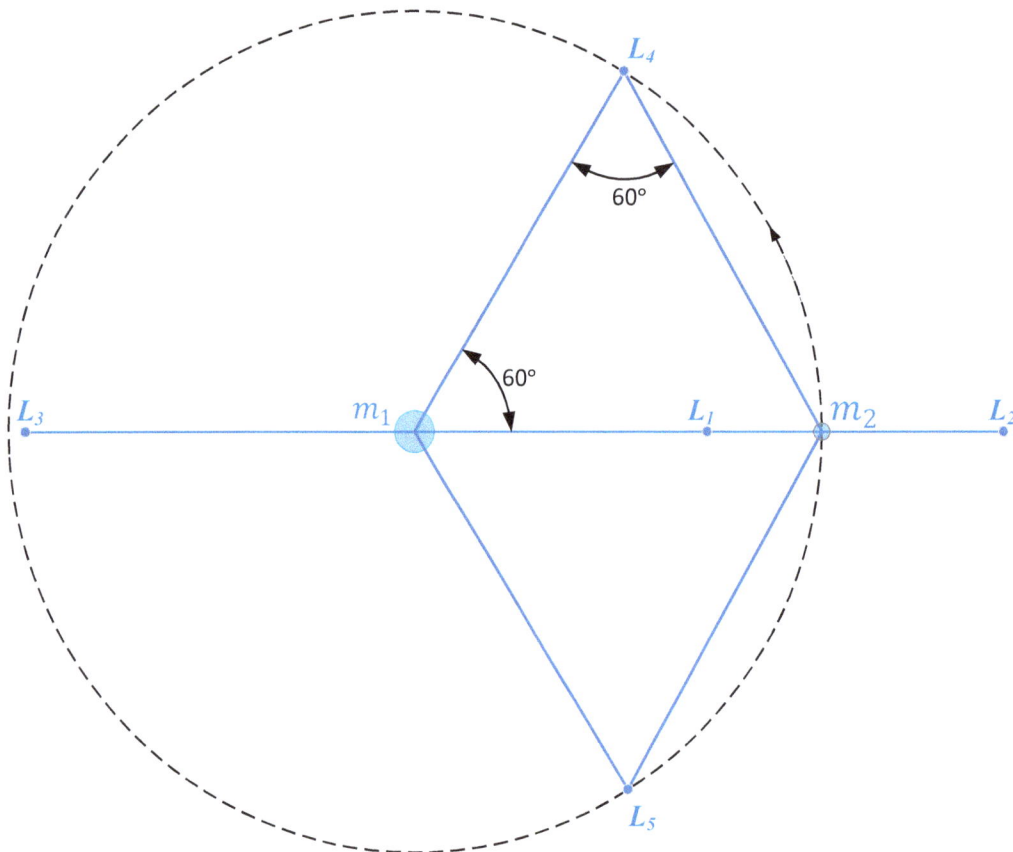

Figure 11-2 Lagrange Point Geometry

Spacecraft can be placed in orbits around Lagrange points. This is a special class called *halo orbits*. In practical applications, orbital maintenance is required even for halo orbits at stable Lagrange points. This is due to the presence of gravitational attractions and other disturbances other than the two celestial bodies in the problem.

Presence of the unstable Lagrange points can be inferred from geometric evaluation by considering both the presence of gravity and the coordinate frame rotation. The presence of L_1 between the two masses can be considered as a circular orbit about the system barycenter. However, the gravitational acceleration from the m_1 celestial body is reduced by an opposing gravitational acceleration from the m_2 celestial body. Hence it is a circular orbit lower than the m_2 orbit around *barycenter*.

The presence of L_2 and L_3 result from the additive gravitational attractions from the m_1 and m_2 celestial bodies. The L_2 circular orbit is outside that of the m_2 celestial body as the result of gravitational acceleration dropping off with the inverse square of the distance. The L_2 circular orbit is at a larger radius resulting from the more massive m_1 body being at a greater distance and the less massive m_1 body being is at a closer distance. The radius of the L_2 circular orbit is such that its period due to the combined gravitational accelerations of m_1 and m_2 equals the period of the m_1 and m_2 bodies' orbits about the barycenter.

The L_3 circular orbit also results from the combined gravitational attraction of the m_1 and m_2 celestial bodies. However, since m_1 is the closer and more massive of the two celestial bodies, the net gravitational effect is closer to that of a slightly more massive m_1 body. Thus, the circular orbital radius is less than that of the m_2 body, such that the L_3 orbital period is the same as that of the m_1 and m_2 bodies.

The presence of the L_4 and L_5 Lagrange points, and particularly the geometry of equilateral triangle in common with the m_1 to m_2 radius is less intuitive by a qualitative explanation. Their presence is best explained by the mathematical evaluation of the three body equations of motion, which are used to establish the radii for each Lagrange point associated with a particular set of celestial bodies and their respective masses.

Key Terms:

Lagrange Points: consist of five discrete locations in the circular restricted three body problem that have a net zero acceleration due to the combined effect of the gravitation from the two celestial bodies and the system's centrifugal rotation.
Stable Lagrange Points: are the L_1, L_2 and L_3 Lagrange points that form a third vertex of an equilateral triangle with the celestial bodies at the other two vertices.
Unstable Lagrange Points: are the L_4 and L_5 Lagrange points that are colinear with two celestial objects. Objects perfectly positioned with zero relative velocity at an unstable Lagrange point will drift away with the slightest disturbance.

➤ **Transition**: *You may continue this section with Lagrange Point Geometry at Level II, or you may skip to the beginning of the next section called Earth-Moon System Lagrange Points.*

11.4.2 Lagrange Point Geometry [Level II – Equations]

The scalar equations of motion can be used to locate the Lagrange equilibrium points. These are points have zero net acceleration ($\ddot{x} = \ddot{y} = \ddot{z} = 0$) in the rotating coordinate frame when an object is placed at their location with zero relative velocity ($\dot{x} = \dot{y} = \dot{z} = 0$). Equation 11.30 expresses the third equation (11.12 and 11.29) of motion with $\ddot{z}$ set to zero.

$$\left(\frac{\mu_1}{r_1^3} + \frac{\mu_2}{r_2^3}\right) z = 0 \qquad\qquad 11.30$$

Since both the μ_1/r_1^3 and μ_2/r_2^3 terms have positive values, the only solution for the third equation is $z = 0$. Consequentially, all Lagrange points must lie within the orbital (x, y) plane.

Equations 11.31 and 11.32 are a simultaneous solution of the x and y equations (11.27 and 11.28).

$$x = \left(\frac{1}{2} - \pi_2\right) r_{12} \qquad\qquad 11.31$$

$$y = \pm\frac{\sqrt{3}}{2} r_{12} \qquad\qquad 11.32$$

The quantities $1/2$ and $\sqrt{3}/2$ in these equations should be recognizable as the trigonometric cosine and sine of $60°$. The $-\pi_2 r_{12}$ offset in the x-direction is the location of the larger celestial body associated with m_1 per equation 11.6. The x, y coordinates therefore position the L_4 and L_5 Lagrange points at locations in which they each are one vertex of an equilateral triangle with the two celestial bodies at the remaining vertices.

The L_1, L_2, and L_3 Lagrange point locations are determined by first recognizing they lie on the x-axis. Thus, the y and z are both zero. What remains is the x equation which is evaluated against r_1 and r_2 along with the π_2 mass ratio from equation 11.9. Equations 11.33 through 11.35, developed by Schaub and Junkins (2003), provide relationships for determining the colinear Lagrange point normalized x positions. Note that the differences for each Lagrange point are in the sign on the various terms.

288

$$L_1: \qquad x - \frac{1 - \pi_2}{(x + \pi_2)^2} + \frac{\pi_2}{(x + \pi_2 - 1)^2} = 0 \qquad\qquad 11.33$$

$$L_2: \qquad x - \frac{1 - \pi_2}{(x + \pi_2)^2} - \frac{\pi_2}{(x + \pi_2 - 1)^2} = 0 \qquad\qquad 11.34$$

$$L_3: \qquad x + \frac{1 - \pi_2}{(x + \pi_2)^2} + \frac{\pi_2}{(x + \pi_2 - 1)^2} = 0 \qquad\qquad 11.35$$

These equations are very non-linear. The x position corresponding to respective equation's zero value should be determined by a numerical search. The search domains are $0.5 \leq x \leq 1.0$ for L_1, $1.0 \leq x \leq 1.5$ for L_2, and $-1.5 \leq x \leq -1.0$ for L_3. The non-linearized x positions are the product of the normalized positions and r_{12}.

11.5 Earth-Moon System Lagrange Points

Figure 11-3 illustrates the Earth-Moon system Lagrange point locations, determined using equations 11.31 through 11.35 and presuming a 384,400 km Earth-Moon distance (as r_{12}). The presumed masses were $5.792 \times 10^{24} kg$ for Earth and $7.349 \times 10^{22} kg$ for the Moon.

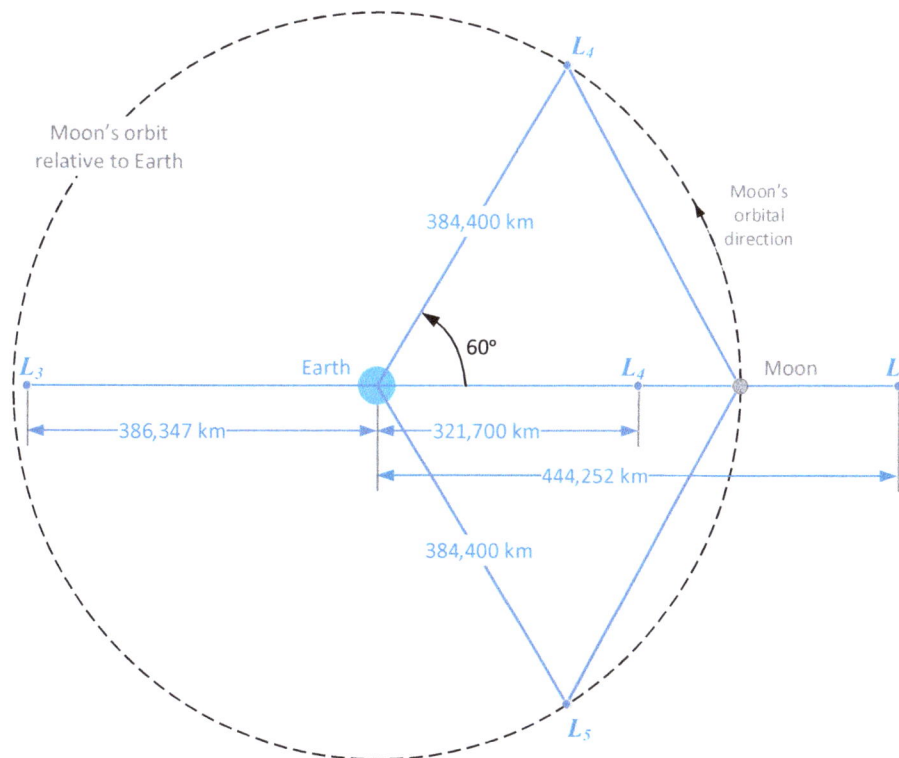

Figure 11-3 Earth-Moon System Lagrange Points

11.6 Jacobi Integral

The Jacobi integral, named for German mathematician Karl Gustav Jakob (1804-1851) helps provide insight into the circular restricted three body problem in terms of the Jacobi constant, expressed as energy per unit mass (km^2/s^2).

11.6.1 Jacobi Constant [Level I – Descriptive]

Jacobi constant evaluation at various energy levels provides perception into the types of motion a spacecraft can experience in the circular restricted three body problem. Figure 11-4 plots zero velocity contours for strategically selected spacecraft specific orbital energies.

The contours indicate the boundary at which the specific orbital energy (expressed by the Jacobi constant) would result in a zero velocity. A spacecraft at a specific energy level can travel in the provided unshaded region but may not cross the contour into or across the shaded region, since this would result in negative speed which is an impossible condition.

A discussion of each of the figure 11-4 plots is provided below.

(a) At this orbital energy level, a spacecraft in the unshaded region that contains the Earth may only access that unshaded region. Likewise, a spacecraft in the unshaded region near the Moon may only access that unshaded region. Spacecraft in either unshaded region could not access any of the Lagrange points.

(b) At this orbital energy level, spacecraft in either unshaded region has a minimal ability to access the L_1 Lagrange point. The slightest energy increase (such as through a velocity change) allows crossover through L_1 between unshaded regions.

(c) At this orbital energy level, a spacecraft has access to the Earth, Moon, and the L_1 Lagrange points. Minimal access is also available to the L_2 Lagrange point. The slightest energy increase allows crossover through L_2 to escape the Earth-Moon system.

(d) At this orbital energy level, a spacecraft has access to the Earth, Moon, and the L_1 and L_2 Lagrange points. Minimal Access is available to the L_3 Lagrange point.

(e) At this orbital energy level, a spacecraft has access to the Earth, Moon, and the L_1, L_2, and L_3 Lagrange points. Minimal Access is available to the L_3 Lagrange points. However, no access is available to the L_4 or L_5 Lagrange points.

(f) At this level, a spacecraft has sufficient orbital energy to access the Earth, Moon, and all five Lagrange points.

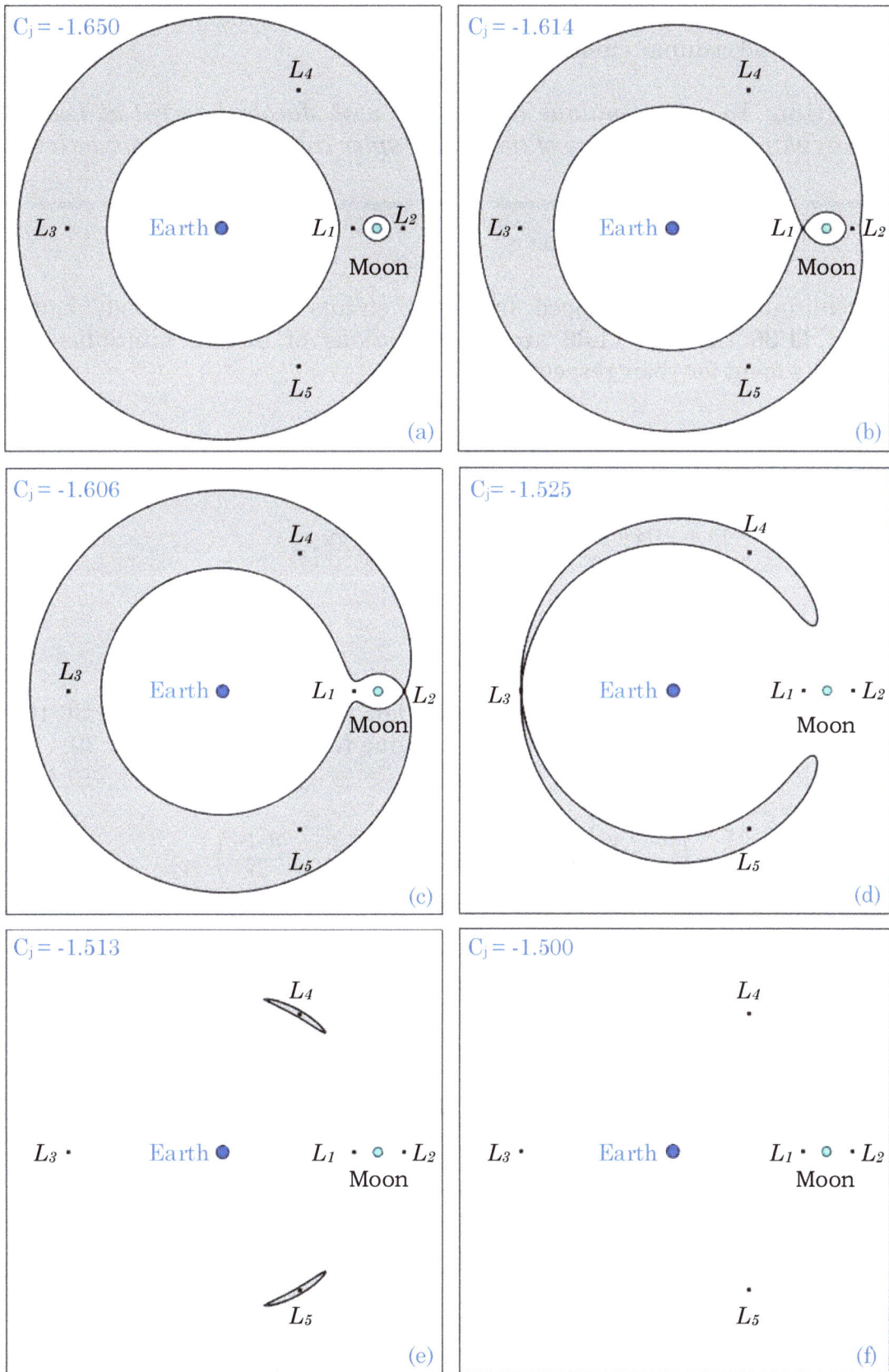

C_J = -1.650 ... C_J = -1.614 ... C_J = -1.606 ... C_J = -1.525 ... C_J = -1.513 ... C_J = -1.500

Figure 11-4 Jacobi Integral Zero Velocity Contours

Understanding the nature of these energy levels is helpful when planning trajectories in the cislunar environment.

➢ **Transition**: *You may continue this section with Jacobi Integral at Level II, or you may skip to the beginning of the next chapter called Lunar Trajectories.*

11.6.1 Jacobi Constant [Level II – Equations]

The Jacobi integral is developed from the circular restricted body equations. Equations 11.36 through 11.38 are the equations of motion multiplied by the velocity component for their respective axis.

$$\dot{x}\ddot{x} - 2\Omega\dot{x}\dot{y} - \Omega^2 x\dot{x} = -\frac{\mu_1}{r_1^3}(x\dot{x} + \pi_2 r_{12}\dot{x}) - \frac{\mu_2}{r_2^3}(x\dot{x} - \pi_1 r_{12}\dot{x}) \qquad 11.36$$

$$\dot{y}\ddot{y} + 2\Omega\dot{x}\dot{y} - \Omega^2 y\dot{y} = -\frac{\mu_1}{r_1^3}y\dot{y} - \frac{\mu_2}{r_2^3}y\dot{y} \qquad 11.37$$

$$\dot{z}\ddot{z} = -\frac{\mu_1}{r_1^3}z\dot{z} - \frac{\mu_2}{r_2^3}z\dot{z} \qquad 11.38$$

Equation 11.39 is the sum of the right and left sides of equations 11.36 through 11.38. Equation 11.40 is the result of rearranging terms in equation 11.39.

$$\dot{x}\ddot{x} + \dot{y}\ddot{y} + \dot{z}\ddot{z} - \Omega^2(x\dot{x} + y\dot{y})$$
$$= -\left(\frac{\mu_1}{r_1^3} + \frac{\mu_2}{r_2^3}\right)(x\dot{x} + y\dot{y} + z\dot{z}) + \left(\frac{\pi_1\mu_2}{r_2^3} - \frac{\pi_2\mu_1}{r_1^3}\right)r_{12}\dot{x} \qquad 11.39$$

$$x\dot{x} + y\dot{y} + z\dot{z} - \Omega^2(x\dot{x} + y\dot{y})$$
$$= -\frac{\mu_1}{r_1^3}(x\dot{x} + y\dot{y} + z\dot{z} + \pi_2 r_{12}\dot{x}) - \frac{\mu_2}{r_2^3}(x\dot{x} + y\dot{y} + z\dot{z} - \pi_1 r_{12}\dot{x}) \qquad 11.40$$

Equation 11.41 relates a parenthetical quantity in 11.40 to the time derivative of v^2 relative to the rotating frame. Equation 11.42 relates the time derivative of the squares of in-plane position to a subset of equation 11.40 quantities. Equation 11.43 relates the spacecraft position relative to the more massive celestial body to another subset of equation 11.40 quantities.

$$x\dot{x} + y\dot{y} + z\dot{z} = \frac{1}{2}\frac{d}{dt}(\dot{x}^2 + \dot{y}^2 + \dot{z}^2) = \frac{1}{2}\frac{d}{dt}v^2 \qquad 11.41$$

$$x\dot{x} + y\dot{y} = \frac{1}{2}\frac{d}{dt}(x^2 + y^2) \qquad 11.42$$

$$r_1^2 = (x + \pi_2 r_{12})^2 + y^2 + z^2 \qquad\qquad 11.43$$

Equation 11.44 is the time derivative of equation 11.43, with equation 11.45 isolating the time derivative of the spacecraft position relative to the more massive celestial body.

$$2r_1 \frac{dr_1}{dt} = 2(x + \pi_2 r_{12})\dot{x} + 2y\dot{y} + 2z\dot{z} \qquad\qquad 11.44$$

$$\frac{dr_1}{dt} = \frac{1}{r_1}(\pi_2 r_{12}\dot{x} + x\dot{x} + y\dot{y} + z\dot{z}) \qquad\qquad 11.45$$

Equations 11.46 and 11.47 are the time derivatives of the reciprocals of r_1 and r_2 respectively.

$$\frac{d}{dt}\frac{1}{r_1} = \frac{1}{r_1^2}\frac{dr_1}{dt} = -\frac{1}{r_1^3}(\pi_2 r_{12}\dot{x} + x\dot{x} + y\dot{y} + z\dot{z}) \qquad\qquad 11.46$$

$$\frac{d}{dt}\frac{1}{r_2} = -\frac{1}{r_2^3}(\pi_1 r_{12}\dot{x} + x\dot{x} + y\dot{y} + z\dot{z}) \qquad\qquad 11.47$$

Equation 11.48 makes substitutions of the above expressions, simplifying equation 11.40. Equation 11.49 is the *Jacobi integral*, with all terms moved to the left side creating a common time derivative.

$$\frac{1}{2}\frac{d}{dt}v^2 - \frac{1}{2}\Omega^2\frac{d}{dt}(x^2 + y^2) = \mu_1\frac{d}{dt}\frac{1}{r_1} + \mu_2\frac{d}{dt}\frac{1}{r_2} \qquad\qquad 11.48$$

$$\frac{d}{dt}\left[\frac{v^2}{2} - \frac{1}{2}\Omega^2(x^2 + y^2) - \frac{\mu_1}{r_1} - \frac{\mu_2}{r_2}\right] = 0 \qquad\qquad 11.49$$

The integration with respect to time leaves the bracketed quantity on the left side. Integration of the right-side results in a constant of integration C, known as the *Jacobi constant*.

The Jacobi constant is the total specific energy (i.e., energy per unit mass) of the spacecraft relative to the rotating coordinate frame. For any value of the Jacobi constant, v^2 is a function of only the spacecraft position in the rotating frame. Equation 11.50 creates a boundary by setting $v^2 = 0$ and forming the Jacobi integral into an inequality. Setting $v^2 = 0$ is the boundary condition since a negative value for v^2 would represent a physically impossible situation (i.e., imaginary speed).

$$\Omega^2(x^2 + y^2) + \frac{2\mu_1}{r_1} + \frac{2\mu_2}{r_2} + 2C \geq 0 \qquad\qquad 11.50$$

The contours in figure 11-4 are generated as the Jacobi integral (x, y) boundary (i.e., v^2) for selectively-chosen Jacobi constant (C) values. The unshaded areas satisfy the inequality, while crossing a contour into a shaded region $(v^2 < 0)$ would violate the boundary constraint.

REFERENCES

1. Curtis, Howard D, *Orbital Mechanics for Engineering Students*, Fourth Edition, © 2020 Elsevier, Ltd., ISBN 978-0-08-102133-0.
2. Cornish, Neil J., *The Lagrange Points*, 1998 NASA technical note for WMAP Education and Outreach.
3. Schaub, Hanspeter and Junkins, John L., *Analytical Mechanics of Space Systems*, © 2003 American Institute of Aeronautics and Astronautics (AIAA), ISBN 1-56347-563-4.
4. Battin, Richard H., *An Introduction to the Mathematics and Methods of Astrodynamics*, © 1987 by author, American Institute of Aeronautics and Astronautics (AIAA), ISBN 0-930403-25-8.
5. Roy, Archie E., *Orbital Motion*, Third Edition Student Text, © 1988 by author, ISBN 0-85274-229-0.
6. Escobal, Pedro R., *Methods of Astrodynamics*, © 1968 John Wiley & Sons, Inc., Library of Congress Catalog Card Number 68-24795.
7. Gurfil, Pini and Seidelmann, P. Kenneth, *Celestial Mechanics and Astrodynamics: Theory and Practice*, © 2016 Springer-Verlag, ISBN 978-3-662-50368-3.
8. Beutler, Gerhard, *Methods of Celestial Mechanics*, Volume I, © 2005 Springer, ISBN 978-3-540-26870-3.
9. MATLAB Monkey, CRTBP Pseudo-Potential and Lagrange Points, https://www.matlab-monkey.com/celestialMechanics/CRTBP/LagrangePoints/LagrangePoints.html, n.d., downloaded 2024-May-12.

12 Lunar Trajectories

Lunar trajectories provide a transition between the Earth and Moon, but unlike interplanetary trajectories they lack the intermediate solar trajectory that governs the planning process. Instead, the circular restricted three body problem provides the primary insight for the planning process.

12.1 The Moon's Orbit

The Earth-Moon system orbits about their common center-of-mass or barycenter, as is seen with the circular restricted three body problem. While not meeting the most accepted qualifier[1], the Earth and Moon behave much like a binary planetary system orbiting the Sun. The system barycenter is approximately 4673 km from the Earth's center and 379727 km from the Moon's center. This positions the barycenter at 73% of an Earth radius, and thus is inside the Earth's volume.

The semimajor axis of the Moon's orbit is 384,400 km with a 0.0549 eccentricity. The Moon's mean orbital speed is 1.023 km/s with a sidereal orbital period of 655.720 hours (27.32167 days), and thus a mean orbital rate of 13.176 degrees per day.

The inclination relative to the ecliptic (i.e., Earth's orbital plane) is $5.145° \pm 0.15°$, varying over 179 days. Since obliquity (inclination) of the ecliptic has a mean value of 23.44°, the Moon's inclination relative to the equator varies between approximately 18.3° and 28.6° over an 18.6-year cycle. The 18.3° inclination (23.44° − 5.145°) occurs when the Earth-Moon descending node is aligned with the Vernal Equinox; the 28.6° inclination (23.4° + 5.145°) occurs when the ascending node is aligned with the equinox. The United States' Apollo program benefitted from the Moon's equatorial inclination being near the 28.6° maximum (versus the 28.5° launch pad latitude) during the years the Moon landings were conducted.

The Earth's mass is approximately $5.799 \times 10^{24} \, kg$ versus $7.346 \times 10^{22} \, kg$ for the Moon, provided a mass ratio of ~81.3. Using the masses and the Moon's 384,400 km semimajor axis in equation 10.31 provides a $66,183 \, km$ lunar sphere of influence radius.

12.2 Lunar SOI Targeting

The fundamental difference between lunar and interplanetary trajectories is direct transition between the Earth and Lunar SOI without an intermediate heliocentric

---

[1] It is commonly accepted that the barycenter of a binary planetary system is outside the volumes of both celestial bodies.

trajectory. Thus, the lunar SOI needs to be targeted directly and at a location that provides a trajectory compatible with mission inside the lunar SOI. Furthermore, and unlike an interplanetary trajectory, the lunar hyperbolic trajectory will generally neither be at the asymptote, nor at the hyperbolic excess speed.

A minimal energy lunar targeting would intercept tangent to the SOI on its leading edge. The tradeoff with trajectory type for human occupied spacecraft is the transit time, which should be fast to minimize depletion of consumables such as breathing air, water, and food. The Apollo program selected a 3-day transit, balancing extra propellant to minimize the need for consumables.

Like interplanetary trajectories, the trajectory must target where the leading edge of the lunar SOI will be at the end of the transit. Departure must be at the location where the lunar antipode will be on arrival. Planning must thus consider the Moon's 13.176 degree per day orbital rate.

12.2.1 Earth-Moon Plane SOI Intercept [Level I – Descriptive]

Targeting the lunar SOI leading edge with a trajectory within the Earth-Moon plane is consistent with how the Apollo missions were executed. The design of the trans-lunar trajectory, which is a challenging task, is complicated by several factors which the interested reader may choose to explore. An analytic solution will only be approximate and must be refined numerically to execute a mission. Figure 12-1 illustrates the targeting geometry.

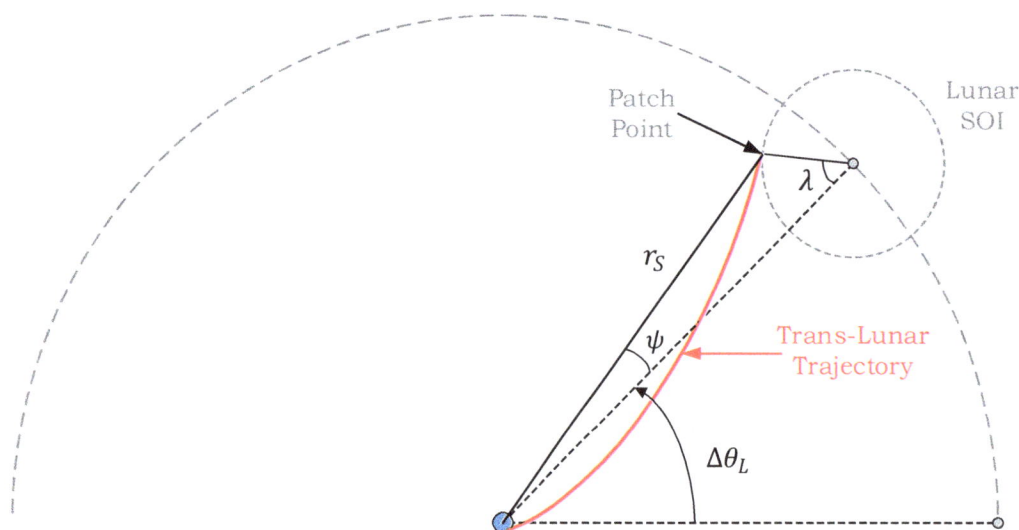

Figure 12-1 Lunar SOI Targeting Geometry

The ranges of values that work for in-plane targeting at a 275 km Earth parking orbit are somewhat like a Hohmann transfer, meaning the semi-major axis value results in an apoapsis somewhat greater than the mean Earth 384400 km distance.

The preliminary design presumes the Trans-Lunar Injection (TLI) maneuver, which transitions from the parking orbit to the trans-lunar trajectory is impulsive and occurs at periapsis. The periapsis speed will be in the 10.85 to 10.90 km/s range and the lunar arrival angle (λ), will typically be in the 45° to 60° range. The trans-lunar trajectory transit time, which is from TLI to the patch point on the leading edge of the lunar SOI typically varies between 2 and 5 days. The translunar departure point will be approximately 125° behind the Earth-Moon line aimed at the patch point arrival. The trans-lunar time affects the Moon's orbital progression ($\Delta\theta_L$), which determines the location of the patch point. The true anomaly difference ($\Delta\theta$) between TLI and the patch point, called the sweep angle, typically varies from 160° to 175°.

The specific parameter choices will affect the trajectory characteristics within the lunar SOI, which can be presumed to be hyperbolic. Given a particular trans-lunar trajectory, the hyperbolic characteristics are computed based on the figure 12-2 geometry.

Key Term:

Lunar Arrival Angle: is the angle between the Earth-Moon line and the lunar SOI patch point.
Patch Point: is the location a trans-lunar trajectory intercepts the lunar SOI.
Trans-Lunar Trajectory: is an Earth-departing trajectory that intercepts the lunar sphere-of-influence.

12.2.2 Earth-Moon Plane SOI Intercept [Level II – Equations]

Given the Moon's semimajor axis ($a_L = 384400\ km$) and the lunar SOI radius ($r_{SOI} = 66183\ km$), equation 12.1 computes the geocentric radius to the SOI patch point (r_S) using the law of cosines. Equation 12.2 computes the angle (ψ) between the moon and SOI patch point using the law of sines.

$$r_S = \sqrt{a_L^2 + r_{SOI}^2 - 2a_L r_{SOI} \cos\lambda} \qquad\qquad 12.1$$

$$\sin\psi = \frac{r_{SOI}}{r_S}\sin\lambda \qquad\qquad 12.2$$

Equation 12.3 specifies the patch point coordinates in the rotating Earth-Moon coordinates used in the circular restricted three-body problem for a specified lunar arrival angle (λ).

$$\bar{r}_S = \begin{bmatrix} a_L - r_{SOI}\cos\lambda \\ r_{SOI}\sin\lambda \\ 0 \end{bmatrix} \qquad\qquad 12.3$$

The patch point coordinates depend only on the lunar arrival angle (λ) for this preliminary analysis. The trans-lunar trajectory has freedom in adjusting its semi-major axis, eccentricity, and TLI departure position to intercept the patch point at the specified trans-lunar transit time. Unless a trans-lunar trajectory is previously specified, these parameters are iteratively adjusted using the provided guidelines until the result produces the desired lunar hyperbolic trajectory. Lambert targeting can be a useful tool for such a design process.

12.2.3 Lunar Trajectory Inside the SOI [Level I – Descriptive]

When the lunar SOI patch point is reached in the patched conic approximation, the trajectory transitions from geocentric to selenocentric (i.e., lunar-centered). The spacecraft hyperbolic lunar velocity ($v_{S,h}$) completes a triangle consisting of the inertial velocities of the Moon (v_M) and the spacecraft (v_S) at the patch point as illustrated in figure 12-2.

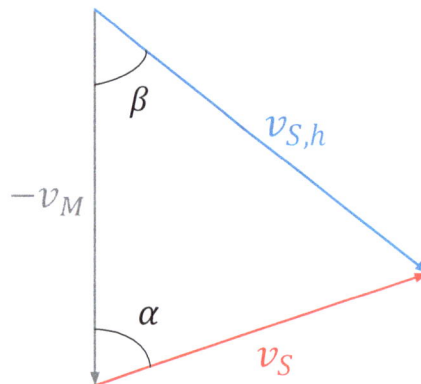

Figure 12-2 Velocity Vector Diagram at SOI Arrival

The spacecraft position in the lunar hyperbolic trajectory at SOI arrival has the SOI radius (66183 km) and is oriented by the lunar arrival angle (λ) relative to the Earth-Moon line. This provides sufficient information to characterize the lunar hyperbolic trajectory.

12.2.4 Lunar Trajectory Inside the SOI [Level II – Equations]

Equation 12.4 is the angle (α) between the Moon's velocity and the trans-lunar velocity at the lunar SOI arrival patch point, which is the difference between the arrival flight path angle (ϕ_S) and the geocentric angle (ψ) between the Moon and the patch point. Note that $v_M = 1.023\ km/s$ is the Moon's orbital speed.

$$\alpha = \phi_S - \psi \qquad\qquad 12.4$$

Equation 12.5 is the hyperbolic speed ($v_{S,h}$) relative to the Moon at the SOI patch point, computed by the law of cosines. Equation 12.6 relates the angle (β) between the lunar hyperbolic velocity and the lunar velocity by the law of sines. Equation 12.7 is the corresponding flight path angle ($\phi_{S,h}$) for the lunar hyperbolic trajectory in radians.

$$v_{S,h} = \sqrt{v_M^2 + v_S^2 - 2v_M v_S \cos\alpha} \qquad\qquad 12.5$$

$$\sin\beta = \frac{v_S}{v_{S,h}}\sin\alpha \qquad\qquad 12.6$$

$$\phi_{S,h} = \pi - \lambda - \beta \qquad\qquad 12.7$$

Equation 12.8 is the hyperbolic lunar trajectory specific energy (ε) using the Moon's gravitational parameter ($\mu_M = 4902.8\ km^3/s^2$) and the speed and radius at the SOI patch point. Equation 12.9 computes the hyperbolic lunar trajectory's specific angular momentum (h). Equations 12.10 and 12.11 use standard relationships to compute the hyperbolic trajectory semi-major axis and eccentricity.

$$\varepsilon = \frac{v_{S,h}^2}{2} - \frac{\mu_M}{r_{SOI}} \qquad\qquad 12.8$$

$$h = r_{SOI}v_{S,h}\cos\phi_{S,h} \qquad\qquad 12.9$$

$$a = -\frac{\mu_M}{2\varepsilon} \qquad\qquad 12.10$$

$$e = \sqrt{1 - \frac{h^2}{\mu_M a}} \qquad\qquad 12.11$$

From these parameters, the hyperbolic trajectory's periapsis radius $r_p = a(1 - e)$ can be computed. The trans-lunar trajectory targeting the SOI and/or the lunar arrival angle (λ) may be iteratively adjusted until a desired periapsis radius and/or other required lunar trajectory characteristics are achieved.

12.3 Free Return Trajectories

Free return trajectories occur within the Earth-Moon system and provide a means of directly transitioning from the Earth's SOI to the Moon and back to the Earth's SOI if no velocity changes ($\Delta v's$) from thrust events are imparted. Speculation as to the existence of such trajectories can be made by analysis of the circular restricted three-body problem and the Jacobi zero velocity contours seen in figure 11-4 case (c).

While nearly every trajectory that lacks the energy to escape the Earth-Moon system, will return to Earth, the timing of such a return can be acute for a human occupied spacecraft with limited consumables. Thus, the free return trajectories of interest are those that given an anomaly that would preclude successful mission execution would return the spacecraft to Earth in a timely manner for crew recovery.

Free Return trajectories were studied by Schwaniger (1963) in preparation for the United States Apollo program. In that NASA Technical Note, free return trajectories are available with relatively low lunar periapsis radii for low inclinations, meaning those in or near the Earth-Moon plane. Another class of higher altitude free return trajectories are also available for inclinations up to lunar polar orbits. The scope herein will use the low inclination trajectories since they are easily understood. Once the concepts are understood, free return trajectories may be developed for the higher inclination cases.

A free return trajectory was first used for a human occupied spacecraft during the Apollo 8 mission. Due to the extremely risky mission profile, the low latitude free return trajectory provided contingencies for safe crew return if lunar orbit could not be established. Figure 12-3 illustrates an in-plane free return trajectory planned and plotted using the NASA General Mission Analysis Tool (GMAT) in the Earth-Moon system used to evaluate the circular restricted three-body problem.

Figure 12-4 illustrates the same trajectory plotted by GMAT in inertial coordinates. In this plot, the Moon's motion begins at the extreme right of the gray arc (i.e., lunar orbit). The Moon's gravity at the encounter point perturbs the trajectory for a timely return to the Earth. The lunar SOI intercept was planned to aim the inbound leg to the reentry corridor in the Earth's atmosphere while still providing the opportunity to establish a lunar orbit.

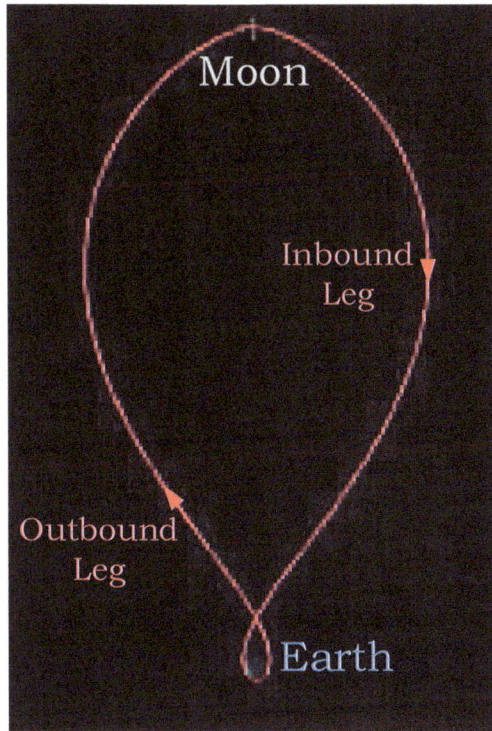

Figure 12-3 Free Return Trajectory in Body Directed Coordinates

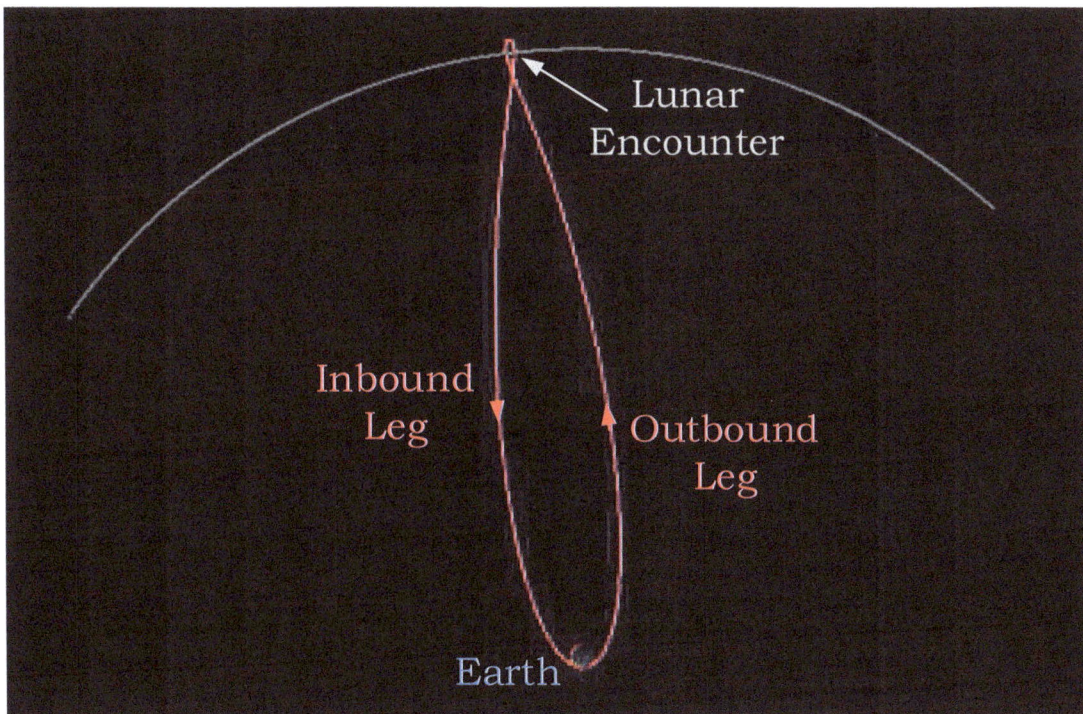

Figure 12-4 Free Return Trajectory in Inertial Coordinates

301

Figure 12-5 provides a close-up of the hyperbolic flyby trajectory inside the lunar SOI. Lunar orbit would be most efficiently established with a braking (i.e., retrograde) executed at perilune (i.e., lunar periapsis).

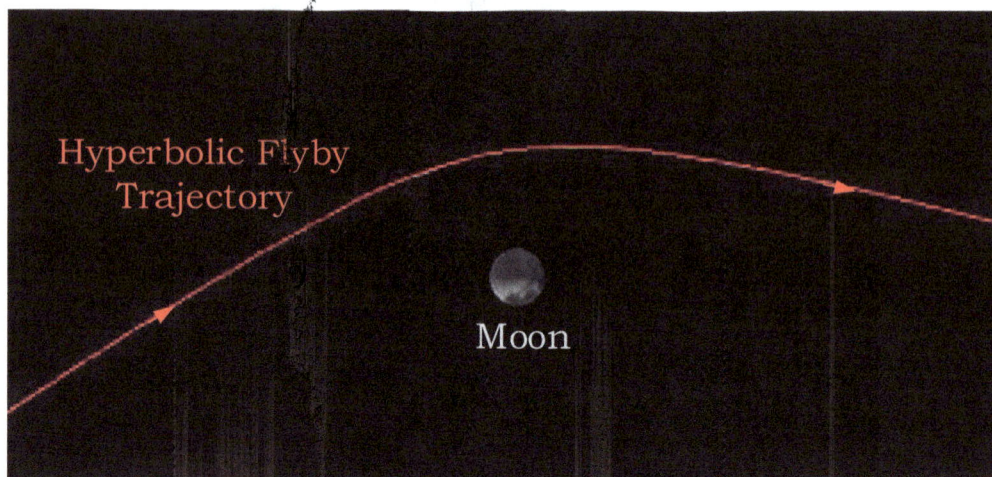

Figure 12-5 Free Return Trajectory Hyperbolic Encounter

Apollo 8 flew a somewhat similar free return trajectory, then successfully entered lunar orbit on 24 December 1968. Unlike this example, Apollo 8 established lunar orbit at a significantly lower (i.e., 70 km) altitude.

Establishing such a low perilune radius was a crucial step in preparation for lunar landing, which first occurred on the Apollo 11 flight eight months later. Apollo missions performed a TLI maneuver to establish a higher altitude a free return trajectory, which was later modified during the trans-lunar timeline to facilitate lunar orbit at an altitude suitable for landing.

Key Term:

> **Free Return Trajectory**: is an Earth-departing trajectory that enters the lunar sphere of influence and is returned to an Earth-bound (i.e., trans-Earth) trajectory without the need for an intervening thrust event.

12.4 Midcourse Corrections

The free return trajectory targets the Earth atmosphere reentry corridor directly, without the need for any additional propulsion events. While this appears ideal in theory, it is more so an artifact of what may be accomplished with a computer simulation. In actual practice, navigation system alignment may be imperfect as is propulsion hardware's ability to exactly execute what is planned. Thus, the limitations of real hardware achieve a trajectory that is a close approximation to what was intended.

Midcourse corrections are therefore planned to fine tune the trajectory within the needed tolerances. The process first requires an OD as done in chapter 9 to measure the spacecraft's actual trajectory. Once the trajectory is accurately known, additional small maneuvers can be planned to adjust the trajectory to the planned orbital profile. It should be noted that midcourse corrections are performed both on the outbound trans-lunar trajectory as well as the return trans-earth trajectory. The return corrections are particularly important to accurately target the spacecraft to the reentry corridor tolerances.

12.5 Lunar Orbital Insertion

Lunar Orbital Insertion (LOI) is generally needed to accommodate a landing and return mission. The spacecraft performs a retrograde thruster firing (i.e., opposite the velocity) to slow the speed, establishing a closed orbit around the Moon as illustrated with a high radius example in figure 12-7. This provides the opportunity to target a corridor aimed at the intended landing site before the spacecraft or a specialized lunar landing spacecraft begins its descent to the surface.

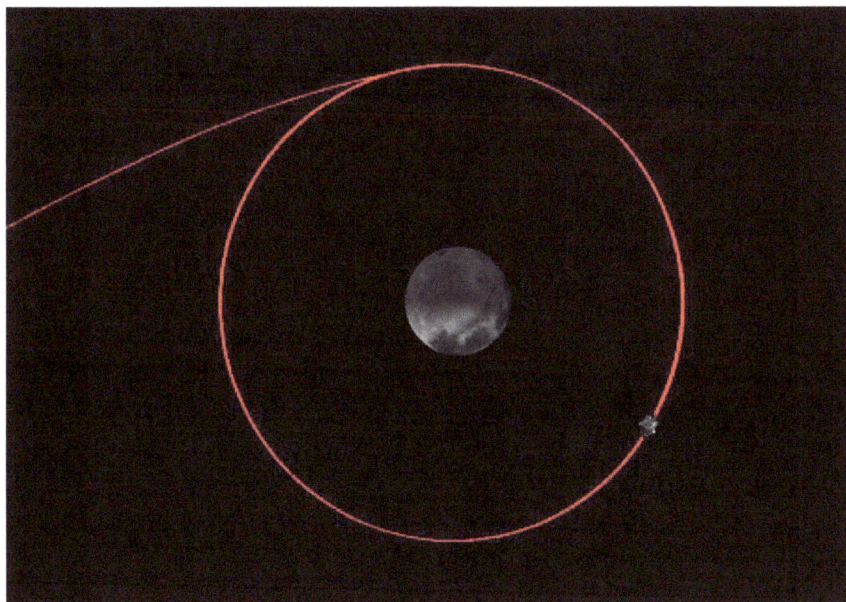

Figure 12-6 Lunar Orbital Insertion

When there is a separate lunar orbiting and landing spacecraft, as was the case with Apollo, the orbiting spacecraft works to cooperatively rendezvous and dock with the lander upon its return from surface activities.

12.6 Trans-Earth Return and Reentry

The Trans-Earth Return is a reversal of the LOI establishing what is a mirror of the trans-lunar trajectory. The Trans-Earth return targets a narrow corridor of the Earth limb for an aerobraking reentry. The Earth limb is targeted to provide a gradual braking effect from the atmosphere. If the reentry angle is too steep, the spacecraft will come in too fast for the heat shield to effectively dissipate the energy and maintain the spacecraft integrity and internal environment. If the reentry angle is too shallow, the spacecraft will skip off the atmosphere much like a horizontally thrown stone skips on a pond surface.

The Apollo Command Module (CM) performed a short skip off the Earth's atmosphere to dissipate part of the kinetic energy from its fast return from lunar altitude. This was facilitated by an off-center center-of-mass with the CM RCS thrusters guiding the process. The short skip blunted the full force of reentry such that the secondary atmospheric entry was within the limits of the CM heat shield.

REFERENCES

1. Brown, Charles D., *Spacecraft Mission Design*, Second Edition, © 1998 American Institute of Aeronautics and Astronautics (AIAA), ISBN 1-56374-262-7.
2. Schwaniger, Arthur J., *Trajectories in the Earth-Moon Space with Symmetrical Free Return Properties*, NASA Technical Note D-1833, Lunar Flight Series: Volume 5, June 1963 George C. Marshall Space Flight Center.
3. Wheeler, Robin, *Apollo Lunar Landing Launch Window: The Controlling Factors and Constraints*, https://www.nasa.gov/history/afj/launchwindow/lw1.html.
4. Curtis, Howard D, *Orbital Mechanics for Engineering Students*, Fourth Edition, © 2020 Elsevier, Ltd., ISBN 978-0-08-102133-0.
5. Chobotov, Vladimir A., *Orbital Mechanics*, © 1991 American Institute of Aeronautics and Astronautics (AIAA), ISBN 1-56347-007-1.
6. Roy, Archie E., *Orbital Motion*, Third Edition Student Text, © 1988 by author, ISBN 0-85274-229-0.
7. Bate, Roger R. et. al, *Fundamentals of Astrodynamics*, © 1971 Dover Publications, Inc., ISBN 0-486-60061-0.
8. NASA, *Apollo Spacecraft Familiarization*, SID 62-435, 1966-December-01, ISBN 978-1-78039-844-0.

13 Spacecraft Attitude Dynamics

Spacecraft attitude dynamics considers the physics of body rotation and its application to orienting a spacecraft to desired celestial references. Processes include measuring the spacecraft attitude, maintaining accurate knowledge of the attitude, and controlling the attitude to establish and maintain desired spacecraft orientations.

13.1 Types of Attitude Stabilization

Spacecraft may have no attitude stabilization or may be stabilized using passive or active methods. The passive methods include gravity gradient stabilization, spacecraft body spin stabilization, and dual spin stabilization. Active stabilization controls the orientation of all three spacecraft axes and is thus referred to as three-axis stabilization.

13.1.1 Gravity Gradient Stabilization

Gravity gradient stabilization is a passive attitude control method. The spacecraft axis corresponding to its minimum moment of inertia naturally aligns vertically to the local gravity gradient.

Since gravitational field strength drops off as an inverse square, a portion of its mass at a slightly lower altitude will have a stronger attraction than an equivalent mass at a slightly higher altitude as shown in figure 13-1. The result is a differential torque that directs the masses toward vertical where the orientation will be stable within approximately 10 degrees.

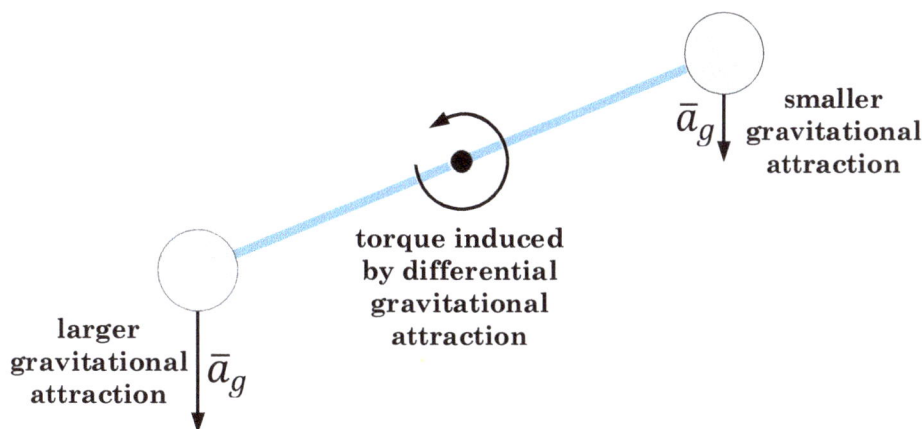

Figure 13-1 Gravity Gradient Differential Torque

While there is a balanced torque if the masses are arranged perfectly horizontally, this arrangement is inherently unstable since the slightest disturbance will cause an imbalance to align the body vertically.

13.1.2 Spin Stabilization

Spin stabilization results from rotating the spacecraft body to maintain a constant inertial direction. Figure 13-2 is an example of the AMSAT Phase 3C spin stabilized spacecraft. This method, as well as all spinning mass-based methods relies on conservation of angular momentum to maintain the spin axis orientation.

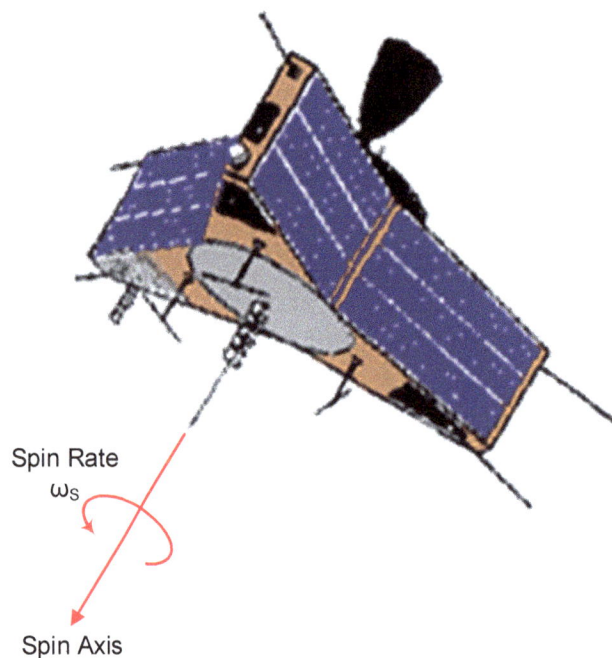

Figure 13-2 Spin Stabilized Spacecraft

Spin stabilization was a commonly-used method before the advent of robust and reliable on-board computers to provide dynamic body pointing. One disadvantage of this method includes the need to occasionally make corrections to the spin axis orientation due to drifts induced by environmental torque disturbances.

This attitude control method is only stable when spinning about the spacecraft's maximum moment of inertia axis. That requires a short and fat shape to the spacecraft body. Such a design is not favorable for fitting large spacecraft in the long and slender cowling that is typical for launch boosters.

When spin stabilization was a prevalent form of attitude control, an innovation was created to allow larger spacecraft to conform to the prevalent long and slender launch booster cowling form. The dual spin stabilized spacecraft has two spinning sections, a primary and a secondary, with each spinning at a different rate.

The platform is the primary spacecraft section that houses the payload. The platform spin rate (ω_P) is set up to accommodate the payload needs. For example, a geosynchronous communications spacecraft would have a primary section spin rate equal to its orbital rate, allowing an antenna to maintain Earth nadir pointing.

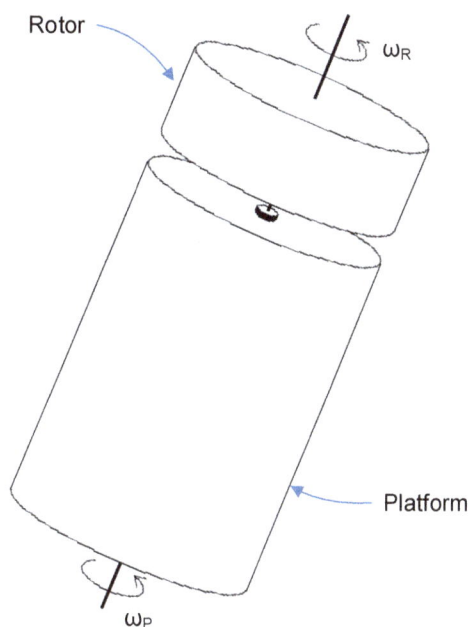

Figure 13-3 Dual Spin Spacecraft Concept

The rotor section spins at a faster rate (ω_R). It has a short and fat profile, causing its spin to be about its major moment of inertia axis. The rotor section provides the dominant stability and causes the spacecraft body to maintain an inertial orientation about an axis aligned with an overall long and slender shape.

307

Active spacecraft attitude control provides three-axis stabilization as illustrated in figure 13-4. This method provides the desired spacecraft orientation by exerting control over two of the three spacecraft axes.

Figure 13-4 Three-Axis Stabilized Spacecraft

Three-axis stabilization is typically specified by pointing one body axis to an *aligned celestial reference* and constraining a second body axis to a plane containing the aligned celestial reference and a *constrained celestial reference*. The aligned and constrained references must be different from each other. The figure 13-4 NAVSTAR GPS spacecraft is Sun-nadir stable, with the spacecraft z-axis aligned to Earth (i.e., nadir) and the x-axis constrained to the plane containing the spacecraft, Sun, and Earth.

Examples of aligned and constrained celestial references include:
- Earth
- Sun
- Moon
- Planets
- Trajectory velocity direction
- Trajectory normal direction
- Defined inertial direction

While the preceding description simplifies the definition to only body axes, it is also possible to align and/or constrain to oblique vectors that are combinations of two or more body axes. But as with the celestial references, the body referenced axis and/or vectors must be unique.

13.2 Spacecraft Attitude Representation

A spacecraft attitude representation is an instantaneous snapshot of the body orientation and its inertial rotation rate at an annotated time. Those with aeronautical experience may relate body orientation to an aircraft yaw, pitch, and roll Euler angles and relate rates to parameters such as rate of turn. Spacecraft attitudes are often provided as an analog to the aircraft representation for the benefit of human operators who more easily relate to this form, as illustrated by the Space Shuttle coordinate system in figure 13-5. However, the underlying attitude representation is in a mathematical form that more efficiently supports determination and control algorithms, while remaining free from gimbal lock and other mathematical issues.

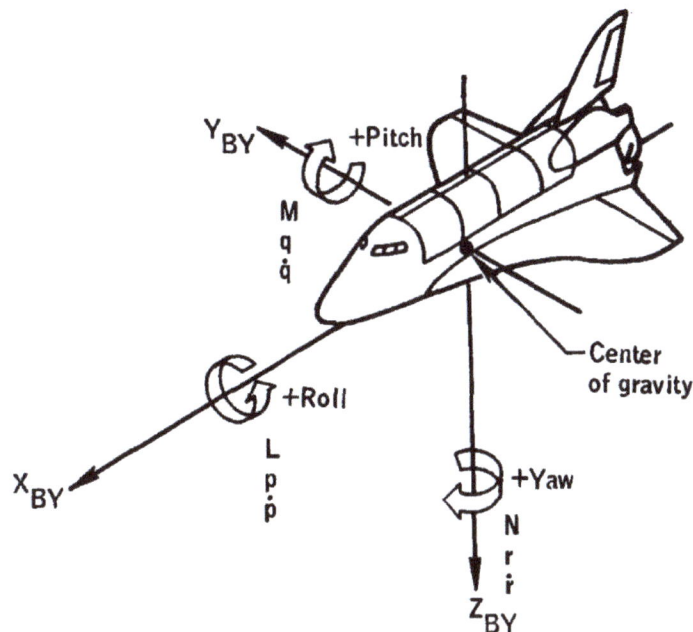

Figure 13-5 Space Shuttle Attitude Coordinates (NASA)

13.2.1 Spacecraft Attitude State [Level I – Descriptive]

The attitude state provides the spacecraft body orientation and inertial rate at a specified time. When humans are actively piloting the spacecraft, the state is represented to provide real time feedback. In such cases, the human interface translates the internal representation to the forms useful to the pilot such as yaw, pitch, roll, and the corresponding time rates of change of these angles. The subparagraphs to this section describe the attitude state's internal representation.

13.2.1.1 Spacecraft Body Orientation [Level I – Descriptive]

Body orientation is specified relative to a celestial reference which is typically an inertial coordinate frame. Since the spacecraft body axes are oriented against a set of inertial axes, a transformation matrix would appear to be a convenient representation. However, a transformation matrix has nine quantities versus the [minimal] yaw, pitch, roll Euler angles, leaving six levels of redundancy. Since the Euler angles also have the gimbal lock issue, which is a realistic concern for a spacecraft, a compromise is reached using a [four parameters] vector called a *quaternion*.

Quaternions are the most common mathematical representation for spacecraft body orientations. They are a four-parameter shorthand for a transformation matrix, having a single level of redundancy. An attitude quaternion is typically an inertial to spacecraft body axis transformation. This form not only orients the spacecraft (via its body axes) to the inertial frame, but also provides a convenient means to convert vectors in the inertial frame to the body frame and vice-versa. Quaternion physical representation and mathematics are covered in depth in appendix D.

Quaternions were introduced in 1843 by Irish mathematician Sir William Rowan Hamilton (1805-65). Hamilton described the quaternion as a *hypercomplex number* (i.e., a representation with one real component and three imaginary components along a set of mutually orthogonal imaginary axes).

Key Term:

> **Quaternion**: is a four parameter coordinate transformation relating a spacecraft's body axes to a reference coordinate system.

13.2.1.2 Spacecraft Body Orientation [Level II – Equations]

Spacecraft body orientations are commonly quaternions representing an inertial to body axis coordinate transformation. Equation 13.1 presents the quaternion ($\bar{q}$) as a hypercomplex number, with a scalar real component (q_s) and a vector imaginary component ($\bar{q}_v$). The vector component may be further decomposed to its three hypercomplex imaginary components (q_i, q_j, q_k).

310

$$\bar{q} = \begin{bmatrix} q_s \\ \bar{q}_v \end{bmatrix} = \begin{bmatrix} q_s \\ q_i \\ q_j \\ q_k \end{bmatrix} \qquad 13.1$$

Equation 13.2 converts the quaternion to its corresponding inertial-to-body DCM (i.e., transformation matrix). Appendix D provides a process to extract the corresponding quaternion from a DCM.

$$C_i^b = \begin{bmatrix} q_s^2 + q_i^2 - q_j^2 - q_k^2 & 2(q_i q_j + q_s q_k) & 2(q_i q_k - q_s q_j) \\ 2(q_i q_j - q_s q_k) & q_s^2 - q_i^2 + q_j^2 - q_k^2 & 2(q_j q_k + q_s q_i) \\ 2(q_i q_k + q_s q_j) & 2(q_j q_k - q_s q_i) & q_s^2 - q_i^2 - q_j^2 + q_k^2 \end{bmatrix} \qquad 13.2$$

The attitude Euler angles use the aircraft yaw (ψ), pitch (θ), and roll (ϕ) convention to provide a standard reference for human-piloted spacecraft. Equation 13.3 computes the attitude inertial-to-body DCM from Euler angles. Equation 13.4 extracts the Euler angles from DCM elements.

$$C_i^b = \begin{bmatrix} \cos\phi\cos\theta & \sin\phi\cos\theta & -\sin\theta \\ \cos\phi\sin\theta\sin\psi - \sin\phi\cos\psi & \sin\phi\sin\theta\sin\psi + \cos\phi\cos\psi & \cos\theta\sin\psi \\ \cos\phi\sin\theta\cos\psi + \sin\phi\sin\psi & \sin\phi\sin\theta\cos\psi - \cos\phi\sin\psi & \cos\theta\cos\psi \end{bmatrix} \qquad 13.3$$

$$\tan\phi = \frac{C_{i_{12}}^b}{C_{i_{11}}^b} \qquad \sin\theta = -C_{i_{13}}^b \qquad \tan\psi = \frac{C_{i_{23}}^b}{C_{i_{33}}^b} \qquad 13.4$$

The pitch angle has a $90° \leq \theta \leq 90°$ range of values. Yaw and roll each have the full range of quadrants (i.e., $0° \leq \psi \leq 360°$ and $0° \leq \phi \leq 360°$). Thus, a two argument inverse tangent function is needed to compute the roll and yaw angles in the correct quadrant.

13.2.1.3 Spacecraft Inertial Rate

The inertial rotation rate ($\bar{\omega}$) is a vector representing the angular rate in the spin axis direction. This may be conveniently visualized as the spacecraft's instantaneous inertial rotation rates about each of its body axes. The attitude quaternion may be used to transform the rates between the instantaneous body inertial ($\bar{\omega}_{bi}$) and the celestial inertial ($\bar{\omega}_i$) reference frames. The inertial rate vector is used for predicting the evolution of the spacecraft orientation over time.

It should be noted that the yaw, pitch, and roll rates (i.e., Euler rates) only equal the inertial body rates when the corresponding yaw, pitch, and roll angles are zero. At any significant yaw, pitch, and/or roll angles there will cause cross coupling that will result with a non-intuitive redistribution of the yaw, pitch, and roll rates amongst the body axes.

Key Term:

Spacecraft Inertial Rate: is a vector representing a spacecraft body's inertial rotation rates about the instantaneous body axes.

13.2.1.4 Spacecraft Inertial Rate [Level II – Equations]

Equation 13.5 transforms the celestial inertial body rate to the body inertial rotation rate. Equation 13.6 computes the yaw, pitch, and roll rates from the corresponding Euler angles and the body inertial rates (i.e., instantaneous inertial rates about the body axes).

$$\bar{\omega}_{bi} = C_i^b \bar{\omega}_i \qquad\qquad 13.5$$

$$\begin{bmatrix} \omega_\psi \\ \omega_\theta \\ \omega_\phi \end{bmatrix} = \begin{bmatrix} 0 & \sin\psi/\cos\theta & \cos\psi/\cos\theta \\ 0 & \cos\psi & -\sin\psi \\ 1 & \sin\psi\tan\theta & \cos\psi\tan\theta \end{bmatrix} \begin{bmatrix} \omega_x \\ \omega_y \\ \omega_z \end{bmatrix} \qquad 13.6$$

Note that the yaw and roll rates are mathematically singular at $\theta = \pm 90°$. This is the gimbal lock case in which the yaw and roll rates are indistinguishable.

13.3 Torque Free Rotation

The rotational motion of a freely-spinning spacecraft in a torque-free environment can be categorized by the orientation of the inertial rate vector relative to the either the body axes or what are known as *principal axes* (i.e., axes of symmetry). Body axes are references defined by spacecraft designers.

The three basic types of rotational motion are pure rotation, coning, and nutation.
- *Pure rotation* occurs when the inertial rate vector ($\bar{\omega}$) is colinear with a spacecraft principal axis ($\hat{P}$). This is the simplest form of torque free rotation. For this case, the angular momentum ($\bar{L}$) is also aligned with the principal axis.
- *Coning* motion occurs for spacecraft when body axes are misaligned from the principal axes. Coning occurs is like a pure rotation in that the inertial rate vector and angular momentum vector are colinear with a principal axis. Since the body axis is misaligned, its motion traces out a cone about the principal axis.

- *Nutation* occurs when the inertial rate vector ($\bar{\omega}$) is not colinear with a principal axis. The nutation angle (θ) is the angle between the principal axis ($\hat{P}$) and the angular momentum ($\bar{L}$). A nutating spacecraft has both the inertial rate vector and the principal axis rotating about the angular momentum.

13.3.1 Spacecraft Mass Properties [Level I – Descriptive]

Basic mass properties include the spacecraft total mass (m) and its center of mass location ($\bar{r}_{cm}$). Total mass is the sum of all mass components (m_i) and center-of-mass also considers component locations ($\bar{r}_i$). Equation 13.7 computes the center of mass position ($\bar{r}_{cm}$) relative to a defined origin in the body frame.

$$\bar{r}_{cm} = \frac{1}{m} \sum_{i=1}^{n} m_i \bar{r}_i \qquad 13.7$$

It is often convenient to distinguish the wet mass (m_{wet}) from dry mass (m_{dry}), with the wet mass being the sum of the dry mass and propellant mass (m_p). When doing so, there are corresponding wet and dry centers-of-mass.

While spacecraft body axes are usually intended to be the axes of symmetry, the most carefully built spacecraft will have some symmetry offsets. Fortunately, all bodies have three mutually-perpendicular axes of symmetry, regardless of how asymmetric they may appear. The distinction between body and principal axes are that the body axes are design references while the principal axes correspond to the spacecraft "as built," and conform to the spacecraft's actual mass arrangement. The principal axes are determined by measuring the "as built" spacecraft mass properties.

Newton's second law describes mass as a resistance to linear acceleration. A *moment of inertia* is a rotational counterpart to mass since it is the resistance to rotational acceleration about a specified axis. A rigid spacecraft's rotational resistance may be represented in a 3x3 inertia matrix (I), consisting of moments and products of inertia. The inertia matrix values may be estimated knowing the masses and locations of each spacecraft component. There are also processes to measure the inertia matrix elements for the "as built" spacecraft that are beyond the scope of this text.

13.3.2 Principal Axes and the Inertia Matrix [Level II – Equations]

Equation 13.8 shows an inertia matrix as a symmetric tensor. The inertia tensor's diagonal elements are the moments of inertia about the body axes. The off-diagonal elements are the *products of inertia* corresponding to a pair of axes. Equations 13.9 through 13.11 compute the moments of inertia about the body axes through the center-of-mass, and equations 13.12 through 13.14 compute the products of inertia for axis pairs through the center-of-mass. In these equations, m_i represents discrete mass elements and x_i, y_i, z_i represent the locations of the discrete mass locations in body coordinates.

$$I = \begin{bmatrix} I_{xx} & I_{xy} & I_{xz} \\ I_{xy} & I_{yy} & I_{yz} \\ I_{xz} & I_{yz} & I_{zz} \end{bmatrix} \qquad\qquad 13.8$$

$$I_{xx} = \sum_{i=1}^{n} m_i(y_i^2 + z_i^2) \qquad\qquad 13.9$$

$$I_{yy} = \sum_{i=1}^{n} m_i(x_i^2 + z_i^2) \qquad\qquad 13.10$$

$$I_{zz} = \sum_{i=1}^{n} m_i(x_i^2 + y_i^2) \qquad\qquad 13.11$$

$$I_{xy} = I_{yx} = -\sum_{i=1}^{n} m_i x_i y_i \qquad\qquad 13.12$$

$$I_{xz} = I_{zx} = -\sum_{i=1}^{n} m_i x_i z_i \qquad\qquad 13.13$$

$$I_{yz} = I_{zy} = -\sum_{i=1}^{n} m_i y_i z_i \qquad\qquad 13.14$$

The inertia matrix products of inertia (equations 13.12 – 13.14) have zero values about the axes of symmetry (i.e., the principal axes). If the inertia matrix is represented in the body coordinate system, the principal axes and principal moments of inertia can be determined by computing the inertia matrix eigenvectors and eigenvalues. The eigenvectors are colinear with the principal axes and the eigenvalues are the principal moments of inertia.

Equation 13.15 is the parallel axis theorem, which is useful for computing the inertia matrix relative to a position ($\bar{r}_p$) offset from the center-of-mass. $[I]_p$ represents the inertia matrix about position p, $[I]_{cm}$ represents the inertia matrix about the center-of-mass, and $\bar{r}_p$ is a column vector representing position p relative to the center-of-mass. The product of $\bar{r}_p\bar{r}_{pT}$ is a vector outer product, producing a 3x3 matrix.

$$[I]_p = [I]_{cm} + m\bar{r}_p\bar{r}_p^T \qquad \qquad 13.15$$

This relationship can relate the wet mass inertia matrix to its dry mass counterpart by applying the effects of propellant mass and location.

Key Term:

> **Inertia Matrix**: is a 3x3 matrix representing a spacecraft or other body's mass distribution.

13.3.3 Euler Torque Free Rotational Dynamics [Level I – Descriptive]

Euler's rotational equations for a rigid body presume rotation about a single body axis. Hence the inertia matrix diagonal elements are principal moments of inertia and the off-diagonal elements (i.e., products of inertia) are all zero. Since the spin is about a single axis in which all points at the same radius from the rotational axis have the same speed, the body is said to be experiencing pure rotation.

The advantage of analyzing rotational motion against the principal axes is the simplicity of eliminating the products of inertia. Since the off-diagonal inertia matrix elements are zero, the cross-axis dependencies are eliminated, and the equations can be evaluated as three separate scalar equations.

13.3.4 Euler Torque Free Rotational Dynamics [Level II – Equations]

Equation 13.16 expresses Euler's torque free rotational equations in terms of a rigid body's principal moments of inertia (I_{xx}, I_{yy}, I_{zz}), the instantaneous inertial rotation rates (ω_x, ω_y, ω_z) about each axis, and the rotational accelerations ($\dot{\omega}_x$, $\dot{\omega}_y$, $\dot{\omega}_z$) about each axis.

$$
\begin{aligned}
I_{xx}\dot{\omega}_x + \left(I_{zz} - I_{yy}\right)\omega_y\omega_z &= 0 \\
I_{yy}\dot{\omega}_y + \left(I_{xx} - I_{zz}\right)\omega_x\omega_z &= 0 \\
I_{zz}\dot{\omega}_z + \left(I_{yy} - I_{xx}\right)\omega_x\omega_y &= 0
\end{aligned}
\qquad 13.16
$$

13.3.5 Euler Torque Free Rotational Dynamics [Level III – Derivation]

The rotational dynamics evaluation initially does not presume torque-free motion. Equation 13.17 presents a torque (or moment) as the time derivative of angular momentum about the body's center-of-mass. Equation 13.18 represents the angular momentum as three-axis components.

$$\overline{M}_{cm} = \dot{\overline{L}}_{cm} \qquad\qquad 13.17$$

$$\overline{L}_{cm} = \begin{bmatrix} L_x \\ L_y \\ L_z \end{bmatrix} \qquad\qquad 13.18$$

Equation 13.19 expresses $\overline{L}_{cm}$ in terms of the inertia matrix ($\boldsymbol{I}$) and the body inertial rate ($\overline{\omega}_{bi}$) in the body-fixed (i.e., rotating) coordinate system. Since in this case the body frame is aligned with the principal axes, the angular momentum components are simplified in equation 13.20 as independent quantities.

$$\overline{L}_{cm} = \boldsymbol{I}\overline{\omega}_{bi} \qquad\qquad 13.19$$

$$\overline{L}_{cm} = \begin{bmatrix} I_{xx}\omega_x \\ I_{yy}\omega_y \\ I_{zz}\omega_z \end{bmatrix} \qquad\qquad 13.20$$

Equation 13.19 is the moment at the center-of-mass expressed as the time derivative of angular momentum in the moving frame, with $\dot{\overline{L}}_{rel}$ as the torque in the rotating frame. Because the body is rigid, the principal moments of inertia are constant, and their time derivatives are zero ($\dot{I}_{xx} = \dot{I}_{yy} = \dot{I}_{zz} = 0$). Equation 13.20 thus expresses the torque in the rotational frame.

$$\overline{M}_{cm} = \dot{\overline{L}}_{rel} + \overline{\omega} \times \overline{L}_{cm} \qquad\qquad 13.19$$

$$\dot{\overline{L}}_{rel} = \boldsymbol{I}\dot{\overline{\omega}} = \begin{bmatrix} I_{xx}\dot{\omega}_x \\ I_{yy}\dot{\omega}_y \\ I_{zz}\dot{\omega}_z \end{bmatrix} \qquad\qquad 13.20$$

$$\overline{M}_{cm} = \begin{bmatrix} M_x \\ M_y \\ M_z \end{bmatrix} = \begin{bmatrix} I_{xx}\dot{\omega}_x \\ I_{yy}\dot{\omega}_y \\ I_{zz}\dot{\omega}_z \end{bmatrix} + \begin{vmatrix} \hat{\imath} & \hat{\jmath} & \hat{k} \\ \omega_x & \omega_y & \omega_z \\ I_{xx}\omega_x & I_{yy}\omega_y & I_{zz}\omega_z \end{vmatrix} \qquad 13.21$$

Equation 13.22 expands and factors the cross-product terms, adding it to the torque in the relative frame.

$$\overline{M}_{cm} = \begin{bmatrix} I_{xx}\dot{\omega}_x + \left(I_{zz} - I_{yy}\right)\omega_y\omega_z \\ I_{yy}\dot{\omega}_y + (I_{xx} - I_{zz})\omega_x\omega_z \\ I_{zz}\dot{\omega}_z + \left(I_{yy} - I_{xx}\right)\omega_x\omega_y \end{bmatrix} \qquad 13.22$$

$$\begin{aligned} M_x &= I_{xx}\dot{\omega}_x + \left(I_{zz} - I_{yy}\right)\omega_y\omega_z \\ M_y &= I_{yy}\dot{\omega}_y + (I_{xx} - I_{zz})\omega_x\omega_z \qquad 13.23 \\ M_z &= I_{zz}\dot{\omega}_z + \left(I_{yy} - I_{xx}\right)\omega_x\omega_y \end{aligned}$$

Equation 13.23 are Euler's moment equations. Setting the net moment (torque) to zero verifies equation 13.16.

13.3.6 Torque-Free Attitude Stability [Level I – Descriptive]

A consequence of Euler's torque-free rotational equations is that a free body can never spin stably about its intermediate moment of inertia axis as will be demonstrated. A body can spin stably about its major moment of inertia axis, and it turns out is marginally stable about its minor moment of inertia axis.

The "marginal" stability about the minor moment of inertia axis is a highly idealized case of a perfectly rigid rotating body with spin that experiences no energy dissipation. It will be shown later that stable spin about the minor moment of inertia axis is not practical in the long term.

13.3.7 Torque-Free Attitude Stability [Level II – Equations]

The stability of a rigid body in a torque-free environment spinning about one axis may be evaluated against Euler's rotational equations. While the z-axis is identified as the rotational axis, the analysis considers each of the principal axes as candidates for the z-axis stability evaluation.

Euler's equations are evaluated for rotation about the z-axis with a slight disturbance to the motion. Equations 13.24 and 13.25 result, which are two second-order linear homogeneous differential equations with constant coefficients. (The $\delta\omega_x$ and $\delta\omega_y$ are the small rotational disturbances about the x- and y-axes.) Equation 13.26 defines an α coefficient in terms of the principal moments of inertia and the spin rate (ω_s) about the z-axis to simplify the mathematics.

$$\ddot{\omega}_x + \alpha\,\delta\omega_x = 0 \qquad\qquad 13.24$$

$$\ddot{\omega}_y + \alpha\,\delta\omega_y = 0 \qquad\qquad 13.25$$

$$\alpha = \frac{(I_{xx} - I_{zz})(I_{yy} - I_{zz})}{I_{xx}I_{yy}}\omega_s^2 \qquad\qquad 13.26$$

Equation 13.27 is the solution for disturbance term ($\delta\omega$) in the two differential equations.

$$\delta\omega = c_1 e^{i\sqrt{\alpha t}} + c_2 e^{-i\sqrt{\alpha t}} \qquad\qquad 13.27$$

Spin stability depends on the α coefficient's sign. If the coefficient is positive (i.e., $\alpha > 0$), equation 13.27's imaginary "i" parameter (i.e., $\sqrt{-1}$) persists in the exponents of both terms which can be evaluated using Euler's equation: $e^{i\theta} = \cos\theta + i\sin\theta$. Thus, for a positive α, both disturbances ($\delta\omega$) vary sinusoidally, and are bounded by the c_1 and c_2 amplitudes. This allows stable spin in the presence of small disturbances.

However, if the α coefficient is negative (i.e., $\alpha < 0$), the equation 13.27 exponents are no longer imaginary, becoming: $-\sqrt{\alpha t}$ and $\sqrt{\alpha t}$ respectively. Regardless of whether $\sqrt{\alpha t}$ is greater or less than 1.0, one of the two exponential functions will experience unbounded growth, and thus instability.

The α coefficient in equation 13.26 can only be positive (and thus allow stable spin) under two sets of conditions:
    a. $I_{zz} < I_{xx}$ and $I_{zz} < I_{yy}$, indicating I_{zz} is the minimum moment of inertia, or
    b. $I_{zz} > I_{xx}$ and $I_{zz} > I_{yy}$, indicating I_{zz} is the maximum moment of inertia.

The α coefficient in equation 13.26 is negative and thus spin is unstable under the remaining two conditions which each indicate I_{zz} is the intermediate moment of inertia:
    a. $I_{xx} < I_{zz} < I_{yy}$, or
    b. $I_{yy} < I_{zz} < I_{xx}$

Thus, Euler's rotational equations in a torque-free environment show stable spin is only possible about the **minimum** or **maximum** moment of inertia axes and can never be stable about the **intermediate** moment of inertia axis.

13.3.8 Torque-Free Attitude Stability [Level III – Derivation]

Derivation of the torque free stability differential equations begins with Euler's torque free rotation as expressed by equation 13.16 with a constant ω_s spin rate about the z-axis. Equation 13.28 defines small rotational disturbances about each axis. Equation 13.29 introduces these disturbances into equation 13.16.

$$\bar{\omega} = \begin{bmatrix} \delta\omega_x \\ \delta\omega_y \\ \omega_s + \delta\omega_z \end{bmatrix} \qquad 13.28$$

$$
\begin{aligned}
I_{xx}\delta\dot{\omega}_x + (I_{zz} - I_{yy})\omega_s\delta\omega_y + (I_{zz} - I_{yy})\delta\omega_y\delta\omega_z &= 0 \\
I_{yy}\delta\dot{\omega}_y + (I_{xx} - I_{zz})\omega_s\delta\omega_x + (I_{xx} - I_{zz})\delta\omega_x\delta\omega_z &= 0 \\
I_{zz}\delta\dot{\omega}_z + (I_{yy} - I_{xx})\delta\omega_x\delta\omega_y &= 0
\end{aligned}
\qquad 13.29
$$

The product of two small disturbances (such as $\delta\omega_y\delta\omega_z$) is an extremely small number that can be neglected. Equation 13.30 simplifies the equations by eliminating the insignificant terms.

$$
\begin{aligned}
I_{xx}\delta\dot{\omega}_x + (I_{zz} - I_{yy})\omega_s\delta\omega_y &= 0 \\
I_{yy}\delta\dot{\omega}_y + (I_{xx} - I_{zz})\omega_s\delta\omega_x &= 0 \\
I_{zz}\delta\dot{\omega}_z &= 0
\end{aligned}
\qquad 13.30
$$

The z-axis equation indicates $\delta\dot{\omega}_z = 0$ and thus $\delta\omega_z$ is constant over time. Equation 13.31 is the derivative of the x-equation (first) in 13.30. Equation 13.32 solves the middle equation 13.30 in terms of $\delta\dot{\omega}_y$.

$$I_{xx}\delta\ddot{\omega}_x + (I_{zz} - I_{yy})\omega_s\delta\dot{\omega}_y = 0 \qquad 13.31$$

$$\delta\dot{\omega}_y = -\frac{(I_{xx} - I_{zz})}{I_{yy}}\omega_s\delta\omega_x \qquad 13.32$$

Equation 13.33 substitutes the right side of equation 13.32 into equation 13.31. Equation 13.34 uses the same process for the y-equation.

$$\delta\ddot{\omega}_x - \frac{(I_{xx} - I_{zz})(I_{zz} - I_{yy})}{I_{xx}I_{yy}}\omega_s^2\delta\omega_x = 0 \qquad 13.33$$

$$\delta\ddot{\omega}_y - \frac{(I_{xx} - I_{zz})(I_{zz} - I_{yy})}{I_{xx}I_{yy}}\omega_s^2\delta\omega_y = 0 \qquad 13.34$$

Equation 13.35 defines an α coefficient that is common to both equations. Equation 13.36 produces the form of the ordinary differential equation.

$$\alpha = \frac{(I_{xx} - I_{zz})(I_{yy} - I_{zz})}{I_{xx}I_{yy}} \qquad 13.35$$

$$\delta\ddot{\omega} + \alpha\,\delta\omega = 0 \qquad 13.36$$

These results confirm equations 13.24 through 13.26.

13.3.9 Attitude Motion with Energy Dissipation Level III – Derivation]

Flexibility (i.e., the tendency to bend) is a characteristic of all rigid bodies, even if their flexure is difficult to measure. Body flexing results in energy dissipation that needs to be considered. This derivation follows the approach by Curtis (2020).

Consider a rotationally symmetric body ($I_{xx} = I_{yy}$) spinning torque free about its principal z-axis. Equation 13.37 describes its angular momentum ($\bar{L}_{cm}$) about its center of mass, with the transverse angular rate ($\bar{\omega}_T = \bar{\omega}_{xy}$) component including the effects of rotation about the x- and y-axes.

$$\bar{L}_{cm} = I_{xx}\bar{\omega}_T + I_{zz}\omega_z \hat{k} \tag{13.37}$$

$$L_{cm}^2 = I_{xx}^2\omega_T^2 + I_{zz}^2\omega_z^2 \tag{13.38}$$

Equation 13.39 is the derivative of 13.38 with respect to time.

$$\frac{dL_{cm}^2}{dt} = I_{xx}^2 \frac{d\omega_T^2}{dt} + 2I_{zz}^2\omega_z\dot{\omega}_z \tag{13.39}$$

Since angular momentum is constant (conserved) $\bar{L}_{cm}$ is constant and thus its derivative is zero.

$$\frac{d\omega_T^2}{dt} = -2\frac{I_{zz}^2}{I_{xx}^2}\omega_z\dot{\omega}_z \tag{13.40}$$

Equation 13.41 is the kinetic energy for a rotationally symmetric body ($I_{xx} = I_{yy}$).

$$T = \frac{1}{2}I_{xx}\omega_x^2 + \frac{1}{2}I_{xx}\omega_x^2 + \frac{1}{2}I_{zz}\omega_z^2 \tag{13.41}$$

$$T = \frac{1}{2}I_{xx}(\omega_x^2 + \omega_x^2) + \frac{1}{2}I_{zz}\omega_z^2 \tag{13.42}$$

$$T = \frac{1}{2}I_{xx}\omega_T^2 + \frac{1}{2}I_{zz}\omega_z^2 \tag{13.43}$$

Equation 13.44 is the time rate of change of rotational kinetic energy.

$$\dot{T} = \frac{1}{2}I_{xx}\frac{d\omega_T^2}{dt} + I_{zz}\omega_z\dot{\omega}_z \tag{13.44}$$

Equation 13.45 solves equation 13.44 in terms of $\dot{\omega}_z$.

$$\dot{\omega}_z = \frac{1}{I_{zz}\omega_z}\left[\dot{T} - \frac{1}{2}I_{xx}\frac{d\omega_T^2}{dt}\right]$$

13.45

Equation 13.46 substitutes the right side of equation 13.45 into equation 13.40 and solves algebraically for $d\omega_T^2/dt$.

$$\frac{d\omega_T^2}{dt} = \frac{2\dot{T}}{I_{xx}}\left[\frac{I_{zz}}{I_{zz} - I_{xx}}\right]$$

13.46

For the spacecraft to spin stably about the z-axis, the transverse rotational velocity (ω_T) may not increase in magnitude. Therefore, $d\omega_T^2/dt$ may only have a negative value for stable rotation.

A spacecraft dissipating energy will have a negative $\dot{T}$ value. The only way for equation 13.46 to produce a negative value in the presence of energy dissipation is for $I_{zz} > I_{xx}$. Therefore, spacecraft spin in the presence of energy dissipation is only stable about the maximum moment of inertia axis. Naturally, as energy dissipates, the body moves toward the minimum rotational energy state.

13.4 Attitude Motion with Torque

In the presence of torques, angular momentum changes, which in turn changes the spacecraft's attitude state. Spacecraft experience torques from environmental (external) sources as well as self-induced torques from on-board actuators.

13.4.1 Environmental Torques

Table 13-1 provides a summary of common environment torques spacecraft experience.

Table 13-1 Environmental Torque Summary		
Torque	Description	Effect
Gravity Gradient	Inverse square variation of distances of different component masses on the spacecraft versus the Earth's center-of mass.	Significant effect in low altitude orbits. Measurable in medium altitude orbits.
Radiation Pressure	Component of radiation (photon impact) pressure	The effect is on all orbit types. It is a dominant

	perpendicular to the offset between spacecraft center-of mass and center of radiation pressure.	torque for high altitude orbits.
Aerodynamic	Component of aerodynamic pressure perpendicular to the offset between spacecraft center-of mass and center of dynamic pressure.	Significant effect only for low altitude orbits (or orbits with a low periapsis altitude) with the presence of an atmosphere.
Magnetic Disturbance	Interaction of magnetic fields induced by on-board electronics with a celestial body's magnetic field.	Significant effect in low and medium altitude orbits about celestial bodies with measurable magnetic fields.

13.4.1.1 Gravity Gradient Torque

A gravity gradient torque tends to cause the spacecraft minimum moment of inertia axis to align vertically to the local gravity gradient. Equation 13.47 is the gravity gradient torque.

$$\bar{\tau}_{gg} = \frac{3\mu}{r^3}(\bar{r} \times I\hat{r}) \qquad\qquad 13.47$$

Vector $\bar{r}$ is the position of the spacecraft center of mass, μ is the centrally attracting body's gravitational parameter, and I is the spacecraft inertia matrix.

13.4.1.2 Radiation Pressure Torque

A radiation pressure torque results from a momentum transfer from photon impacts. The Sun is the most common radiation source in the solar system, but radiation pressure can also come from visible light reflection off a celestial body (i.e., albedo) or from infrared radiation emitted from a planet or other celestial body.

Equation 13.48 computes a differential force component due to the radiation pressure. It depends on the momentum flux (P) from the radiating source acting on a spacecraft surface. It also depends on the surface's specular reflection (C_s) and diffuse reflection (C_d) coefficients for the radiation type.

$$\overline{df}_{total} = -P \int \left[(1 - C_s)\hat{S} + 2\left(C_s \cos\theta + \frac{1}{3}C_d \right)\hat{N} \right] \cos\theta \, dA \qquad 13.48$$

In this equation, $\hat{S}$ is the direction of the radiation from the source to the spacecraft. $\hat{N}$ is normal to the surface elemental area dA. θ is the angle between $\hat{N}$ and $\hat{S}$. Equation 13.39 is the imparted radiation pressure torque.

$$\bar{\tau}_{rad} = \int \bar{\rho} \times \overline{df}_{total} \qquad 13.49$$

Vector $\bar{\rho}$ is the offset between the spacecraft center-of-mass and the differential surface area center of radiation pressure.

While radiation pressure is presented in integral form, practical implementations use summations of finite elements.

13.4.1.3 Aerodynamic torque

Aerodynamic torque is a function of atmospheric density (ρ) and the spacecraft body drag coefficient (C_D). The strength is proportional to the square of the velocity ($\bar{v}_a$) relative to the atmosphere. Equation 13.50 computes a differential aerodynamic force.

$$\overline{df}_{aero} = -\frac{1}{2}C_D \rho v_a^2 (\hat{N} \cdot \hat{v}_a)\hat{v}_a dA \qquad 13.50$$

Equation 13.51 is the aerodynamic torque.

$$\bar{\tau}_{aero} = \int \bar{\rho} \times \overline{df}_{aero} \qquad 13.51$$

Vector $\bar{\rho}$ is the offset between the spacecraft center-of-mass and the differential surface area center of aerodynamic pressure.

While aerodynamic pressure is presented in integral form, practical implementations use summations of finite elements.

13.4.1.4 Magnetic Disturbance Torque

A magnetic disturbance torque is a function of the spacecraft net effective magnetic moment ($\bar{m} \ A \cdot m^2$) and the planetary magnetic flux density ($\bar{B} \ Wb/m^2$). Equation 13.52 is the magnetic disturbance torque.

$$\bar{\tau}_{mag} = \bar{m} \times \bar{B} \qquad\qquad 13.52$$

13.4.2 Attitude Actuators

Attitude actuators are mechanisms on a spacecraft used to change its orientation and/or inertial rotation rate. While actuators are devices that deliberately change the attitude, on-board mechanisms that impart incidental torques that change the spacecraft angular momentum may also be considered actuators.

13.4.2.1 Thrusters

Thrusters can impart torques on a spacecraft body to change its angular momentum. Section 4.2.3 showed opposing thruster pairs in an RCS being used to impart a torque to change the attitude without a translational acceleration that would change its trajectory.

13.4.2.2 Gyroscopic Attitude Stabilization

Gyroscopic mechanisms are spinning masses that store and/or change spacecraft angular momentum by maintaining or changing rotational speed or changing the spinning mass's rotational axis.

13.4.2.2.1 Spinning Spacecraft Body

Spinning spacecraft bodies are a traditional method of attitude stabilization that is now only used in rare circumstances. A spacecraft body rotating about its maximum moment of inertia axis will maintain a stable spin attitude. This stabilization method is included since it remains practical for some spacecraft missions (such as communications satellites).

13.4.2.2.2 Momentum Wheels

A momentum wheel is a single, fixed-speed angular momentum storage device that, like a spinning spacecraft body, provides attitude stabilization along a single axis. While secondary actuators may change body orientation, momentum wheels have most of the limitations as a spinning spacecraft body. This stabilization device is presented primarily from a historical perspective.

13.4.2.2.3 Reaction Wheels

Reaction wheels are currently the most common spinning mass device used for non-human occupied spacecraft. Reaction wheels are mounted in fixed orientations within the spacecraft body. Spacecraft angular momentum is changed by speeding up, slowing down, and reversing wheel speeds. According to the action-reaction law, a change to wheel speed will result in a reactive rotational torque on the spacecraft body in the opposite direction. The rotational acceleration imparted to the spacecraft ($\dot{\omega}_s$) about the wheel spin axis is proportional to the wheel acceleration ($\dot{\omega}_w$), the wheel moment of inertia (I_w), and the spacecraft moment of inertia (I_s) about the wheel axis. Equation 13.53 has the relationship as a net zero momentum change, expressing conservation of angular momentum.

$$I_w \dot{\omega}_w + I_s \dot{\omega}_s = 0 \qquad\qquad 13.53$$

$$\dot{\omega}_s = -\frac{I_w}{I_s} \dot{\omega}_w \qquad\qquad 13.54$$

A three-axis stabilized spacecraft requires at least three reaction wheels for maintaining attitude control. A three-wheel system is normally configured orthogonally. Another common arrangement uses four or more wheels in which all wheels have a component along a common axis. Such a system provides redundancy in the event of a wheel failure. Equally important, such an arrangement allows re-distribution of angular momentum, balancing the wheel speeds without changing the spacecraft body angular momentum.

One consideration when using reaction wheels, like any mechanical device, is that a wheel has a maximum speed at which it can spin. As the wheel approaches its maximum speed, its ability to apply torque to the spacecraft diminishes. A secondary actuator is used to apply torque opposing the wheel. The applied torque causes the attitude control system to induce the opposing corrective torque by causing the wheel to slow down (desaturate) in the process. Such wheel desaturation (i.e., momentum management) is commonly performed using RCS thrusters or electromagnetic torquers.

13.4.2.2.4 Control Moment Gyroscopes

Control Moment Gyroscopes (CMGs) are gimbaled, tilting the spinning mass axis (and in some cases changing rotational speed) to impart a torque on a spacecraft. CMGs are commonly used on large human occupied spacecraft such as space stations for attitude control.

13.4.2.2.5 Articulating Solar Arrays

Articulating a solar array orients the solar panels normal to the Sun direction for maximum energy collection. But since solar arrays have a moment of inertia about their articulation axis, reorienting the arrays imparts a proportional counter torque to the spacecraft. The attitude control algorithms consider such commanded torques and counteracts them automatically while orienting solar arrays.

13.4.2.3 Magnetic Torquers

Magnetic torquers are electromagnets designed to interact with the planetary magnetic fields to impart a torque on the spacecraft. These are only useful for spacecraft orbiting a planet with a significant magnetic field (such as Earth). Since magnetic field strength drops off rapidly with altitude, magnetic torquers are usually only effective for spacecraft in which at least a portion of the orbit is at low altitude.

13.4.2.4 Nutation Dampers

Nutation dampers are passive attitude control devices used in spin stabilized spacecraft. Nutation dampers induce energy dissipation in a spinning spacecraft, causing the body to move toward the minimum rotational energy state.

Angular momentum is the product of moment of inertia and inertial rate ($\bar{L} = I\bar{\omega}$) and rotational energy is proportional to the square of the rotation rate ($\varepsilon = I\omega^2/2$). For angular momentum to be conserved (i.e., remain constant), the $I\bar{\omega}$ product must remain constant.

With energy dissipation the inertial spin rate (ω) is decreasing. Having $I\bar{\omega}$ remain constant simultaneously with a decreasing spin rate requires that the moment of inertia (I) increase. The physical manifestation of these constraints is the body realigning its spin axis toward the maximum moment of inertia axis.

13.4.3 Attitude Sensors

Attitude sensors provide feedback to attitude prediction algorithms and are critical in closing the loop and estimating the actual spacecraft attitude.

13.4.3.1 Star Trackers

Star trackers measure the spacecraft inertial orientation with modern trackers producing attitude quaternions. Star trackers compare detected star locations against the expected coordinates specified by a star catalog.

Since the star catalogs reference the star locations to a standard epoch time and generally from the solar system barycenter point of view, corrections need to be made to the catalog coordinates to predict the expected star directions from the spacecraft point of view. The IAU SOFA libraries provide the functions needed to determine the expected star directions.

13.4.3.2 Inertial Measurement Units

An Inertial Measurement Unit (IMU) is an integrated three axis assembly of inertial sensors. IMUs typically consist of a three-axis set of rate gyroscopes with orthogonal alignment. Aligned with each gyroscope axis is a linear accelerometer. Both gyroscopes and accelerometers are inertial instruments since the measurements are made relative to inertial space.

13.4.3.2.1 Gyroscopes

Gyroscopes (or "gyros" for short) as inertial instruments can be used to measure or maintain orientation or measure angular rates of change to orientation. Traditional gyroscopes are spinning masses that maintain an inertial orientation due to conservation of angular momentum.

Spinning mass gyroscopes are still used extensively in aircraft attitude and direction-indicating instruments. Spacecraft more commonly use photonic-based gyroscopes that drift less than spinning mass gyroscopes. The two common photonic gyroscopes are the Ring Laser Gyro and the Fiber Optic Gyros.

13.4.3.2.1.1 Ring Laser Gyroscopes

Ring Laser Gyroscopes (RLGs) detect changes in angular rate through interferometry. Two lasers are aimed in opposite directions over a closed path created by reflective elements, that is perpendicular to the sense axis. The interferometric effect causes a phase shift in the form of an interference pattern, which is translated to an angular rate about the sense axis.

RLGs require countermeasures at slow rotation speeds to avoid measurement confusion. Thus, a certain amount of additional complexity is needed to get accurate measurements at slow rotation rates.

13.4.3.2.1.2 Interferometric Fiber Optic Gyroscopes

Fiber Optic Gyroscopes (FOGs), like RLGs, detect changes in angular rate through interferometry. Two lasers are injected in opposite directions at opposite ends of an extremely long fiber (hundreds or thousands of meters long). The fiber is wound on a spool perpendicular to the sense axis. The interferometric effect causes a phase shift in the form of an interference pattern, which is translated to an angular rate about the sense axis.

Since photons are the only moving parts in the sensing mechanism, FOGs provide extremely accurate angular rate measurements and are usually insensitive to vibration and acceleration. FOGs are a suitable choice for precise space applications.

13.4.3.3 Sun Sensors

There are two basic types of Sun sensors: conical and slit sensors. Conical sensors measure illumination when the Sun is within the conical field-of-view. Illumination is diminished when the Sun disk is only partially within the cone. Conical Sun sensors are useful for directing solar arrays.

Slit Sun sensors have a narrow opening that restricts the ability to sense the Sun in one angular dimension. It has an array of detectors that determines the Sun's angle along the slit direction. Slit Sun sensors are useful for determining when a particular axis crosses the Sun in a spinning spacecraft. Slit Sun sensors may also be used in pairs with the slits arranged perpendicularly to precisely determine the Sun's direction in two dimensions.

13.4.3.4 Limb Sensors

Limb sensors are used to align to the outer edge of a planetary disk. They are often two rings of concentric detector arrays in which the planetary disk should fully illuminate the inner ring to be centered, but not illuminate the outer detectors. Any sensing in the outer detectors indicates a drift off center, prompting the attitude control system to direct the orientation away from the illuminated outer detector.

13.5 Attitude Determination and Control

Spacecraft attitude control is predicated on having knowledge of the current attitude. The desired orientation and rate are compared with what is measured. A closed loop control system such as a Proportional-Integral-Derivative (PID) controller is used to drive momentum-changing actuators to correct the attitude. The task is to drive the attitude error, as measured by the feedback sensors, as close to zero as practical.

13.5.1 Attitude Determination

Attitude Determination (AD) is a process that estimates the spacecraft attitude (orientation and/or inertial rate) using measurements from the available attitude sensors. AD is fully analogous for attitudes as OD is for determining a spacecraft's trajectory, although AD tends to be more difficult and less straight forward.

AD may consist of batch least squares differential corrections and/or extended Kalman filters. While there are numerous successful approaches, the author prefers a Multiplicative Extended Kalman Filter (MEKF) for determining the body orientation. The MEKF solves for the attitude quaternion (or its equivalent) indirectly by estimating a corrective coordinate transformation in the form of another quaternion or a minimal parameter set such as Modified Rodrigues Parameters. The updated estimate of the body orientation is the product of the current quaternion with the corrective coordinate transformation.

13.5.2 Attitude Control

Attitude control is performed by a closed loop servo system in which the error signal is the difference between the commanded and the measured attitude. Depending on the level of sophistication, the attitude may be measured either directly by the attitude sensors or consist of the attitude estimated by the AD process.

The error loops are closed by commanding the applicable attitude actuators to apply torques that drive the attitude toward the commanded state, cancelling the attitude errors.

The imparted torques change the spacecraft's net angular momentum in a controlled manner. Momentum management processes can reduce excessive angular momentum in mechanisms such as reaction wheels, allowing them to operate within the design limits and/or their most effect speed ranges.

REFERENCES

1. Schaub, Hanspeter and Junkins, John L., *Analytical Mechanics of Space Systems*, AIAA Education Series, © 2003 American Institute of Aeronautics and Astronautics (AIAA), ISBN 1-56347-563-4.
2. Wertz, James R. (editor), *Spacecraft Attitude Determination and Control*, © 1987 Kluwer Academic Publishers, ISBN 90-277-1204-2.
3. Markley, F. Landis and Crassidis, John L., *Fundamentals of Spacecraft Attitude Control*, © 2014 Springer, ISBN 978-1-4939-0801-1.
4. Kuipers, Jack B., *Quaternions and Rotation Sequences*, A Primer with Applications to Orbits, Aerospace, and Virtual Reality, © 1999 Princeton University Press, ISBN 0-691-10298-8.

5. Agrawal, Brij N., *Design of Geosynchronous Spacecraft*, © 1986 by author, Prentice-Hall, ISBN 0-13-200114-4.

6. Curtis, Howard D, *Orbital Mechanics for Engineering Students*, Fourth Edition, © 2020 Elsevier, Ltd., ISBN 978-0-08-102133-0.

7. Sidi, Marcel J., *Spacecraft Dynamics & Control*, A Practical Engineering Approach, © 1997 Cambridge University Press, ISBN 0-521-78780-7.

8. Hughes, Peter C., *Spacecraft Attitude Dynamics*, © 1986 by author, Dover Publications, Inc., ISBN 0-486-43925-9.

9. Kane, Thomas R. et al., *Spacecraft Dynamics*, © 1983 McGraw-Hill Book Company, ISBN 0-07-037843-6.

10. Kaplan, Marshall H., *Modern Spacecraft Dynamics & Control*, © 1976 John Wiley & Sons, Inc., ISBN 0-471-45703-5.

11. Kreyszig, Erwin, *Advanced Engineering Mathematics*, ©1962 John Wiley & Sons, Inc., ISBN 0-471-15496-2.

12. Beer, Ferdinand P. and Johnston Jr., E. Russell, *Vector Mechanics for Engineers, Statics and Dynamics*, © 1962 McGraw-Hill, Inc., ISBN 0-07-079923-7.

13. Wells, D.A., *Schaum's Outline of Theory and Problems of Lagrangian Dynamics, with a treatment of Euler's Equations of Motion, Hamilton's Equations and Hamilton's Principal*, © 1967 McGraw-Hill, Inc., ISBN 07-069258-0.

14. NASA Technical Memorandum, *Coordinate Systems for the Space Shuttle Program*, October 1974, NASA TM X-58153.

15. Ernandes, Kenneth J., Joseph, Benjamin E., and Cefola, Paul J., *Implementation of a Multiplicative Extended Kalman Filter for Spinning Spacecraft Attitude Determination in the Astrodynamics Environment (ADE)*, AAS-07-337, American Astronautical Society, 2007.

14 Space System Simulation

Spacecraft simulation models the physical processes governing a mission. This typically involves not only orbital and attitude dynamics, but also the interaction of onboard spacecraft components and subsystems with each other, and corresponding on-board and ground-based command and control systems. Simulations are used in both a spacecraft's conceptual and design phases and during operations.

During the design phase, various candidate tradeoffs are considered and run through exercises to simulate mission operations. These simulations are used to evaluate the viability and performance of various alternatives through various design iterations. Simulations of typical and the most extreme of mission environments are conducted to gain knowledge under a full set of conditions. This leads to design maturation with a degree of confidence that the spacecraft and command and control assets will be capable of performing the mission at the specified levels.

Operational simulators model the spacecraft and command and control assets as built. These simulations emulate specific anticipated conditions or sets of conditions as scenarios. The purpose is usually for training and mission rehearsal. This can include activities such as rehearsing various sequences of events including thruster firings, attitude maneuvers, and interactions with other spacecraft and ground command and control systems. Such simulations are usually prerequisites to approving mission plans.

14.1 Systems Being Simulated

The first stages of a simulator design ask the fundamental questions of: (1) what is being simulated and (2) to what level of fidelity should the simulated systems emulate what is expected in the real mission. The answers to these questions are fundamental to determining the requirements that drive the simulator design. This includes whether the simulation includes only spacecraft flight dynamics, or it also models collateral and environmental effects on all or a subset of spacecraft subsystems.

Equally important to defining the simulation scope (i.e., what is being simulated) is the degree (or fidelity) of each simulation component. This includes how closely (or exactly) hardware components match the real world or how closely a software model emulates reality. A tradeoff usually needs to be made particularly for software models to balance the requirements against computational overhead and throughput.

Simulations can range from testing a single component, such as a single spacecraft or ground control terminal to a full up spacecraft with human occupants interacting with a mission control segment and human operators. Figure 14-1 is a notional example of an elaborate simulation system with both mission control and a human occupied spacecraft simulator.

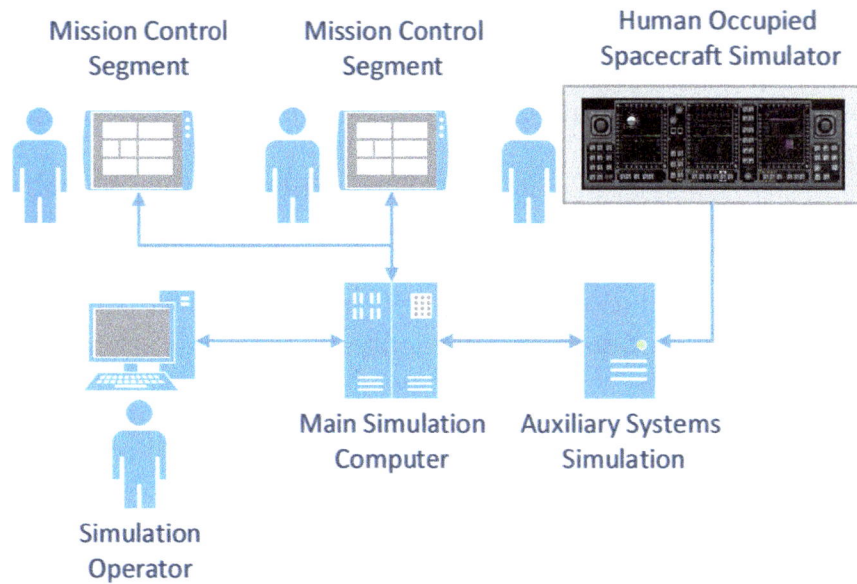

Figure 14-1 Example Simulation System Hardware Layout

14.1.1 Spacecraft Subsystems

Simulators involving space flight dynamics usually involve some degree spacecraft simulation. In some cases, the simulation involves a single spacecraft, in many others multiple spacecraft are involved. The following sections provide an example top-level description of the tasks performed by the software-simulated subsystems. Figure 14-2 is a notional block diagram of a spacecraft simulator.

14.1.1.1 Spacecraft Structure

The spacecraft structure is commonly a rigid or semi-rigid frame to which other subsystems are mounted. The spacecraft structure carries the spacecraft body coordinate system. In a spacecraft simulator, the structure model typically maintains the mass properties, including mass management, center-of-mass cognizance, and inertia matrix maintenance. The spacecraft structure is often modeled as the thermal backbone, providing the thermal control system one conduit for heat transfer.

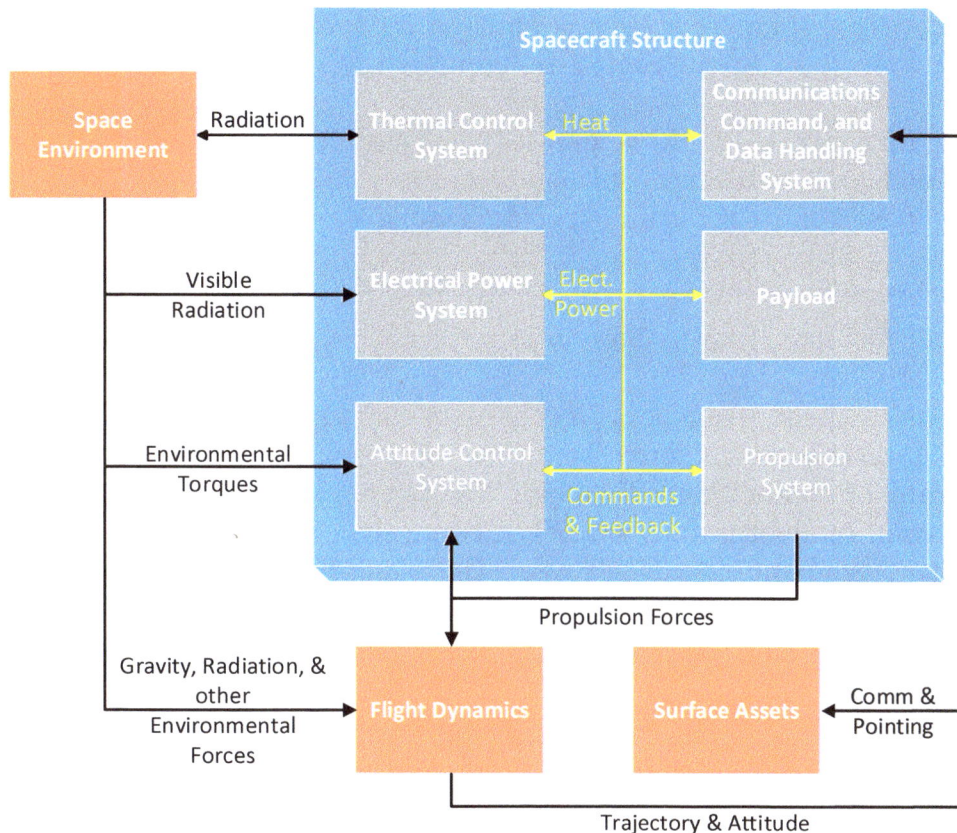

Figure 14-2 Spacecraft Simulator Block Diagram

14.1.1.2 Thermal Control System [Level I – Descriptive]

The thermal control system is involved with any spacecraft activities that collect, generate, distribute, or expel heat (i.e., thermal energy). The thermal control system manages the mutual heat transfer between the spacecraft components and the space environment. Thermal modeling activities consider heat generation and transport in terms of the following:

- <u>Heat Generation</u> – is typically the result of electrical inefficiency or chemical reaction. Examples include the inherent inefficiency of electrically-powered components expressed as heat energy and the inefficiency of a thruster in producing pure thrust.
- <u>Heat Conduction</u> – is the movement of heat energy between spacecraft components based on the temperature difference and characteristics of the thermal interface (connection). Conductive transfer cause heat to flow from higher to lower temperatures at a rate proportional to the temperature difference.
- <u>Radiative Absorption</u> – is energy absorbed by spacecraft components through radiative transfer.

- Radiative Emission – is energy emitted by spacecraft components through radiative transfer by the Stefan-Boltzmann law.

A flow model or network should be created that simultaneously models the mutual heat generation and transfer of all components. The component temperature is then updated based on its mass and specific heat capacity.

14.1.1.2.1 Heat Conduction [Level II – Equations]

Conduction is a heat transport mechanism between spacecraft components. Equation 14.1 describes conductive heat flow, as provided by Fourier, with Q being the linear rate of conduction (W) along the temperature gradient, K being the material thermal conductivity ($W/m \cdot K$), A being the cross-sectional area normal to the temperature gradient, and dT/dx being the temperature gradient (K/m).

$$Q = -KA\frac{dT}{dx}$$

14.1

Table 14-1 is the thermal conductivity (K) for common spacecraft materials as provided by Agrawal (1986).

Table 14-1 Thermal Conductivities of Common Spacecraft Materials	
Material	**K ($W/m \cdot K$) at 25°C**
Aluminum	210
Aluminum Alloys	117-175
Magnesium	157
Magnesium Alloys	52-111
Titanium	21
Stainless Steel	16.2
Teflon	0.25

Heat transfer can be estimated between spacecraft components across their mutual surface contact area. Equation 14.2 determines the thermal resistivity (R_c) in K/W of a component with a case wall thickness τ across a contact area A.

$$R_c = \frac{\tau}{KA}$$

14.2

Equation 14.3 is the heat conduction across the contact interface, with T_1 as the temperature of the warmer component and T_2 being the temperature of the cooler component.

$$Q = \frac{T_1 - T_2}{R_c}$$ 14.3

This method provides a reasonable estimate of the heat flow if the component cases carry most of the thermal load. Higher fidelity simulations should use heat flow values measured from actual components.

14.1.1.2.2 Absorbed versus Emitted Radiation [Level I – Descriptive]

Bodies absorb and emit radiation equally at any given wavelength according to Kirchoff's law. A black body is an ideal absorber and emitter of radiation, meaning its absorptivity (α) and emissivity (ε) are both unity ($\alpha = \varepsilon = 1$) at all wavelengths. A gray body is one in which the absorptivity and emissivity are a constant between zero and one ($0 < \{\alpha = \varepsilon\} < 1$) at all wavelengths.

Much of the radiation incident on a spacecraft is in the visible spectrum. In contrast, spacecraft operating temperatures cause them to emit primarily in the infrared spectrum (i.e., what humans sense as heat). Most surfaces have a different absorptivity and emissivity at visible and infrared wavelengths. Thus, the characteristics of spacecraft components surface materials should be characterized to understand their absorptive and radiative properties.

Generally, light surfaces (such as white paint) or polished metal surfaces have low visible absorptivity (0.1 to 0.2) and have high emissivity (0.8 to 0.9). In contrast, dark surfaces (such as black paint) have high absorptivity (0.8 to 0.9) and high emissivity (0.8 to 0.9). Materials such as gold have both low absorptivity and emissivity.

14.1.1.2.3 Stefan-Boltzmann Law [Level II – Equations]

The Stefan-Boltzmann Law states that the energy emitted per unit area of a black body is proportional to the absolute temperature to the fourth power as indicated in equation 14.4. The parameter $\sigma = 5.670374419 \times 10^{-8}\, W/(m^2 K^4)$ is the Stefan-Boltzmann constant.

$$E = \sigma T^4$$ 14.4

Most surfaces have an effective emissivity (ε) for a given radiative wavelength (λ), which is a coefficient that modifies equation 14.5.

$$E = \varepsilon \sigma T^4$$ 14.5

The rate at which a black body at absolute temperature T in Kelvins radiates energy by wavelength (λ) is characterized by in equation 14.6. A characteristic black body curve results from plotting this equation versus wavelength.

$$E(\lambda) = \frac{2\pi hc^2}{(e^{hc/\lambda kT} - 1)\lambda^5} \qquad 14.6$$

In equation 14.6, h is Planck's constant, c is the speed of light, and k is the Boltzmann constant.

14.1.1.3 Electrical Power System

The electrical power system is responsible for generating, regulating, storing, and distributing electrical power to the remainder spacecraft components that use electricity. The electrical generation model requires inputs from flight dynamics (trajectory and attitude) for solar array energy collection. In that process, it must also be cognizant of any eclipse entry and exit.

The modeling considers the electrical inefficiency from powered components, converting the associated energy to heat for management by the thermal control system. Another management function modeled is the energy budget in terms of generation, storage, and usage. The system may also model load shedding in the case of extreme electrical power deficits.

14.1.1.4 Communications, Command, and Data Handling System

The communications, command, and data handling system is responsible for communicating with entities external to the spacecraft and well as for internal communications. It is also responsible for on-board commanding, processing, storage, and data distribution.

The simulation functions model the radio and/or optical communications link budgets between the spacecraft and other assets (such as other spacecraft and surface sites). Radio communications links consider the antenna gain and frequencies, the free space path loss, and pointing offsets. In the case of a marginal communications link, the models compute the probabilities of successful (voice or data) bit transfers and communications errors are modeled by inverting bit states according to the probabilistic frequency as discerned by a pseudo random number generator.

When operational command and control terminals are in the loop, the simulated communications, command, and data handling system decodes commands received from surface terminals. The models must also encode applicable on-board parameters into the telemetry data stream sent to the surface terminals.

14.1.1.5 Attitude Determination and Control System

The attitude determination and control system is responsible for measuring the orientation of the spacecraft body coordinate system and the inertial rotation rate against established reference coordinate systems. It is also responsible for managing on board momentum transfer actuators that change the body orientation and rotation rate to what is commanded to facilitate mission execution.

The simulation can model the AD process at levels as simple as applying the appropriate statistical levels of error to the true attitude state, all the way to embedding the flight AD software and processing the measurements from the simulated attitude sensors. The simulation models drive the attitude actuators to correct the differences between commanded and actual attitude states.

14.1.1.6 Propulsion System

The propulsion system provides thrusters, tanks, and associated hardware (manifolds, feed lines, valves, and regulators) that provide thrust for orbital translation and attitude control.

The models determine the net effects of thrust events on the vehicle. This includes net torques to the flight dynamics attitude truth model for integration into the vehicle angular momentum. The momentum changes result in the associated updates to the attitude state.

The model also provides net translational accelerations to the flight dynamics trajectory model for integration into the vehicle net acceleration. The acceleration applies the appropriate change to the spacecraft trajectory.

14.1.1.7 Spacecraft Payload

The spacecraft payload is the system that performs the intended mission. Since there is a multitude of potential spacecraft missions, the design and activities of a simulated spacecraft payload and its interaction with the spacecraft bus must be considered on a case-by-case basis.

14.1.2 Body-Based Assets

Body-based assets are those associated with celestial bodies. These can include fixed or mobile surface-based assets as well as airborne assets (i.e., those operating above the surface but confined to the body's atmosphere).

The communications between two widely separated antennas, such as on a spacecraft and a body-based asset, experience a light time delay (δt). Equation 14.7 computes the light time delay based on separation distance ($\Delta \rho$) and the speed of light (c).

$$\delta t = \frac{\Delta \rho}{c} \qquad\qquad 14.7$$

The relative speed between two antennas causes a Doppler frequency shift. Equation 14.8 is the Doppler shift (Δf) on a transmitted frequency (f_0) resulting from the radial speed difference ($\dot{\rho} = \delta \bar{v} \cdot \hat{\rho}$) or range rate between two antennas. Equation 14.9 is the frequency (f) observed at the receiving antenna.

$$\Delta f = f_0 \frac{\dot{\rho}}{c} \qquad\qquad 14.8$$

$$f = f_0 \left(1 + \frac{\dot{\rho}}{c} \right) \qquad\qquad 14.9$$

14.1.2.1 Control Centers

Control centers usually involve persons-in-the loop, interacting with the simulation by operating at system control consoles. However, it is possible to simulate the control center operations with interactions that would follow standard checklists.

14.1.2.2 Tracking and Communications Antennas [Level I – Descriptive]

Spacecraft tracking and communications is often accomplished through tracking antennas positioned at the surface of a celestial body.

The concepts of *antenna gain* and *isotropic radiators* are fundamental to conceptualizing signal link strengths and margins. An isotropic radiator is a device (such as an antenna) that radiates equally in all directions. Signal strength for an isotropic radiator is the same in all directions, providing a spherical surface at radius r an equal signal strength. Practical antennas focus the signal power in the preferred direction(s), at the expense of those directions that are less preferred.

Thus, the signal power is allocated strongly in the preferred direction and reduced elsewhere.

An antenna's gain is in reference to its direction(s) of maximum power output. The gain value, expressed in decibels (dB), is computed from the ratio of maximum power intensity to the power intensity of an isotropic radiator. It should be noted that antenna gain follows a reciprocity principle for reception. That means the antenna functions equally in establishing a preferred direction in which the antenna can more efficiently receive a signal, with weaker reception capability in other directions.

Key Terms:

> **Antenna Gain**: results from antenna design characteristics in which transmitted energy is reallocated to favored directions. Gain likewise gives reciprocal sensitivity enhancement for received signals in the favored directions.
> **Isotropic Radiator**: is an idealized antenna that transmits equal power in all directions over a concentric sphere.

14.1.2.3 Tracking and Communications Antennas [Level II – Equations]

Equation 14.10 is the general expression for determining gain based on an antenna's maximum power intensity (I_{MP}) in its favored direction versus the isotropic radiator's power intensity (I_0). The use of the common logarithm (i.e., log_{10}) produces the gain in dB. Equation 14.11 computes the gain of an ideal parabolic reflecting antenna in dB for a given signal wavelength (λ) and antenna area (A). Equation 14.11 also computes the parabolic antenna gain equivalently for a given frequency (f), with c being the speed of light.

$$G = 10 \, log_{10} \left(\frac{I_{MP}}{I_0} \right) \qquad\qquad 14.10$$

$$G = 10 \, log_{10} \left(\frac{4\pi A}{\lambda} \right) = 10 \, log_{10} \left(\frac{4\pi A f}{c} \right) \qquad\qquad 14.11$$

A parabolic reflecting antenna has maximum gain along the paraboloid's axis of symmetry toward the focus. Minimum gain is also along the axis of symmetry, but opposite the focus direction.

14.1.2.4 Signal Link Strength [Level I – Descriptive]

A communication signal's maximum link strength may be determined by considering the gain between the transmitting antenna, the gain of the receiving antenna, the transmission power, and the distance between the two antennas. Higher gain in the transmitting and/or receiving antenna facilitates a stronger signal as does a higher transmission power. However, the signal power drops off as an inverse square of the distance between the two antennas.

$$P = G_T + 10\ log_{10}(P_T) + G_R - 20\ log_{10}\left(\frac{4\pi r f}{c}\right) \qquad 14.12$$

Equation 14.12 is the first two terms of equation 14.13, which is known as the Effective Isotropic Radiated Power (EIRP) in dBW.

$$EIRP = G_T + 10\ log_{10}(P_T) \qquad 14.13$$

Equation 14.14 uses a mathematical property of logarithms to separate frequency-dependent term of the free space path loss (P_L) from the distance-dependent term.

$$P_L = -20\ log_{10}\left(\frac{4\pi r f}{c}\right) = -20\ log_{10}\left(\frac{4\pi f}{c}\right) - 20\ log_{10}(r) \qquad 14.14$$

When the simulation includes antenna pointing errors, a reduction in the antennas' gain values is modeled to reduce the signal strength.

Key Term:

> **Effective Isotropic Radiated Power**: is the power radiated in an antennas most favored direction .

14.1.2.5 Surface Observational Subjects

Surface observational subjects, when present, consist of what a spacecraft payload is observing, such as weather phenomena or natural resources. These may be simulated, but sometimes include providing information from a real database.

14.1.3 Space Environment

The space environment defines the physical aspects to which spacecraft and other elements of the simulation are subjected.

14.1.3.1 Celestial Body Ephemerides

Celestial body ephemerides provide the positions of celestial bodies at any given time. Ephemerides sources include planetary orbital elements, Jet Propulsion Laboratory (JPL) planetary ephemerides, and those available from the IAU SOFA functions.

14.1.3.2 Radiation Environment

The radiation environment primarily consists of visible and infrared electromagnetic radiation. Most visible radiation emits from the Sun and is reflected by the celestial bodies. Celestial bodies reflect a fraction of incoming sunlight on their surface facing the Sun. The fraction of reflected radiation is called *albedo*. Infrared radiation is emitted by celestial bodies based on their absolute temperature by the Stefan-Boltzmann law.

High energy (ionizing) radiation may also be important to model for certain simulations. Such modeling requirements may be investigating the effects on electronics or human occupants.

14.1.3.3 Magnetic Fields and Effects

Magnetic fields can cause a torque on a spacecraft attitude if the vehicle has a net magnetic field from the flow of onboard electrical current. Magnetic fields also trap and have a ducting effect on high energy radiation.

14.1.4 Software Simulation versus Hardware in the Loop

A fundamental aspect of simulator design is which subset of components or systems are simulated by software versus having real (or facsimile) hardware in the simulation loop. Similarly, and based on the simulation goals, is whether human operators and/or spacecraft occupants participate in the simulation.

14.2 Simulation versus Scripting

An important distinction needs to be made between simulating and scripting. By its nature, scripting has a finite set of interactions that represent well-known outcomes to a set of input stimuli. As such, its usefulness is limited to operator training to a defined scenario set.

In contrast, a simulator is an open-ended physical representation of a system limited primarily by the fidelity of the models and capabilities of the hosting hardware. The strength of a simulation is that the results of unknown outcomes to various conditions and interactions can be inferred. Simulation provides robust operator training to the extent that scenarios tend to be very open-ended and responsive to the provided inputs.

14.3 Event versus Time-Driven Simulators

There are fundamental architectural differences between event-driven and time-driven simulators. Event-driven simulators have functions scheduled when needed by events and called when needed. Time-driven simulators call processes on the scenario timeline as scheduled updates, ensuring all information is synchronized to the current scenario time.

First impressions might tempt one to opt for the time-driven architecture due to its apparently logical structuring and the fact that all information is always current. The disadvantages, however, are that some computational cycles may be wasted in updating information at a time that does not affect the scenario. Depending on the simulation complexity, this can add a computational burden that might prevent the simulation from being sped up or even prevent it from being able to even run in real time.

More importantly, not all processes produce a closed-loop, deterministic result. Processes that need to run at shorter time steps than the main simulation or must numerically solve for a value need to be capable of running asynchronously from the rest of the simulation. This would not be compatible with a time-driven architecture. Thus, it is advisable that simulators be designed as event-driven.

14.4 Truth Data versus Presented Data

A robust simulation needs to keep "truth" data separate from data presented to the components participating in the simulation. Truth data is the underlying reality in the simulator's universe which is the basis for various responses, but is knowledge kept at a private or protected level and only available to the simulation operator and other trusted agents.

Data presented to components participating in the simulation is based on the truth data but deviates from the truth by application of random noise. The noise application represents the outcome, modeling things such as the inaccuracy of measurement devices or the ability of processes to estimate the truth when given inputs with a suitable level of measurement noise.

14.5 Random Noise Generation and Usage

Proper use of pseudo-random noise is key to simulator realism. The types of random noise to be discussed are uniform noise, Gaussian noise, and Brownian noise. Noise applied is "pseudo-random" due to its underlying deterministic characteristic and thus the development of realistic pseudo-random number generators is a branch of mathematics related to probabilities and statistics.

Pseudo-random number generators are deterministic to the degree that they have a sequence in which the same number will always be produced from its predecessor in the sequence. However, a robust random number generator has not only the appearance of randomness by lacking patterns that makes the order of numbers generated appear independent, but also the statistical probabilities of numbers produced conform to a specified probabilistic distribution. There are mathematical tests to evaluate various random number generators that are beyond the scope of this work. Instead, the three types of pseudo-random number generators presented have been proven to be sufficiently robust for most applications and may be used when the programming environment does not offer suitable random number generators.

One strength of pseudo-random number generators is sometimes considered an inherent weakness. That is, if the random sequence is always the same, the simulation should always produce the same outcome. The strength of that feature is reproducibility. The same scenario may be run multiple times under a variety of starting conditions and/or operator actions taken. Having the same random sequence separates the effects of starting conditions and/or operator stimuli from randomness that might otherwise confuse cause and effect.

However, the scenario need not repeat using the same starting point in the pseudo-random number sequence. That is because random number generators have a parameter called a *seed* that determines what is the first number in the sequence to be produced. When no input is provided, the default seed is used, and the pseudo random sequence will always be the same. But when an alternative seed is provided, the sequence starts from a different location. Randomization of the seed itself can be either a controlled configured input, or could incorporate an arbitrary reference such as a read of the operating system's current seconds and milliseconds at the execution time or an early operator-initiated event.

14.5.1 Uniform Random Noise

Uniform random numbers have a limited range of outcomes, of which any outcome has an equal probability of occurring on any draw. Most programming languages include a uniform pseudo random number generator. The method below is presented to provide a uniform pseudo random number for systems that do not include such a capability.

The algorithm below uses the linear congruential method from an initial integer seed ($u_0 = seed$) value. It preliminarily generates a 32-bit integer (u_i) from the preceding 32-bit integer (u_{i-1}) in the sequence. The uniform pseudo random number is a floating-point normalization of the current 32-bit integer such that the generated floating point pseudo random number falls in the range: $0.0 \leq rnd_i \leq 1.0$. Any number returned has the same statistical likelihood as any other.

$$u_i = mod(\lfloor a + b \cdot u_{i-1} \rfloor, m)$$

14.15

$$rnd_i = \frac{u_i}{m}$$

14.16

In this algorithm, $m = (2^{31} - 1) = 2147483647$, $a = 7^5 = 16807$, and $b = (2^{22} - 3) = 4194301$. The $\lfloor \Box \rfloor$ brackets are the greatest integer function, sometimes called the floor function (needed only in systems that lack an integer arithmetic capability), while $mod()$ is the modulus function.

14.5.2 Gaussian Random Noise

Gaussian random numbers have an unlimited range of outcomes, with outcomes increasingly more statistically likely to occur on a Gaussian distribution. Many programming languages include a pseudo random number generator in which the probability of any value returned follows a normalized Gaussian distribution. The method presented is for systems that do not include such a capability.

The algorithm below is the Box-Muller method which returns pseudo random numbers centered on an input mean (μ) with a user-specified standard deviation (σ). The algorithm returns pseudo random numbers on a *normal distribution* for a zero mean ($\mu = 0$) and unity standard deviation ($\sigma = 1$).

The first step in the computation cycle computes two pseudo random numbers between -1 and $+1$ using a uniform random number generator. These two random numbers are treated as planar orthogonal vector components.

$$u_1 = 2 \cdot rand() - 1$$
$$u_2 = 2 \cdot rand() - 1$$

14.17

Equation 14.18 computes the square of the vector radius. The algorithm requires the vector to lie within a unit circle. Thus, if the radius squared is 1.0 or greater the two random numbers are reduced with the help of a new uniform random draw.

$$r^2 = u_1^2 + u_2^2$$

14.18

$$if \ r^2 \geq 1 \ then: \begin{cases} r = \sqrt{r^2 + rand()} \\ u_1 = \frac{u_1}{r} \\ u_2 = \frac{u_2}{r} \end{cases}$$

14.19

The Gaussian radius variable is computed using the natural logarithm function.

$$\rho = \sqrt{-2\frac{ln(r)}{r}}$$

14.20

Two Gaussian pseudo random numbers are generated from the uniform random numbers.

$$rnd_{G1} = (\sigma \cdot \rho)u_1 + \mu$$
$$rnd_{G2} = (\sigma \cdot \rho)u_2 + \mu$$

14.21

The Box-Muller method generates two pseudo-Gaussian random numbers per computational cycle. Since it is more desirable to have the function return a single random number, the recommended design is to maintain the second random number within the function (*static* variable for C/C++, *persistent* variable in MATLAB) to return at the next call.

Figure 14-3 is a histogram of the frequencies of more than 18 billion Gaussian random numbers generated by the Box-Muller algorithm with a 0.001σ bin width. The clean "bell curve" shape is indicative of the robustness of the Box Muller algorithm for Generating Gaussian pseudo random numbers.

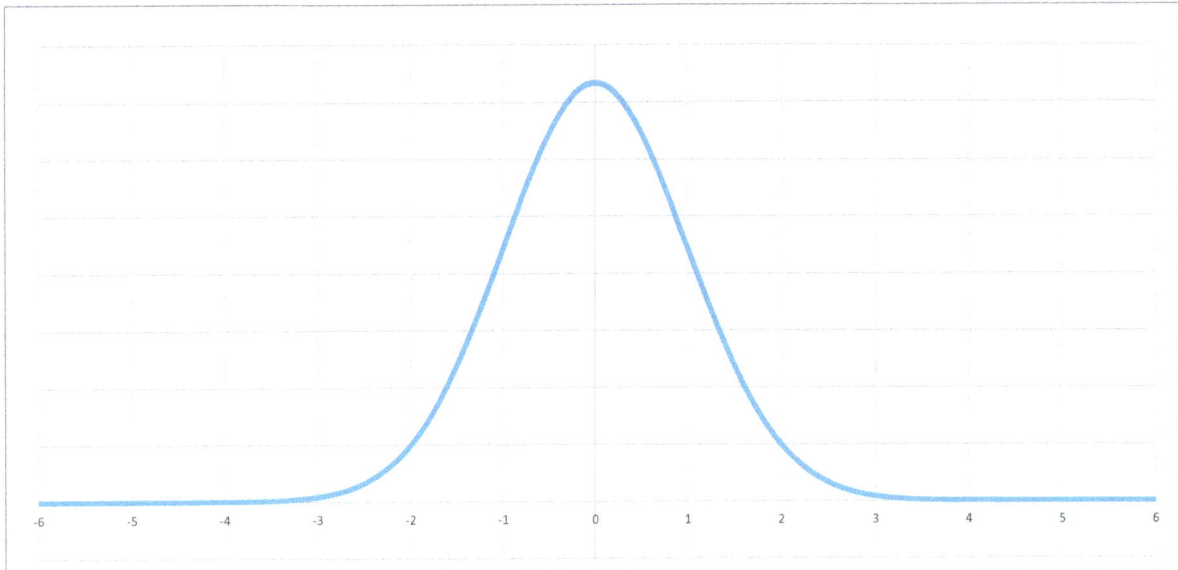

Figure 14-3 Gaussian Random Number Histogram

14.5.3 Brownian Noise

Brownian noise is characterized by random drift. It is useful for simulating a drifting bias in measurements or certain physical parameters. Such noise is typically present as a drift over time in instruments such as transducers, clock drift, and gyroscope senso axis orientations. Thus, it is useful in applicable instances for applying noise to truth parameters when generating simulated presentation values for telemetry or spacecraft on-board displays.

Few programming languages have Brownian noise generators. Fortunately, effective pseudo random Brownian noise may be readily generated using a Gaussian random number generator. This requires maintaining the previously-generated Brownian random number and adding it to a currently-generated Gaussian random number.

$$rnd_B = rnd_{B,i-1} + rnd_G \qquad\qquad 14.22$$

Figure 14-4 is an example of a series of 10000 Brownian pseudo random numbers.

Brownian random noise should be scaled to a level characteristic of the noise in the measuring instrument. In some instances, it may be necessary to bound the upper and lower limits. When bounds are employed, the sign on the Gaussian random number in equation 14.22 is reversed at any time the Brownian random number would violate the bounds.

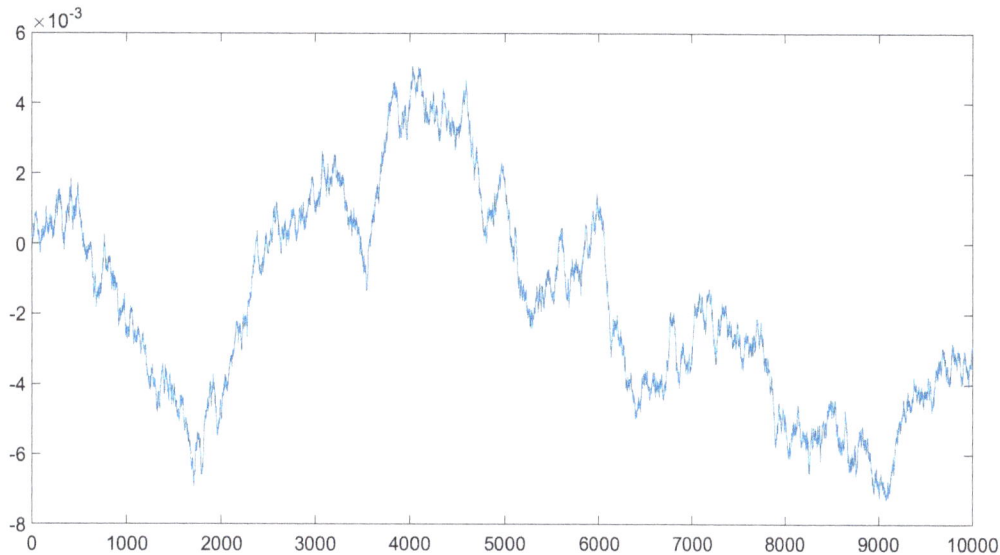

Figure 14-4 Brownian Random Noise

Key Terms:

Brownian Random Noise: is random noise characterized by progressive random drift.

Gaussian Random Noise: is random noise with an unlimited domain of outcomes, with the probability of an outcome highest near the mean of the domain's associated Gaussian distribution.

Uniform Random Noise: is random noise that has a limited domain of outcomes, with each possible value having an equal probability of occurrence.

REFERENCES

1. Zipfel, Peter H., *Modeling and Simulation of Aerospace Vehicle Dynamics*, AIAA Education Series, © 2000 American Institute of Aeronautics and Astronautics (AIAA), ISBN 1-56347-456-5.

2. Agrawal, Brij N., *Design of Geosynchronous Spacecraft*, © 1986 by author, Prentice-Hall, ISBN 0-13-200114-4.

3. Rogers, Robert M., *Applied Mathematics in Integrated Navigation Systems*, AIAA Education Series, © 2000 American Institute of Aeronautics and Astronautics (AIAA), ISBN 1-56347-397-6.

4. Nickey, Galen, Black, Jonathan, Ernandes, Kenneth J., and Johnson, W. Joel D, *Human-in-the-Loop Space System Simulation*, January 2020, AIAA SciTech Forum, Orlando, FL.

5. Henderson, David M., *Applied Cartesian Tensors for Aerospace Simulations*, © 2006 American Institute of Aeronautics and Astronautics (AIAA), ISBN 1-56347-793-9.

6. Griffin, Michael D. and French, James R., *Space Vehicle Design*, Second Edition, © 2004 American Institute of Aeronautics and Astronautics (AIAA), ISBN 1-56347-539-1.

7. Fortescue, Peter et al., *Spacecraft Systems Engineering*, Fourth Edition, © 2011 John Wiley & Sons, Ltd., ISBN 978-0-470-75012-4.

8. Kreyszig, Erwin, *Advanced Engineering Mathematics*, ©1962 John Wiley & Sons, Inc., ISBN 0-471-15496-2.

9. Chetty, PRK, *Satellite Technology and Its Applications*, © 1988 Tab Books, Inc., ISBN 0-8306-2931-9.

10. Press, William H. et al, *Numerical Recipes In C, The Art of Scientific Computing*, © 1988 Cambridge University Press, ISBN 0-521-34565-X.

11. Hillier, Frederick S. and Lieberman, Gerald J., *Introduction to Operations Research*, Fifth Edition, © 1990 McGraw-Hill, Inc.

Appendices

The appendices provide supplements to the text. Appendix A provides values for astronomical and physical constants for use in various computations.

Appendices B, C, and D are references for topics that are typically prerequisite mathematics for space flight dynamics (i.e., vector mathematics, matrix mathematics, and coordinate transformations). These topics were separated into appendices to avoid interrupting the flow of the main chapter presentations for those already familiar, and to organize them in easily accessible handbook format. These are obviously far from being comprehensive in their respective subject areas. Instead, a sufficient level is provided to accommodate a foundational presentation of the space flight mechanics subject matter. The interested reader may refer to the cited references for more comprehensive treatment of the individual topics.

Appendix E provides introductory information for numerical integration, specifically of a second order ordinary differential equation. While this is applicable to a variety of practical applications, its primary scope is for predicting trajectories and spacecraft attitudinal motion. In that regard, it focuses on Runge-Kutta methods. The interested reader may refer to the cited references for information regarding other numerical integration methods.

A.1 Astronomical and Physical Constants

A.1.1 Solar System Body Properties

Table A-1 Celestial Body Physical Properties

Body	Mass (kg)	$\mu\ (km^3/s^2)$	Equatorial Radius (km)	Polar Radius (km)	Flattening	Geometric Albedo
Sun	1.9884×10^{30}	132712000000	695712	695677	0.00005	–
Mercury	3.301×10^{23}	22032	2440.5	2483.3	0.0009	0.142
Venus	4.8673×10^{24}	324860	6051.8	6051.8	0	0.689
Earth	5.9722×10^{24}	398600.4415	6378.1	6356.8	0.00335	0.434
Luna	7.346×10^{22}	4900	1738.1	1736	0.0012	0.12
Mars	6.4169×10^{23}	42828	3396.2	3376.2	0.00589	0.17
Jupiter	1.89813×10^{27}	126687000	71492	66854	0.06487	0.538
Saturn	5.8632×10^{26}	37931000	60268	54364	0.09796	0.499
Uranus	8.6811×10^{25}	5794000	25559	24973	0.02293	0.488
Neptune	1.02409×10^{26}	6835100	24764	24341	0.01708	0.442
Pluto	1.303×10^{22}	870	1188	1188	0	0.52

Table A-2 Solar System Orbital Parameters

Body	Semi-Major Axis (km)	Period (days)	Inclination (deg)	Eccentricity
Mercury	5.7909×10^7	87.869	7.004	0.2056
Venus	1.0821×10^8	224.701	3.395	0.0068
Earth	1.49598×10^8	365.256	0	0.0167
Luna	3.844×10^5	27.3217	20.43 ± 2.15	0.0549
Mars	2.27956×10^8	686.98	1.848	0.0935
Jupiter	7.78479×10^8	4332.589	1.304	0.0487
Saturn	1.432041×10^9	10759.22	2.486	0.052
Uranus	2.867043×10^9	30685.4	0.77	0.0469
Neptune	4.514953×10^9	60189	1.77	0.0097
Pluto	5.869656×10^9	90560	17.16	0.2444

Table A-3 Celestial Body Rotation

Body	Obliquity (deg)	Sidereal Rotation Period (hours)	Length of Day (hours)
Sun	7.25	609.12	–
Mercury	0.034	1407.6	4222.6
Venus	2.64	−5832.6	2802
Earth	23.44	23.9345	24
Luna	6.68	665.72	665.72
Mars	25.19	24.6229	24.6597
Jupiter	3.13	9.925	9.9259

Table A-3 Celestial Body Rotation			
Body	**Obliquity (deg)**	**Sidereal Rotation Period (hours)**	**Length of Day (hours)**
Saturn	26.73	10.656	10.656
Uranus	97.77	−17.24	17.24
Neptune	28.32	16.11	16.11
Pluto	119.51	−153.2928	153.828

A.1.2 Physical Constants

Universal Gravitational constant
$G = 6.6743 \times 10^{-11} \, m^3/(kg \cdot s^2)$

Universal gas constant
$R = 8.31446261815324 \, J \cdot K^{-1} mol^{-1}$

Planck's Constant
$h = 6.626176 \times 10^{-34} \, J/Hz$

Boltzmann constant
$k = 1.38064852 \times 10^{-23} \, J/K$

Stefan-Boltzmann constant
$\sigma = 5.670374419 \times 10^{-8} \, W/(m^2 K^4)$

Speed of light
$c = 299792458 \, m/s$

B.1   Vector Mathematics

Mastery of vector mathematics provides a powerful set of tools and problem-solving methods. But vectors are often intimidating to the uninitiated. An effective way to reduce the intimidation factor is by objectifying the vector – i.e., initially visualizing the vector as a "thing," without undue focus on its constituents.

B.2   Vector Properties

A *vector* is a directed quantity, having both a value (i.e., its *magnitude* or *norm*) and a *direction*. In contrast, a *scalar* is a simple quantity describable by a number, but having no specific direction.

The convention typically used to symbolize an entity as a vector is a flat bar (or arrow) accent over the symbol (such as an alphabetic character) identifying the vector. For example, a vector using the symbol "a" would be denoted as $\bar{a}$, with the symbol a (without the overlying accent bar) indicating its corresponding value or magnitude.

Vectors are usually free-floating, meaning they are positioned conveniently in a diagram. One notable exception is position vectors since they indicate a particular location relative to a reference (or origin). In such cases, the vector's beginning (or tail) is tied to that reference origin.

B.3   The Vector Magnitude

A vector's magnitude (or norm) is a scalar that quantifies the vector in its units of measure. For example, the magnitude of a position vector would be a length or distance with kilometers (km) being an example unit of measure. Similarly, the magnitude of a velocity vector is a speed with kilometers per second (km/s) being an example unit of measure.

A vector's magnitude may be symbolized by surrounding the vector by absolute value bars $|\bar{a}|$, or by the vector's symbol without the overlying accent bar, as previously indicated. For example, the magnitude of vector $\bar{a}$ could be denoted either as $|\bar{a}|$ or simply by a. The former representation with the absolute value bars is typically used to indicate the process of computing the vector's magnitude, while the latter symbol it normally used to represent the magnitude's value, without an indication of when or how it was computed.

B.4 Unit Vectors

Unit vectors are a special form since they indicate only a direction, with no quantity or units of measure. As such, the unit vector's magnitude always equals 1.0, and there is no associated unit of measure. Thus, unit vectors are dimensionless.

Due to its dimensionless status, unit vectors have an overlying caret ($\hat{\Box}$) accent instead of the standard overlying flat bar accent. For example, the direction of vector $\bar{a}$ would be symbolized by the $\hat{a}$ unit vector.

The unit vector provides a convenient means to separate the vector's magnitude and direction in mathematical expressions. For example, vector $\bar{a}$ can be separated into its magnitude and direction constituents in the following equations:

$$\bar{a} = a\hat{a}$$

$$\hat{a} = \frac{\bar{a}}{a}$$

While these may appear trivial, this separation is useful for manipulating algebraic expressions involving vectors.

B.5 Cartesian Vectors

The associated mathematics is conveniently performed when vectors are expressed in *right-handed* Cartesian coordinate systems. A Cartesian coordinate system consists of three mutually perpendicular axes, typically labeled x, y, and z that intersect at an origin. These axes are also accompanied by unit vectors, that are typically labeled $\hat{\imath}$, $\hat{\jmath}$, and $\hat{k}$ and aligned with the x, y, and z directions.

A mutually perpendicular coordinate system is also called an *orthogonal* system. An orthogonal coordinate system with unit vectors ($\hat{\imath}$, $\hat{\jmath}$, and $\hat{k}$) defining the axis directions is called an *orthonormal* system.

The equations have a right-handed flow of rotation, from $x \to y \to z \to x$. Right-handed means, for example, if the straightened fingers of the right hand were directed toward x and subsequently closed toward y, your thumb direction dictates the relative orientation of z. The same sequence of y and z would result in the thumb orienting x, and the sequence of x and z would orient y.

In a Cartesian system, vector $\bar{a}$ is represented by components (a_x, a_y, a_z) projected along each of the coordinate axes. Thus, $\bar{a}$ is the summation of the individual components, multiplied by their respective unit vectors as follows:

$$\bar{a} = a_x\hat{\imath} + a_y\hat{\jmath} + a_z\hat{k}$$

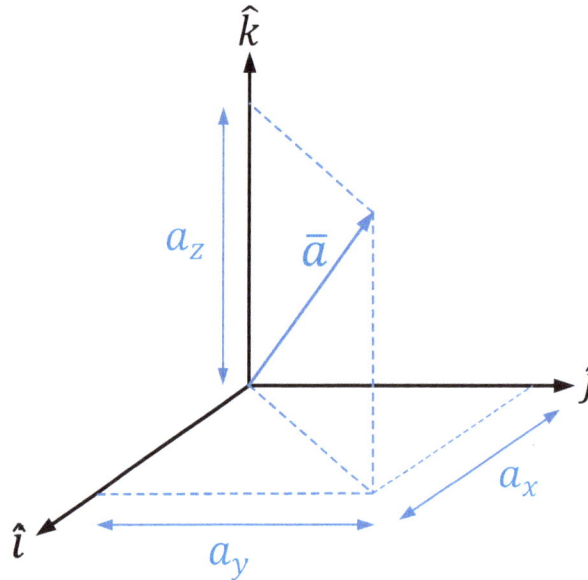

Since the three axes are mutually perpendicular, vector $\bar{a}$ is a three-dimensional hypotenuse to the three Cartesian components. With the Cartesian representation, the vector magnitude is computed using the Pythagorean Theorem, albeit in three dimensions:

$$a = \sqrt{a_x^2 + a_y^2 + a_z^2}$$

B.6   Vector Addition and Subtraction

Vectors can be viewed as free floating. When adding two vectors, the tail of one vector may be placed at the head of the vector to which it is being added in a diagrammatic view.

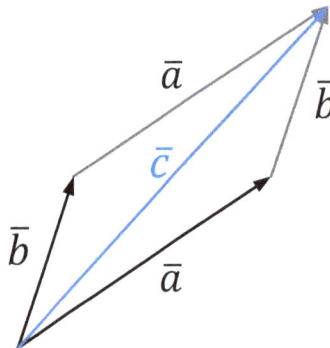

Vector addition and subtraction are mathematically straight forward processes in the Cartesian representation. The respective components along each coordinate axis are added or subtracted as applicable. Thus, given vectors $\bar{a}$ and $\bar{b}$ the sum and difference are computed by:

$$\bar{a} + \bar{b} = (a_x + b_x)\hat{\imath} + (a_y + b_y)\hat{\jmath} + (a_z + b_z)\hat{k}$$

$$\bar{a} - \bar{b} = (a_x - b_x)\hat{\imath} + (a_y - b_y)\hat{\jmath} + (a_z - b_z)\hat{k}$$

Given the above equations, the vector sum is commutative, meaning: $\bar{a} + \bar{b} = \bar{b} + \bar{a}$. Conversely, vector subtraction is anti-commutative, meaning: $\bar{a} - \bar{b} = -(\bar{b} - \bar{a})$.

B.7 Direction Cosines

Direction cosines relate the angles $(\theta_x, \theta_y, \theta_z)$ between vector $\bar{a}$ and the corresponding coordinate axes. For example, consider the relationship between $\bar{a}$ and its projection on the x-axis (a_x):

$$\cos \theta_x = \frac{a_x}{a}$$

Using this convention, unit $\hat{a}$ may also be defined in terms of the direction cosines:

$$\hat{a} = \cos \theta_x \, \hat{\imath} + \cos \theta_y \, \hat{\jmath} + \cos \theta_z \, \hat{k}$$

B.8 The Dot Product

The dot product (or inner product) is the multiplication of two vectors that produces a scalar result. The result of the dot product is the product of the magnitudes of the two vectors, multiplied by the cosine of the angle between them. Thus, the dot product of vectors $\bar{a}$ and $\bar{b}$ with angle θ between them is:

$$\bar{a} \cdot \bar{b} = ab \cos \theta$$

Notice the dot (·) multiplication operator used in equations to denote the dot product. By extension, the dot product of vector $\bar{a}$ with a unit vector $\hat{u}$ with angle θ between them is:

$$\bar{a} \cdot \hat{u} = a \cos \theta$$

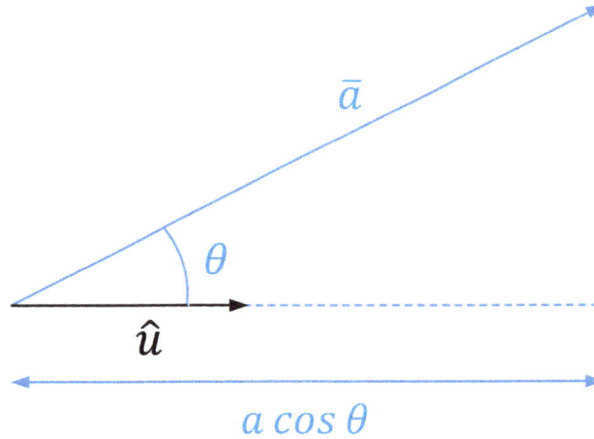

$$a \cos \theta$$

This result is the projection (or component) of vector $\bar{a}$ along the $\hat{u}$ direction. This could alternatively be represented as $\bar{a} \cdot \hat{u} = a_u$. Comparing this result with the direction cosine definition indicates the direction cosines could be defined using the dot product as follows:

$$\cos \theta_x = \hat{a} \cdot \hat{\imath}$$

$$\cos \theta_y = \hat{a} \cdot \hat{\jmath}$$

$$\cos \theta_z = \hat{a} \cdot \hat{k}$$

From the above results, the conclusion is that the dot product of two unit vectors is a direction cosine, or more simply the cosine of the angle between the two unit vectors. This leads to the following set of rules:

- The dot product of two parallel unit vectors equals 1.0.
- The dot product of two perpendicular unit vectors equals 0.0.
- The dot product of two opposing unit vectors equals -1.0.

These rules can be further applied to the coordinate axes unit vectors:

$$\hat{\imath} \cdot \hat{\imath} = \hat{\jmath} \cdot \hat{\jmath} = \hat{k} \cdot \hat{k} = 1$$

$$\hat{\imath} \cdot \hat{\jmath} = \hat{\imath} \cdot \hat{k} = \hat{\jmath} \cdot \hat{k} = 0$$

$$(-\hat{\imath}) \cdot \hat{\imath} = (-\hat{\jmath}) \cdot \hat{\jmath} = (-\hat{k}) \cdot \hat{k} = -1$$

The dot product in the Cartesian system is the sum of the products of the respective components along each axis. Thus, given vectors $\bar{a}$ and $\bar{b}$ the dot product is computed by:

$$\bar{a} \cdot \bar{b} = a_x b_x \hat{\imath} + a_y b_y \hat{\jmath} + a_z b_z \hat{k}$$

The vector dot product is commutative, meaning: $\bar{a} \cdot \bar{b} = \bar{b} \cdot \bar{a}$.

B.9   The Cross Product

The cross product (or vector product) is the multiplication of two vectors that produces a vector result. The result of the cross product is a vector that is perpendicular to both input vectors. The direction of the vector resulting from the cross product uses the same "right hand rule" that defines the orientations of the coordinate system axes relative to each other.

The cross-product magnitude is the product of the magnitudes of the two vectors, multiplied by the sine of the angle α between them. The cross product is denoted by use of the cross ($\times$) multiplication operator in equations.

$$\left| \bar{a} \times \bar{b} \right| = ab \sin \theta$$

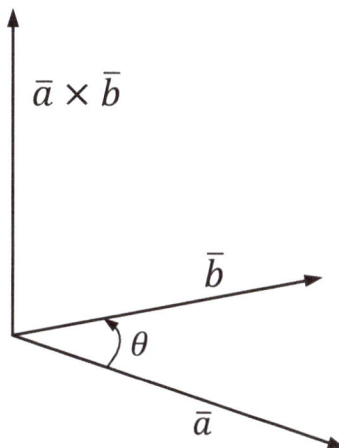

The rules regarding the cross product are as follows:
- The cross product of two colinear vectors equals zero. (The term "colinear" can mean either in the same direction or opposite directions.) This rule reflects the fact that there is *no unique vector* that is mutually perpendicular to two colinear vectors.
- The cross product of two perpendicular unit vectors is a unit vector perpendicular to both input vectors.

These rules can be further applied to the coordinate axes unit vectors:

$$\hat{\imath} \times \hat{\imath} = 0 \quad \hat{\jmath} \times \hat{\jmath} = 0 \quad \hat{k} \times \hat{k} = 0$$

$$\hat{\imath} \times \hat{\jmath} = \hat{k} \quad \hat{\jmath} \times \hat{k} = \hat{\imath} \quad \hat{k} \times \hat{\imath} = \hat{\jmath}$$

The cross product in the Cartesian system is a particular set of products and differences of perpendicular axis components. Thus, given vectors $\bar{a}$ and $\bar{b}$ the cross product is computed by:

$$\bar{a} \times \bar{b} = \left(a_y b_z - a_z b_y\right)\hat{\imath} + \left(a_z b_x - a_x b_z\right)\hat{\jmath} + \left(a_x b_y - a_y b_x\right)\hat{k}$$

The cross product may also be evaluated using a matrix determinant (see appendix C for more information on matrices):

$$\bar{a} \times \bar{b} = \begin{vmatrix} \hat{\imath} & \hat{\jmath} & \hat{k} \\ a_x & a_y & a_z \\ b_x & b_y & b_z \end{vmatrix} = \begin{bmatrix} a_y b_z - a_z b_y \\ a_z b_x - a_x b_z \\ a_x b_y - a_y b_x \end{bmatrix}$$

The vector cross product is anti-commutative, meaning: $\bar{a} \times \bar{b} = -\bar{b} \times \bar{a}$.

B.10 Cross Product Distributive Property

The cross product distributive property is provided below.

$$\bar{a} \times \left(\bar{b} + \bar{c}\right) = \bar{a} \times \bar{b} + \bar{a} \times \bar{c}$$

B.11 Vector Triple Cross Product Identity

The vector triple cross product identity is:

$$\bar{a} \times \left(\bar{b} \times \bar{c}\right) = (\bar{a} \cdot \bar{c})\bar{b} + \left(\bar{a} \cdot \bar{b}\right)\bar{c}$$

B.12 Dot and Cross Product Interchange Identity

The dot and cross product interchange identity is provided below.

$$\bar{a} \cdot \left(\bar{b} \times \bar{c}\right) = \left(\bar{a} \times \bar{b}\right) \cdot \bar{c}$$

B.13 The Outer Product

The vector outer product is the multiplication of two vectors that produces a two-dimensional array (i.e., a matrix) result. The mathematical representation is the multiplication of a column vector ($\bar{a}$) by a row vector ($\bar{b}^T$). The meaning of the "T" superscript as a "transpose" is described more generally in the matrix mathematics appendix.

$$\bar{a} \otimes \bar{b} = \bar{a}\bar{b}^T = \begin{bmatrix} a_x b_x & a_x b_y & a_x b_z \\ a_y b_x & a_y b_y & a_y b_z \\ a_z b_x & a_z b_y & a_z b_z \end{bmatrix}$$

B.14 An Important Vector Consideration

It should be emphasized that vector mathematics is only valid between vectors represented in the same coordinate system. Thus, cognizance of the coordinate system that the vector is represented must be maintained.

References

1. Harper, Charlie, *Introduction to Mathematical Physics*, © 1976 Prentice-Hall, Inc., ISBN 0-13-487538-9.
2. Beer, Ferdinand P. and Johnston Jr., E. Russell, *Vector Mechanics for Engineers, Statics and Dynamics*, Fifth Edition, © 1988 McGraw-Hill, Inc., ISBN 0-07-079923-7.
3. Kreyszig, Erwin, *Advanced Engineering Mathematics*, ©1962 John Wiley & Sons, Inc., ISBN 0-471-15496-2.
4. Hibbeler, R.C., *Engineering Mechanics, Statics*, Thirteenth Edition, © 2013 by author, Pearson Prentice Hall, ISBN 978-0-13-291554-0.
5. Ying, Shuh-Jing, *Advanced Dynamics*, © 1997 American Institute of Aeronautics and Astronautics, ISBN 1-56347-224-4.

C.1 Matrix Mathematics and Linear Algebra

Matrices are rectangular arrays, typically consisting of numeric values. Two-dimensional matrices with n rows and m columns – i.e., $(n \times m)$ – are discussed herein. The discussion includes basic arithmetic and other operations, limited in scope to that which is most useful for space flight dynamics.

C.1.1 Matrix Elements

Standard matrix notation has the elements indicated by the matrix name, with a row and column subscript as shown for the $n \times m$ matrix below. If the number of rows and columns exceeds single digits, there is a comma (,) delimiter to distinguish row from column.

$$A = \begin{bmatrix} a_{11} & a_{12} & \cdots & a_{1m} \\ a_{21} & a_{22} & \cdots & a_{2m} \\ \vdots & \vdots & \ddots & \vdots \\ a_{n1} & a_{n2} & \cdots & a_{nm} \end{bmatrix}$$

C.2 Special Matrix Types

While there are numerous special matrix types, this section focuses on those pertinent to the space flight dynamics topics herein.

C.2.1 Square Matrix

A square matrix is one with an equal number of rows and columns (i.e., $n = m$).

C.2.2 Row and Column Vectors

Matrices with either the row or column count equaling one are called vectors. A matrix with a single row is a row vector; a matrix with a single column is a column vector. Thus, Cartesian vectors may be represented as row or column matrices. This provides some advantages in being able to interact mathematically with matrices to perform certain operations such as coordinate transformations.

C.2.3 Zero or Null Matrix

A matrix with all zero values is called a null or zero matrix.

C.2.4 Identity Matrix

An identity matrix is a square matrix with values equal to one on the main diagonal and values equal to zero off the main diagonal. The identity matrix typically uses I as its symbol. Below is an example of a 3×3 identity matrix.

$$I = \begin{bmatrix} 1 & 0 & 0 \\ 0 & 1 & 0 \\ 0 & 0 & 1 \end{bmatrix}$$

The identity matrix is a multiplicative identity for matrices, just as "1" is the multiplicative identity for scalars. Thus, multiplying a matrix A by a properly dimensioned identity matrix returns the matrix A.

C.2.5 Orthogonal Matrix

An orthogonal matrix is a special square matrix that has row and column independence. If the product of the matrix with its transpose (to be defined below) is an identity matrix (i.e., $AA^T = I$), then the matrix is orthogonal.

C.3 Basic Matrix Operations

This section provides basic mathematical and manipulative operations applicable to matrices.

C.3.1 Matrix Transpose

The matrix transpose operation, typically denoted by a "T" superscript, interchanges rows and columns as shown below for $n \times m$ matrix A.

$$A^T = \begin{bmatrix} a_{11} & a_{21} & \cdots & a_{n1} \\ a_{12} & a_{22} & \cdots & a_{n2} \\ \vdots & \vdots & \ddots & \vdots \\ a_{1m} & a_{2m} & \cdots & a_{nm} \end{bmatrix}$$

C.3.2 Trace of a Square Matrix

The trace of a square matrix is the sum of its main diagonal elements. In shorthand summation notation, the matrix trace is computed as below.

$$Tr(A) = \sum_{i=1}^{n} a_{ii}$$

In the above summation, the parameter n represents the number of rows and columns in the square matrix A.

C.3.3 Matrix Addition and Subtraction

To add or subtract two matrices, the two matrices must have equal numbers of rows and columns. Thus, two matrices with equal dimensions are said to be *conformable* for addition (or subtraction).

The sum of two matrices is a matrix of equal dimension to the two being added. The matrix sum is thus a matrix in which each element is the sum of the corresponding elements as shown below.

$$C = A + B$$

$$C = \begin{bmatrix} a_{11} + b_{11} & a_{12} + b_{12} & \cdots & a_{1m} + b_{1m} \\ a_{21} + b_{21} & a_{22} + b_{22} & \cdots & a_{2m} + b_{2m} \\ \vdots & \vdots & \ddots & \vdots \\ a_{n1} + b_{n1} & a_{n2} + b_{n2} & \cdots & a_{nm} + b_{nm} \end{bmatrix}$$

In shorthand notation, the i,j element of the matrix sum can be characterized as below.

$$c_{ij} = a_{ij} + b_{ij}$$

Matrix subtraction is analogous to addition, except that the corresponding elements are subtracted as shown below.

$$C = A - B$$

$$C = \begin{bmatrix} a_{11} - b_{11} & a_{12} - b_{12} & \cdots & a_{1m} - b_{1m} \\ a_{21} - b_{21} & a_{22} - b_{22} & \cdots & a_{2m} - b_{2m} \\ \vdots & \vdots & \ddots & \vdots \\ a_{n1} - b_{n1} & a_{n2} - b_{n2} & \cdots & a_{nm} - b_{nm} \end{bmatrix}$$

In shorthand subscript notation, the i,j element of the matrix difference can be characterized as below.

$$c_{ij} = a_{ij} - b_{ij}$$

Matrix addition is commutative. Matrix subtraction is anti-commutative.

C.3.4 Matrix Multiplication

To multiply two matrices, the number of columns in the first matrix must equal the number of rows in the second matrix. Given this case, the two matrices are considered *conformable* for multiplication.

The product of two matrices is a matrix of with the number of rows in the first matrix and the number of columns in the second matrix. The product of two matrices is a matrix in which each i, j element is the sum of the products of each element in row i with its corresponding element in column j. As such, every element of the matrix product is the n-dimensional vector dot product of row i of the first matrix with column j of the second matrix.

In shorthand summation notation, the i, j element of the product of matrix A with matrix B can be characterized as below.

$$c_{ij} = \sum_{k=1}^{n} a_{ik} b_{kj}$$

In the above summation, the parameter n represents the number of columns in matrix A, which also equals the number of rows in matrix B to meet the conformability constraint.

Matrix multiplication is not commutative. However, matrix multiplication is associative.

C.3.5 Matrix Division / Inversion

As with standard algebra, matrix division is equivalent to the multiplicative inverse. Thus, the matrix inverse is analogous to the reciprocal of a scalar value. Thus, to remain consistent with scalar algebra, the matrix multiplicative inverse is denoted with a negative one exponent (i.e., A^{-1}).

The key property of a matrix inverse is that any matrix multiplied by its inverse yields an identity matrix as shown below. This is completely analogous to a scalar being multiplied by its reciprocal yielding the number one (multiplicative identity) as the result.

$$AA^{-1} = I$$

Not all matrices have an inverse. Matrices with no inverse are said to be *singular*. This is analogous to the number zero not having a multiplicative inverse.

Determining a matrix inverse is far more complicated since the process must invert the multiplication process. There are a variety of general techniques for inverting matrices, such as lower upper (LU) decomposition and singular value decomposition (SVD), which are beyond the scope of this text. The interested reader is referred to a text dedicated to linear algebra.

The special case of the orthogonal matrix will, however, be covered. Recall that multiplying an orthogonal matrix by its transpose yields an identity matrix ($AA^T = I$). This is the same result that is achieved when multiplying any matrix by its inverse ($AA^{-1} = I$). Therefore, by equivalence a relationship for an orthogonal matrix is:

$$AA^T = AA^{-1}$$

If both sides are multiplied by the inverse of A (and recognizing I as the multiplicative identity):

$$A^{-1}AA^T = A^{-1}AA^{-1}$$

$$IA^T = IA^{-1}$$

$$A^T = A^{-1}$$

Therefore, the *inverse of an orthogonal matrix is its transpose*.

C.3.6 Solving Systems of Linear Equations

The utility of the matrix inverse can be illustrated by solving a simultaneous set of n linear equations in n unknowns. Consider a set of four equations in four unknowns illustrated below in which the desire is to solve for the $a, b, c,$ and d coefficients. For this example, the u_i, v_i, w_i, x_i and y_i values are all known.

$$au_1 + bv_1 + cw_1 + dx_1 = y_1$$
$$au_2 + bv_2 + cw_2 + dx_2 = y_2$$
$$au_3 + bv_3 + cw_3 + dx_3 = y_3$$
$$au_4 + bv_4 + cw_4 + dx_4 = y_4$$

Using the principles of matrix multiplication, the above can be set into a factored matrix/vector equation as follows:

$$\begin{bmatrix} u_1 & v_1 & w_1 & x_1 \\ u_2 & v_3 & w_2 & x_2 \\ u_3 & v_3 & w_3 & x_3 \\ u_4 & v_4 & w_4 & x_4 \end{bmatrix} \begin{bmatrix} a \\ b \\ c \\ d \end{bmatrix} = \begin{bmatrix} y_1 \\ y_2 \\ y_3 \\ y_4 \end{bmatrix}$$

The square matrix will be defined as A, the column vector ($\bar{z}$) containing the a, b, c, d coefficients to be solved will be defined as $\bar{z}$, and the vector on the right side will be defined as $\bar{y}$. Thus, the equation can be re-written in shorthand linear algebra as:

$$A\bar{z} = \bar{y}$$

From algebra, the equation remains in balance by multiplying both sides by the same value. In this case, the multiplication with be by the multiplicative inverse of A (i.e., A^{-1}). In doing so, the equation becomes:

$$A^{-1}A\bar{z} = A^{-1}\bar{y}$$

The product of matrix A with its inverse is the identity matrix (I), which is also the matrix multiplicative identity. Hence:

$$I\bar{z} = A^{-1}\bar{y}$$

$$\bar{z} = A^{-1}\bar{y}$$

Thus, matrix inversion and multiplication provide the solution for the original set of four equations in four unknowns. This may be extended to an arbitrary number of n equations in n unknowns.

C.4 Matrix Determinant

A determinant is a special quantity associated with a square matrix denoted by an absolute value $|\ldots|$ symbol. The determinant's order (n) is that of the square matrix. The determinant is computed by the equation below:

$$|A| = \sum_{i=1}^{n} a_{ij}c_{ij}$$

In the above expression, c_{ij} is the i, j element of the cofactor matrix of A, which is constructed from what are referred to as minors (m_{ij}) of matrix A.

$$c_{ij} = (-1)^{i+j}m_{ij}$$

The minors are the determinants of the submatrices of C, indicating a nested recursion. The submatrix of the i, j element of A is the matrix that results from eliminating A's ith row and jth column. The expansion by minors can continue until a 1x1 submatrix is left, with the determinant of a 1x1 matrix being that single 1x1 value.

Since this process has complicated nesting, the example below demonstrates the process by producing the determinant of a 3x3 matrix by first computing its cofactor matrix.

$$A = \begin{bmatrix} a_{11} & a_{12} & a_{13} \\ a_{21} & a_{22} & a_{23} \\ a_{31} & a_{32} & a_{33} \end{bmatrix}$$

$$C = \begin{bmatrix} +\begin{vmatrix} a_{22} & a_{23} \\ a_{32} & a_{33} \end{vmatrix} & -\begin{vmatrix} a_{21} & a_{23} \\ a_{31} & a_{33} \end{vmatrix} & +\begin{vmatrix} a_{21} & a_{22} \\ a_{31} & a_{32} \end{vmatrix} \\ -\begin{vmatrix} a_{12} & a_{13} \\ a_{32} & a_{33} \end{vmatrix} & +\begin{vmatrix} a_{11} & a_{13} \\ a_{31} & a_{33} \end{vmatrix} & -\begin{vmatrix} a_{11} & a_{12} \\ a_{31} & a_{32} \end{vmatrix} \\ +\begin{vmatrix} a_{12} & a_{13} \\ a_{22} & a_{23} \end{vmatrix} & -\begin{vmatrix} a_{11} & a_{13} \\ a_{21} & a_{23} \end{vmatrix} & +\begin{vmatrix} a_{11} & a_{12} \\ a_{21} & a_{22} \end{vmatrix} \end{bmatrix}$$

$$C = \begin{bmatrix} +(a_{22}a_{33} - a_{23}a_{32}) & -(a_{21}a_{33} - a_{23}a_{31}) & +(a_{21}a_{32} - a_{22}a_{31}) \\ -(a_{12}a_{33} - a_{13}a_{32}) & +(a_{11}a_{33} - a_{13}a_{31}) & -(a_{11}a_{32} - a_{12}a_{31}) \\ +(a_{12}a_{23} - a_{13}a_{22}) & -(a_{11}a_{23} - a_{13}a_{21}) & +(a_{11}a_{22} - a_{12}a_{21}) \end{bmatrix}$$

The determinant can then be produced using the first row of A and its cofactor matrix C.

$$|A| = a_{11}(a_{22}a_{33} - a_{23}a_{32}) + a_{12}(a_{23}a_{31} - a_{21}a_{33}) + a_{13}(a_{21}a_{32} - a_{22}a_{31})$$

While it was not necessary to compute any cofactors other than the first row for the determinant, the cofactor matrix is also useful for computing the matrix inverse. The matrix inverse may be computed by:

$$A^{-1} = \frac{C^T}{|A|}$$

This analytic matrix inversion method would be difficult to implement for an arbitrary matrix size as a computer algorithm. Hence numerical matrix inversion methods are suggested that are easier to implement. Nevertheless, this provides the insight that a matrix with a zero determinant is singular (i.e., not invertible).

C.5 Matrix Pseudoinverse

The Moore-Penrose pseudoinverse is the inverse of a rectangular $(n \times m)$ matrix and is also called the general inverse. An important application for the pseudoinverse in space flight dynamics is with over-determined least-squares problems, particularly in trajectory and attitude determination.

Important properties of the pseudoinverse for real matrices are as follows for an A matrix (with matrix A^+ being its pseudoinverse):

$$AA^+A = A$$
$$A^+AA^+ = A^+$$

The pseudoinverse (A^+) is computed by:

$$A^+ = (A^T A)^{-1} A^T$$

The $(A^T A)$ term produces an $(m \times m)$ square matrix that must be invertible for the pseudoinverse to exist. When the pseudoinverse exists, the least squares problem may be posed as the solution to a system of linear equations as follows:

$$\bar{x} = A^+ \bar{y}$$

In the above expression, $\bar{x}$ is the desired solution state vector and $\bar{y}$ is the measurements uncertainties (or residuals) vector.

C.5 Skew Symmetric Matrix Operator

The skew symmetric matrix operator is a tool that can be employed for computing vector cross products. Given column vectors $\bar{a}$ and $\bar{b}$, the cross product can be equivalently represented by matrix multiplication.

$$\bar{a} = \begin{bmatrix} a_x \\ a_y \\ a_z \end{bmatrix} \quad \bar{b} = \begin{bmatrix} b_x \\ b_y \\ b_z \end{bmatrix}$$

$$\bar{a} \times \bar{b} = \begin{bmatrix} 0 & -a_z & a_y \\ a_z & 0 & -a_x \\ -a_y & a_x & 0 \end{bmatrix} \begin{bmatrix} b_x \\ b_y \\ b_z \end{bmatrix} = \begin{bmatrix} a_y b_z - a_z b_y \\ a_z b_x - a_x b_z \\ a_x b_y - a_y b_x \end{bmatrix}$$

Vector $\bar{a}$ is represented in the 3×3 square matrix. The matrix's arrangement and usage to multiply by column vector $\bar{b}$ results in the cross product. Given that $\bar{b}$ stands separately in the process, the skew symmetric operator, denoted as $[a \times]$ is defined by separating $\bar{b}$ from the expression.

$$[a \times] = \begin{bmatrix} 0 & -a_z & a_y \\ a_z & 0 & -a_x \\ -a_y & a_x & 0 \end{bmatrix}$$

Thus, the $[a \times]$ skew symmetric operator arranges the vector in matrix form, with the meaning of the cross product of $\bar{a}$ (with whatever vector follows).

C.6 Vector Outer Product as a Matrix Multiplication

The vector outer product is produced by multiplying the vectors represented as a column and a row matrix, respectively.

$$\bar{a} = \begin{bmatrix} a_x \\ a_y \\ a_z \end{bmatrix} \quad \bar{b} = \begin{bmatrix} b_x & b_y & b_z \end{bmatrix}$$

$$\bar{a} \otimes \bar{b} = \begin{bmatrix} a_x b_x & a_x b_y & a_x b_z \\ a_y b_x & a_y b_y & a_y b_z \\ a_z b_x & a_z b_y & a_z b_z \end{bmatrix}$$

References

1. Harper, Charlie, *Introduction to Mathematical Physics*, © 1976 Prentice-Hall, Inc., ISBN 0-13-487538-9.
2. Strang, Gilbert, *Introduction to Linear Algebra*, Third Edition, © 2003 by author, Wellesley-Cambridge Press, ISBN 0-9614088-9-8.
3. Golub, Gene H. and Van Loan, Charles F., *Matrix Computations*, 4th Edition, © 1983 The Johns Hopkins University Press, ISBN 978-1-4214-0794-4.
4. Demmel, James W., *Applied Numerical Linear Algebra*, © 1997 Society for Industrial and Applied Mathematics, ISBN 978-0-898713-89-3.
5. Press, William H., et al., *Numerical Recipes in C, The Art of Scientific Computing*, © 1988 Cambridge University Press, ISBN 0-521-35465-X.
6. Bronson, Richard, *Schaum's Outline of Theory and Problems of Matrix Operations*, © 1989 McGraw-Hill Companies, Inc., ISBN 0-07-007978-1.
7. Kreyszig, Erwin, *Advanced Engineering Mathematics*, 8th Edition, © 1999 John Wiley & Sons, Inc., ISBN 0-471-15496-2.
8. Lipschutz, Seymour and Lipson, Marc Lars, *Schaum's Outline of Theory and Problems of Linear Algebra*, Third Edition, © 1968 McGraw-Hill Companies, Inc., ISBN 0-07-136200-2.

D.1 Coordinate Transformations

Coordinate transformations are mathematical tools to express objects, such as vectors, in a desired coordinate system. The Cartesian vector representation inherently represents a vector as components against a specified coordinate frame. The transformation represents the vector in a different desired coordinate frame, effectively changing the point of view.

It should be emphasized that what the vector represents physically or otherwise is invariant. Coordinate systems provide the means of expressing the vector as components in a specific coordinate system. Transformations are mathematical tools that change the presentation from the point of view of one coordinate system to another.

D.2 Direction Cosine Matrices

Direction Cosine Matrices (DCMs) (or transformation matrices) are the most straightforward way of transforming a vector from one coordinate system to another. DCMs are 3×3 normalized orthogonal (or orthonormal) matrices and thus do no change the magnitude of the vector they are transforming.

Consider two right-handed orthogonal coordinate systems with coordinate axis basis vectors $(\hat{\imath}, \hat{\jmath}, \hat{k})$ for the initial (or "from") system and basis vectors $(\hat{\imath}', \hat{\jmath}', \hat{k}')$ for the final (or "to") system. From vector mathematics (Appendix B), the axes of the primed system can be projected on to the unprimed system using dot products, with the result of each dot product being a direction cosine (i.e., the cosine of the angle between the primed axis and unprimed axis unit vectors).

$$\hat{\imath}' = c_{11}\hat{\imath} + c_{12}\hat{\jmath} + c_{13}\hat{k}$$
$$\hat{\jmath}' = c_{21}\hat{\imath} + c_{22}\hat{\jmath} + c_{32}\hat{k}$$
$$\hat{k}' = c_{31}\hat{\imath} + c_{32}\hat{\jmath} + c_{33}\hat{k}$$

The coefficients are the direction cosines:

$$
\begin{array}{lll}
c_{11} = \hat{\imath}' \cdot \hat{\imath} = \cos\theta_{\hat{\imath}',\hat{\imath}} & c_{12} = \hat{\imath}' \cdot \hat{\jmath} = \cos\theta_{\hat{\imath}',\hat{\jmath}} & c_{13} = \hat{\imath}' \cdot \hat{k} = \cos\theta_{\hat{\imath}',\hat{k}} \\
c_{21} = \hat{\jmath}' \cdot \hat{\imath} = \cos\theta_{\hat{\jmath}',\hat{\imath}} & c_{22} = \hat{\jmath}' \cdot \hat{\jmath} = \cos\theta_{\hat{\jmath}',\hat{\jmath}} & c_{23} = \hat{\jmath}' \cdot \hat{k} = \cos\theta_{\hat{\jmath}',\hat{k}} \\
c_{31} = \hat{k}' \cdot \hat{\imath} = \cos\theta_{\hat{k}',\hat{\imath}} & c_{32} = \hat{k}' \cdot \hat{\jmath} = \cos\theta_{\hat{k}',\hat{\jmath}} & c_{33} = \hat{k}' \cdot \hat{k} = \cos\theta_{\hat{k}',\hat{k}}
\end{array}
$$

The relationship may be represented by a matrix multiplication, with the transformation matrix, as shown below:

$$\begin{bmatrix} \hat{\imath}' \\ \hat{\jmath}' \\ \hat{k}' \end{bmatrix} = \begin{bmatrix} C_{11} & C_{12} & C_{13} \\ C_{21} & C_{22} & C_{23} \\ C_{31} & C_{32} & C_{33} \end{bmatrix} \begin{bmatrix} \hat{\imath} \\ \hat{\jmath} \\ \hat{k} \end{bmatrix}$$

The inverse transformation may be formed by redistributing the direction cosines as shown in the figure.

$$\begin{bmatrix} \hat{\imath} \\ \hat{\jmath} \\ \hat{k} \end{bmatrix} = \begin{bmatrix} C_{11} & C_{21} & C_{31} \\ C_{12} & C_{22} & C_{32} \\ C_{13} & C_{23} & C_{33} \end{bmatrix} \begin{bmatrix} \hat{\imath}' \\ \hat{\jmath}' \\ \hat{k}' \end{bmatrix}$$

It should be apparent that the two direction cosine matrices are transposes of each other. This can be further developed by properties of the dot products of parallel and perpendicular unit vectors, since the dot products of parallel unit vectors equals one ($\hat{\imath} \cdot \hat{\imath} = \hat{\jmath} \cdot \hat{\jmath} = \hat{k} \cdot \hat{k} = 1$) and the dot products of perpendicular unit vector equals zero ($\hat{\imath} \cdot \hat{\jmath} = \hat{\jmath} \cdot \hat{k} = \hat{\imath} \cdot \hat{k} = 0$). Multiplying both sides of the primed to the unprimed transformation by a row vector of the unprimed system yields the following expression:

$$\begin{bmatrix} \hat{\imath}' \\ \hat{\jmath}' \\ \hat{k}' \end{bmatrix} [\hat{\imath} \quad \hat{\jmath} \quad \hat{k}] = \begin{bmatrix} C_{11} & C_{12} & C_{13} \\ C_{21} & C_{22} & C_{23} \\ C_{31} & C_{32} & C_{33} \end{bmatrix} \begin{bmatrix} \hat{\imath} \\ \hat{\jmath} \\ \hat{k} \end{bmatrix} [\hat{\imath} \quad \hat{\jmath} \quad \hat{k}]$$

The left side is the matrix outer product of the primed and unprimed vector; the outer product on the right side produces the identity matrix:

$$\begin{bmatrix} \hat{\imath}' \cdot \hat{\imath} & \hat{\imath}' \cdot \hat{\jmath} & \hat{\imath}' \cdot \hat{k} \\ \hat{\jmath}' \cdot \hat{\imath} & \hat{\jmath}' \cdot \hat{\jmath} & \hat{\jmath}' \cdot \hat{k} \\ \hat{k}' \cdot \hat{\imath} & \hat{k}' \cdot \hat{\jmath} & \hat{k}' \cdot \hat{k} \end{bmatrix} = \begin{bmatrix} C_{11} & C_{12} & C_{13} \\ C_{21} & C_{22} & C_{23} \\ C_{31} & C_{32} & C_{33} \end{bmatrix} \begin{bmatrix} 1 & 0 & 0 \\ 0 & 1 & 0 \\ 0 & 0 & 1 \end{bmatrix}$$

The same process may be repeated for the unprimed to primed transformation. Thus, the direction cosines of the constituent dot products are:

$$C^{primed}_{unprimed} = \begin{bmatrix} C_{11} & C_{12} & C_{13} \\ C_{21} & C_{22} & C_{23} \\ C_{31} & C_{32} & C_{33} \end{bmatrix} = \begin{bmatrix} \hat{\imath}' \cdot \hat{\imath} & \hat{\imath}' \cdot \hat{\jmath} & \hat{\imath}' \cdot \hat{k} \\ \hat{\jmath}' \cdot \hat{\imath} & \hat{\jmath}' \cdot \hat{\jmath} & \hat{\jmath}' \cdot \hat{k} \\ \hat{k}' \cdot \hat{\imath} & \hat{k}' \cdot \hat{\jmath} & \hat{k}' \cdot \hat{k} \end{bmatrix}$$

$$C^{unprimed}_{primed} = \begin{bmatrix} C_{11} & C_{21} & C_{31} \\ C_{12} & C_{22} & C_{32} \\ C_{13} & C_{23} & C_{33} \end{bmatrix} = \begin{bmatrix} \hat{\imath} \cdot \hat{\imath}' & \hat{\imath} \cdot \hat{\jmath}' & \hat{\imath} \cdot \hat{k}' \\ \hat{\jmath} \cdot \hat{\imath}' & \hat{\jmath} \cdot \hat{\jmath}' & \hat{\jmath} \cdot \hat{k}' \\ \hat{k} \cdot \hat{\imath}' & \hat{k} \cdot \hat{\jmath}' & \hat{k} \cdot \hat{k}' \end{bmatrix}$$

The shorthand notion for the direction cosine matrix ($C^{superscript}_{subscript}$) was introduced. The *subscript* represents the coordinate system being transformed *from*; the *superscript* represents the coordinate system being transformed *to*. The subscript (*from*) is also called the DCM's *base* coordinate system; the superscript (*to*) is also called the *destination* coordinate system.

The dot product relationships between the primed vectors are listed below. These equations recognize that the dot product of parallel unit vectors equals one and the dot product of perpendicular unit vectors equals zero.

$$\hat{\imath}' \cdot \hat{\imath}' = c_{11}^2 + c_{12}^2 + c_{13}^2 = 1$$
$$\hat{\jmath}' \cdot \hat{\jmath}' = c_{21}^2 + c_{22}^2 + c_{23}^2 = 1$$
$$\hat{k}' \cdot \hat{k}' = c_{31}^2 + c_{32}^2 + c_{33}^2 = 1$$

$$\hat{\imath}' \cdot \hat{\jmath}' = c_{11}c_{21} + c_{12}c_{22} + c_{13}c_{23} = 0$$
$$\hat{\jmath}' \cdot \hat{k}' = c_{21}c_{31} + c_{22}c_{32} + c_{23}c_{33} = 0$$
$$\hat{\imath}' \cdot \hat{k}' = c_{11}c_{31} + c_{12}c_{32} + c_{13}c_{33} = 0$$

Since the dot product is commutative, the product of the two transformations may be simplified by:

$$C_{unprimed}^{primed} C_{primed}^{unprimed} = \begin{bmatrix} c_{11}^2 + c_{12}^2 + c_{13}^2 & c_{11}c_{21} + c_{12}c_{22} + c_{13}c_{23} & c_{11}c_{31} + c_{12}c_{32} + c_{13}c_{33} \\ c_{11}c_{21} + c_{12}c_{22} + c_{13}c_{23} & c_{21}^2 + c_{22}^2 + c_{23}^2 & c_{21}c_{31} + c_{22}c_{32} + c_{23}c_{33} \\ c_{11}c_{31} + c_{12}c_{32} + c_{13}c_{33} & c_{21}c_{31} + c_{22}c_{32} + c_{23}c_{33} & c_{31}^2 + c_{32}^2 + c_{33}^2 \end{bmatrix}$$

$$C_{unprimed}^{primed} C_{primed}^{unprimed} = \begin{bmatrix} \hat{\imath}' \cdot \hat{\imath}' & \hat{\imath}' \cdot \hat{\jmath}' & \hat{\imath}' \cdot \hat{k}' \\ \hat{\imath}' \cdot \hat{\jmath}' & \hat{\jmath}' \cdot \hat{\jmath}' & \hat{\jmath}' \cdot \hat{k}' \\ \hat{\imath}' \cdot \hat{k}' & \hat{\jmath}' \cdot \hat{k}' & \hat{k}' \cdot \hat{k}' \end{bmatrix} = \begin{bmatrix} 1 & 0 & 0 \\ 0 & 1 & 0 \\ 0 & 0 & 1 \end{bmatrix}$$

The product of the two transformations (which are inverses of each other) not surprisingly produces the identity matrix. Since the two transformations are also transposes of each other, this result verifies that the transpose of a direction cosine matrix is also its inverse.

D.2.1 Row-Column Coordinate Axes Representation

The transformation from the unprimed to the primed system may be viewed by constructing the transformation matrix as either row or column vectors. Specifically, the matrix may be viewed as either:
- Three row vectors $(\hat{\imath}', \hat{\jmath}', \hat{k}')$ represented in the $(\hat{\imath}, \hat{\jmath}, \hat{k})$ frame, or
- Three column vectors $(\hat{\imath}, \hat{\jmath}, \hat{k})$ represented in the $(\hat{\imath}', \hat{\jmath}', \hat{k}')$ frame.

$$C^{primed}_{unprimed} = \begin{bmatrix} [\leftarrow & \hat{\imath}'_{ijk} & \rightarrow] \\ [\leftarrow & \hat{\jmath}'_{ijk} & \rightarrow] \\ [\leftarrow & \hat{k}'_{ijk} & \rightarrow] \end{bmatrix} = \begin{bmatrix} \begin{bmatrix} \uparrow \\ \hat{\imath}_{i'j'k'} \\ \downarrow \end{bmatrix} & \begin{bmatrix} \uparrow \\ \hat{\jmath}_{i'j'k'} \\ \downarrow \end{bmatrix} & \begin{bmatrix} \uparrow \\ \hat{k}_{i'j'k'} \\ \downarrow \end{bmatrix} \end{bmatrix}$$

Viewing the transformation as row vectors provides the connection to the vector dot product. The transformation rows are unit vectors aligned to the three axes of the coordinate system being transformed to, represented in the coordinate system being transformed from. This representation illustrates the transformation as three vector dot products, when considering the matrix multiplication process.

Recall that the dot product of a vector with any unit vector provides the projection of that vector along the unit vector's direction. Since all vector operations must be with vectors represented in the same coordinate frame, the row vectors must be represented in the coordinate frame of the vector being transformed (i.e., the "from" frame). To represent the vector as components in the frame being transformed to, the matrix multiplication becomes three dot products against the new (i.e., the "to") coordinate axes.

The column representation harkens to the relationship with the inverse transformation. When the transformation matrix is transposed, the columns become rows and the superscript and subscript invert. Thus, the transpose of the transformation matrix maintains the dot product projection paradigm by inverting the to/from transformation relationship.

D.2.2 Rotations About Single Axes

Standard transformations exist for a right-handed rotation by a specified angle (θ), about a single coordinate axis. These rotations are referred to as an R_1 transformation for rotations about the first axis (such as the x-axis), an R_2 transformation for rotations about the second axis (such as the y-axis), and an R_3 transformation for rotations about the third axis (such as the z-axis). Each of these rotation types are listed below:

$$R_1(\theta) = \begin{bmatrix} 1 & 0 & 0 \\ 0 & \cos\theta & \sin\theta \\ 0 & -\sin\theta & \cos\theta \end{bmatrix}$$

$$R_2(\theta) = \begin{bmatrix} \cos\theta & 0 & -\sin\theta \\ 0 & 1 & 0 \\ \sin\theta & 0 & \cos\theta \end{bmatrix}$$

$$R_3(\theta) = \begin{bmatrix} \cos\theta & \sin\theta & 0 \\ -\sin\theta & \cos\theta & 0 \\ 0 & 0 & 1 \end{bmatrix}$$

Notice in each case that the axis about which the rotation occurs is left unaltered.

D.2.3 Chaining Successive DCM Transformations

The transformation subscript/superscript (to/from) shorthand is used commonly since it facilitates the chaining of successive coordinate transformations in the correct sequence. For example, the transformations below, when sequenced from right to left, produce a composite transformation:

$$C_A^F = C_E^F C_D^E C_C^D C_B^C C_A^B$$

The mnemonic for arranging the transformations requires agreement between a transformation's superscript and the subscript of the transformation immediately to the left. Subscripts cancel superscripts with lower left/upper right agreement, provided they are in immediate proximity in the sequence.

To illustrate, consider a vector ($\bar{a}$) represented in the "A" coordinate frame, with the need to transform its representation to the "F" coordinate frame. The vector's subscript indicates the frame in which it is currently represented. If the above coordinate transformations are available, the vector can be represented in the "F" frame by a nested series of transformations:

$$\bar{a}_F = C_E^F \left(C_D^E \left(C_C^D \left(C_B^C (C_A^B \bar{a}_A) \right) \right) \right)$$

The innermost product transforms $\bar{a}$ from being represented in the "A" coordinate frame to the "B" coordinate frame. The next level transforms from the "B" coordinate frame to the "C" coordinate frame, and so on until the vector is ultimately represented in the "F" frame. But since matrix multiplication is associative, individual transformations can be moved outside the parentheses, leaving the net product as the composite transformation:

$$\bar{a}_F = C_A^F \bar{a}_A$$

In practical problems, it is common to only have a transformation available that is the inverse of what is needed for lower left/upper right subscript cancellation. This is remedied using the transpose inversion. For example, if in the above composite transformation chain, C_D^C was available instead of C_C^D, the transformation chain is computed by embedding the transpose of the inverse quaternion as follows:

$$C_A^F = C_E^F C_D^E (C_D^C)^T C_B^C C_A^B$$

D.2.4 Roll-Pitch-Yaw DCM

The DCM representing a traditional aircraft roll-pitch-yaw (Euler angle) attitude is determined by three successive single axis rotations: (1) a rotation about the z-axis by the yaw (ψ) angle, (2) a rotation about the y-axis by the negative pitch ($-\theta$) angle, and (3) rotation about the x-axis by the roll (φ) angle.

$$C_{\varphi\theta\psi} = R_1(\varphi)R_2(-\theta)R_3(\psi)$$

$$C_{\varphi\theta\psi} = \begin{bmatrix} \cos\theta\cos\psi & \cos\theta\sin\psi & \sin\theta \\ -\cos\varphi\sin\psi - \sin\varphi\sin\theta\cos\psi & \cos\varphi\cos\psi - \sin\varphi\sin\theta\sin\psi & \sin\varphi\cos\theta \\ \sin\varphi\sin\psi - \cos\varphi\sin\theta\cos\psi & -\sin\varphi\cos\psi - \cos\varphi\sin\theta\sin\psi & \cos\varphi\cos\theta \end{bmatrix}$$

The sign conventions for this sequence are:
- **Roll (φ)**: positive right, negative left.
- **Pitch (θ)**: positive up, negative down.
- **Yaw (ψ)**: positive left, negative right

The Euler angles may also be recovered from the DCM:

$$\tan\varphi = \frac{c_{23}}{c_{33}} \quad \sin\theta = c_{13} \quad \tan\psi = \frac{c_{12}}{c_{11}}$$

Since both roll and yaw can have values from 0 to 2π radians, the correct quadrant must be resolved based on the individual signs of the coefficients before the ratios are computed. While this may be done by individual cases, it may also be automated using a two-argument inverse tangent function.

D.3   Quaternions

Quaternions are an alternative form coordinate transformations and thus possess all the properties of DCMs. Quaternions remove all but one level of the inherent redundancy in DCMs, while remaining free from singularities. As a coordinate transformation, quaternions are the standard means of representing the orientations of spacecraft body axes, relative to a standard (typically inertial) coordinate frame. A quaternion consisting of an inertial-to-body transformation may be denoted as $\bar{q}_{ib}$. This is commonly known as an attitude quaternion.

D.3.1 Quaternion Representation

Sir William Rowan Hamilton presented quaternions to the Royal Irish Academy in 1843 as hypercomplex numbers (i.e., complex numbers with three imaginary axes). The three complex axes ($i, j,$ and k) each represent $\sqrt{-1}$ offsets from the real component and behave as a right-hand Cartesian coordinate system. Thus, a

374

quaternion ($\bar{q}$) may be presented as a four-element array or alternatively by scalar (q_s) and vector ($\bar{q}_v$) constituents.

$$\bar{q} = \begin{bmatrix} q_s \\ \bar{q}_v \end{bmatrix} = \begin{bmatrix} q_s \\ q_i \\ q_j \\ q_k \end{bmatrix}$$

Euler's Principal Rotation Theorem states that a rigid body or coordinate reference frame can be transformed between two orientations by a single rotation about a principal axis ($\hat{e}$) through a principal angle (θ). Consistent with Euler's theorem, the quaternion's scalar and vector constituents may be represented as:

$$q_s = \cos\frac{\theta}{2}$$

$$\bar{q}_v = \hat{e}\sin\frac{\theta}{2}$$

From the above representation, a quaternion having a zero value for its rotation angle ($\theta = 0$) has a value of one for its scalar component ($q_s = 1$) and zeros as its vector component ($\bar{q}_v = \bar{0}$). Since this represents no rotation, such a quaternion is analogous to the identity matrix.

Important Note:

Presenting the quaternion's scalar constituent first (as is done here) is consistent with Hamilton's hypercomplex number convention. However, NASA adopted a convention in which the scalar constituent follows the vector constituent. Most spacecraft attitude quaternions follow the NASA convention.

D.3.2 Quaternion Shadow Forms

With one exception, every quaternion has a twin (or shadow form) that describes the equivalent transformation in a slightly different, but recognizable form. There are two differences between a quaternion and its shadow form that result in the transformation rotating through either the *short path* or the *long path*.

One difference is that the short path transformation has a rotation less than π radians ($\theta_{short} < \pi$), while the long path has a rotation angle greater than π radians ($\theta_{long} > \pi$). (Rotation angles of π radians represent a special exception case.) The short and long path rotation angles always add up to 2π radians. Thus, the long path rotation is equivalent to the short path rotation, but in the *opposite direction*.

$$\theta_{short} = 2\pi - \theta_{long}$$

The second difference ensures the long and short path transformations represent equivalent transformations. Since the rotation angles are in opposite directions for the long and the short paths, the sign on the principal axis must be opposite to maintain right-handed rotation.

$$\hat{e}_{short} = -\hat{e}_{long}$$

Now consider the following trigonometric relationship (using the cosine of angular differences identity):

$$\cos\left(\frac{2\pi - \theta}{2}\right) = \cos\left(\pi - \frac{\theta}{2}\right) = -\cos\frac{\theta}{2}$$

This result, coupled with the sign flip on the principal rotation vector demonstrates that the form of the long and short path quaternions can be characterized as only a difference in sign.

$$\bar{q}_{short} = -\bar{q}_{long}$$

Thus, inverting the sign on all four quaternion components switches from one form to its shadow form. Furthermore, the short and long path forms are readily distinguishable. The short path transformation always has a positive value for its scalar constituent, while the long path transformation always has a negative value for its scalar constituent.

D.3.3 Quaternion to DCM Conversion

Since they are both coordinate transformations, a quaternion may be readily converted to a DCM as follows:

$$C = \begin{bmatrix} q_s^2 + q_i^2 - q_j^2 - q_k^2 & 2(q_i q_j + q_s q_k) & 2(q_i q_k - q_s q_j) \\ 2(q_i q_j - q_s q_k) & q_s^2 - q_i^2 + q_j^2 - q_k^2 & 2(q_j q_k + q_s q_i) \\ 2(q_i q_k + q_s q_j) & 2(q_j q_k - q_s q_i) & q_s^2 - q_i^2 - q_j^2 + q_k^2 \end{bmatrix}$$

Extracting a quaternion from a DCM may be accomplished by algebraic manipulation of matrix components. There is more than one possible formulation, but some may be better conditioned numerically in a computer solution. The Stanley method seeks the best conditioned numerical approach.

The Stanley algorithm may be implemented as follows:
1. Compute the trace of the DCM: $T = Tr(C)$.
2. Compute four evaluation quantities:
    a. $Q_s^2 = (1 + T)/4$

b. $Q_i^2 = (1 + 2c_{11} - T)/4$
c. $Q_j^2 = (1 + 2c_{22} - T)/4$
d. $Q_k^2 = (1 + 2c_{33} - T)/4$

3. Determine the maximum value of the evaluation quantities: $Q_{max} = max(Q_s^2, Q_i^2, Q_j^2, Q_k^2)$.

    a. If $Q_{max} = Q_s^2$

$$q_s = \sqrt{Q_i^2}$$

$$d = 4q_s$$

$$q_i = (c_{23} - c_{32})/d$$
$$q_j = (c_{31} - c_{13})/d$$
$$q_k = (c_{12} - c_{21})/d$$

    b. If $Q_{max} = Q_i^2$

$$q_i = \sqrt{Q_i^2}$$

$$d = 4q_i$$

$$q_s = (c_{23} - c_{32})/d$$
$$q_j = (c_{12} - c_{21})/d$$
$$q_k = (c_{31} - c_{13})/d$$

    c. If $Q_{max} = Q_j^2$

$$q_j = \sqrt{Q_j^2}$$

$$d = 4q_j$$

$$q_s = (c_{31} - c_{13})/d$$
$$q_i = (c_{12} - c_{21})/d$$
$$q_k = (c_{31} - c_{13})/d$$

    d. If $Q_{max} = Q_k^2$

$$q_k = \sqrt{Q_k^2}$$

$$d = 4q_k$$

$$q_s = (c_{12} - c_{21})/d$$
$$q_i = (c_{31} - c_{13})/d$$

$$q_j = (c_{23} - c_{32})/d$$

D.3.4 Quaternion Multiplication

Quaternion multiplication may be performed using several methods. One method uses a matrix multiplication operator $[\boldsymbol{Q} \times]$.

$$[\boldsymbol{Q} \times] = \begin{bmatrix} q_s & -q_i & -q_j & -q_k \\ q_i & q_s & -q_k & q_j \\ q_j & q_k & q_s & -q_i \\ q_k & -q_j & q_i & q_s \end{bmatrix}$$

Quaternion $\bar{q}_a$ multiplication by quaternion $\bar{q}_b$ is performed using the operator as follows:

$$\bar{q}_a \bar{q}_b = [\boldsymbol{Q} \times]_a \bar{q}_b$$

$$\bar{q}_a \bar{q}_b = \begin{bmatrix} q_s & -q_i & -q_j & -q_k \\ q_i & q_s & -q_k & q_j \\ q_j & q_k & q_s & -q_i \\ q_k & -q_j & q_i & q_s \end{bmatrix}_a \begin{bmatrix} q_s \\ q_i \\ q_j \\ q_k \end{bmatrix}_b$$

The values inside the $[Q \times]_a$ matrix correspond to the constituents of $\bar{q}_a$; the values inside the column vector correspond to the constituents of $\bar{q}_b$.

D.3.5 Quaternion Inverse

Quaternions have a straightforward inverse, (consistent with a DCM having its transpose as a straightforward inverse). The inverse of a quaternion is its conjugate $(\bar{q}^*)$.

$$(\bar{q})^{-1} = \bar{q}^*$$

$$\bar{q}^* = \begin{bmatrix} q_s \\ -\bar{q}_v \end{bmatrix} = \begin{bmatrix} q_s \\ -q_i \\ -q_j \\ -q_k \end{bmatrix}$$

D.3.6 Quaternion Transformation of a Vector

A quaternion may transform a vector between its base (i.e., *"from"*) coordinate frame and its destination (i.e., *"to"*) coordinate frame. Consider a quaternion that transforms from base coordinate frame A to destination coordinate frame B called $\bar{q}_{AB}$. A vector expressed in frame A may be expressed in frame B using the quaternion frame rotation:

$$\bar{v}_B = \bar{q}_{AB}^* \bar{v}_A \bar{q}_{AB}$$

An efficient equation for implementing the frame rotation is:

$$\bar{v}_B = 2(q_s^2 - 1)\bar{v}_A + 2[(\bar{v}_A \cdot \bar{q}_v)\bar{q}_v + q_s(\bar{v}_A \times \bar{q}_v)]$$

While the equation above appears complicated, a computer implementation requires fewer floating-point operations than would be needed to convert the quaternion to a DCM and transform through matrix multiplication.

D.3.7 Chaining Successive Quaternion Transformations

Successive quaternion transformations may be chained in a manner like that done with DCMs. The one difference is the chaining direction. Quaternions are chained from left-to-right for successive transformations, while the DCMs are chained from right-to-left.

D.3.7.1 Transformational Quaternion Computation

Given two quaternions $(\bar{q}_o, \bar{q}_f)$ representing an initial and final quaternion, the interim transformation $(\delta \bar{q})$ can be determined. The transformational quaternion represents an intermediate transformation, such that the final quaternion $(\bar{q})$ can be computed as the product of the initial quaternion $(\bar{q}_o)$ and the transformational quaternion $(\delta \bar{q})$ as follows:

$$\bar{q}_f = \bar{q}_o \delta \bar{q}$$

The transformational quaternion is recovered recognizing that the quaternion's conjugate $(\bar{q}^*)$ is its inverse.

$$\bar{q}_0^* \bar{q}_f = \bar{q}_0^* \bar{q}_o \delta \bar{q}$$

$$\delta \bar{q} = \bar{q}_0^* \bar{q}_f$$

D.3.7.2 Rotation Rate from Two Successive Quaternions

If a constant rotation rate $(\bar{\omega})$ can be presumed, the rotation vector can be recovered using the transformational quaternion.

Recall that a quaternion has both scalar (q_s) and a vector $(\bar{q}_v)$ components. Thus, for the transformational quaternion from $\bar{q}_o$ to $\bar{q}_f$:

$$\delta\bar{q} = \begin{bmatrix} \delta q_s \\ \delta\bar{q}_v \end{bmatrix}$$

Recall also, that the scalar and vector components have magnitudes related to the rotation angle (θ) and that transformation is about the principal rotation vector ($\delta\hat{q}$) as follows:

$$\delta q_s = \cos\frac{\theta}{2}$$

$$\delta q_v = \sin\frac{\theta}{2}\,\delta\hat{q}$$

A constant rotation rate (ω) over the time interval (t), may be computed as:

$$\bar{\omega} = \frac{\theta}{t}\delta\hat{q}$$

Direct recovery of the rotation rate vector is computed by:

$$\bar{\omega} = \frac{2\delta q_v\,\sin^{-1}|\delta\bar{q}_v|}{|\delta\bar{q}_v|t}$$

The constituents are computed from:

$$|\delta\bar{q}_v| = \sqrt{\delta\bar{q}_v \cdot \delta\bar{q}_v} = \sin\frac{\theta}{2}$$

$$\delta\hat{q} = \frac{\delta\bar{q}_v}{|\delta\bar{q}_v|}$$

D.3.8 Quaternion Time Derivative

The quaternion's time derivative may be used for kinematic computations. The time derivative is computed from the quaternion or its inverse and inertial body rotation rate vector, depending on which is most conveniently available. An angular rate matrix (Ω) is arranged in a skew symmetric:

$$\Omega_i = \begin{bmatrix} 0 & -\omega_i & -\omega_j & -\omega_k \\ \omega_i & 0 & \omega_k & -\omega_j \\ \omega_j & -\omega_k & 0 & \omega_i \\ \omega_k & \omega_j & -\omega_i & 0 \end{bmatrix}$$

The Ω matrix, populated from components represented in the inertial coordinate system, is denoted as Ω_i. The time derivative of the quaternion from the inertial-to-body frame is computed as:

$$\dot{\bar{q}}_{ib} = \frac{1}{2}\Omega_i \bar{q}_{ib}$$

D.3.9 Quaternion Extrapolation

Quaternions ($\bar{q}$) experiencing an inertial rotation ($\bar{\omega}$) may be extrapolated forwarded in time (Δt). The updated quaternion ($\bar{q}'$) is computed as follows:

$$\bar{q}' = \left[I\cos(\omega\,\Delta t) + \Omega\,\frac{\sin(\omega\,\Delta t)}{\omega} \right]\bar{q}$$

The Ω matrix is constructed from the $\bar{\omega}$ vector's components as shown below and I is a 4×4 identity matrix.

$$\Omega = \begin{bmatrix} 0 & -\omega_i & -\omega_j & -\omega_k \\ \omega_i & 0 & \omega_k & -\omega_j \\ \omega_j & -\omega_k & 0 & \omega_i \\ \omega_k & \omega_j & -\omega_i & 0 \end{bmatrix} \qquad I = \begin{bmatrix} 1 & 0 & 0 & 0 \\ 0 & 1 & 0 & 0 \\ 0 & 0 & 1 & 0 \\ 0 & 0 & 0 & 1 \end{bmatrix}$$

This formulation presumes the Hamiltonian convention with the scalar as the first quaternion component, followed by the vector components. Note that the $\bar{\omega}$ vector is in the coordinate system as the quaternion principal rotation vector.

D.4 Modified Rodrigues Parameters

Modified Rodrigues Parameters (MRPs) are another coordinate transformation that leverage Euler's Principal Rotation Theorem. MRPs are a useful alternative to the quaternion when it is desirable to represent the body orientation without redundancy while avoiding singularities.

D.4.1 MRP Representation

The MRP ($\bar{\sigma}$) is a three-element vector consisting of the principal rotation vector ($\hat{e}$) scaled to the tangent of one quarter the principal rotation angle (θ).

$$\bar{\sigma} = \hat{e}\tan\left(\frac{\theta}{4}\right) = \begin{bmatrix} \sigma_i \\ \sigma_j \\ \sigma_k \end{bmatrix}$$

It is noteworthy that the MRP has singularities at principal rotation angles that are multiples of $\pm 2\pi$ radians. The singularities are easily averted while maintaining a full range of rotations by only using the short path $(-\pi \le \theta \le \pi)$ shadow form.

D.4.2 MRP Shadow Forms

The MRP magnitude determines whether it represents a short path $(\sigma < 1)$ or long path $(\sigma > 1)$ rotation. The shadow form of an MRP $(\bar{\sigma}')$ is computed by dividing the MRP by the negative square of its magnitude.

$$\sigma = \sqrt{\bar{\sigma} \cdot \bar{\sigma}}$$

$$\bar{\sigma}' = -\frac{\bar{\sigma}}{\sigma^2}$$

D.4.3 MRP to DCM Conversion

The MRP may be converted to a DCM as follows:

$$C = \frac{1}{(1+\sigma^2)^2} \begin{bmatrix} 4\left(\sigma_i^2 - \sigma_j^2 - \sigma_k^2\right) + (1-\sigma^2)^2 & 8\sigma_i\sigma_j + 4\sigma_k(1-\sigma^2) & 8\sigma_i\sigma_k - 4\sigma_j(1-\sigma^2) \\ 8\sigma_i\sigma_j - 4\sigma_k(1-\sigma^2) & 4\left(-\sigma_i^2 + \sigma_j^2 - \sigma_k^2\right) + (1-\sigma^2)^2 & 8\sigma_j\sigma_k + 4\sigma_i(1-\sigma^2) \\ 8\sigma_i\sigma_k + 4\sigma_j(1-\sigma^2) & 8\sigma_j\sigma_k - 4\sigma_i(1-\sigma^2) & 4\left(-\sigma_i^2 - \sigma_j^2 + \sigma_k^2\right) + (1-\sigma^2)^2 \end{bmatrix}$$

D.4.4 MRP to Quaternion Conversion

MRPs may be converted to quaternions as shown below:

$$q_s = \frac{1 - \sigma^2}{1 + \sigma^2} \qquad \bar{q}_v = \frac{2\bar{\sigma}}{1 + \sigma^2}$$

Conversely, quaternions may be converted to MRPs by:

$$\bar{\sigma} = \frac{\bar{q}_v}{1 + q_s}$$

D.4.5 MRP Time Derivatives

The MRP's time derivative may be used for kinematic computations. The time derivative is computed from the MRP and inertial body rotation rate vector.

$$\dot{\bar{\sigma}} = \frac{1}{4} \begin{bmatrix} 1 - \sigma^2 + 2\sigma_i^2 & 2\left(\sigma_i\sigma_j - \sigma_k\right) & 2\left(\sigma_i\sigma_k + \sigma_j\right) \\ 2\left(\sigma_i\sigma_j + \sigma_k\right) & 1 - \sigma^2 + 2\sigma_j^2 & 2\left(\sigma_j\sigma_k - \sigma_i\right) \\ 2\left(\sigma_i\sigma_k - \sigma_j\right) & 2\left(\sigma_j\sigma_k + \sigma_i\right) & 1 - \sigma^2 + 2\sigma_k^2 \end{bmatrix} \begin{bmatrix} \omega_i \\ \omega_j \\ \omega_k \end{bmatrix}$$

The matrix portion may be abbreviated as $\mathbf{\Sigma}$. This leads to a shorthand equation for the MRPs first time derivative:

$$\dot{\bar{\sigma}} = \frac{1}{4}\mathbf{\Sigma}\bar{\omega}$$

The MRP's second time derivative is therefore:

$$\ddot{\bar{\sigma}} = \frac{1}{4}\left(\mathbf{\Sigma}\dot{\bar{\omega}} + \dot{\mathbf{\Sigma}}\bar{\omega}\right)$$

$$\dot{\mathbf{\Sigma}} = \begin{bmatrix} 4\sigma_i\dot{\sigma}_i - 2\bar{\sigma}\cdot\dot{\bar{\sigma}} & 2\left(\sigma_i\dot{\sigma}_j + \dot{\sigma}_i\sigma_j - \dot{\sigma}_k\right) & 2\left(\sigma_i\dot{\sigma}_k + \dot{\sigma}_i\sigma_k + \dot{\sigma}_j\right) \\ 2\left(\sigma_i\dot{\sigma}_j + \dot{\sigma}_i\sigma_j + \dot{\sigma}_k\right) & 4\sigma_j\dot{\sigma}_j - 2\bar{\sigma}\cdot\dot{\bar{\sigma}} & 2\left(\sigma_j\dot{\sigma}_k + \dot{\sigma}_j\sigma_k - \dot{\sigma}_i\right) \\ 2\left(\sigma_i\dot{\sigma}_k + \dot{\sigma}_i\sigma_k - \dot{\sigma}_j\right) & 2\left(\sigma_j\dot{\sigma}_k + \dot{\sigma}_j\sigma_k + \dot{\sigma}_i\right) & 4\sigma_k\dot{\sigma}_k - 2\bar{\sigma}\cdot\dot{\bar{\sigma}} \end{bmatrix}$$

D.4.6 Body Rate from MRP Time Derivative

The inertial body rate vector ($\bar{\omega}$) in global coordinates may be recovered from the MRP ($\bar{\sigma}$) and its first time derivative ($\dot{\bar{\sigma}}$):

$$\bar{\omega} = 4\mathbf{\Sigma}^{-1}\dot{\bar{\sigma}}$$

References

1. Schaub, Hanspeter and Junkins, John L., *Analytical Mechanics of Space Systems*, AIAA Education Series, © 2003 American Institute of Aeronautics and Astronautics (AIAA), ISBN 1-56347-563-4.
2. Kuipers, Jack B., *Quaternions and Rotation Sequences, A Primer with Applications to Orbits, Aerospace*, and Virtual Reality, © 1999 Princeton University Press, ISBN 0-691-10298-8.
3. Hamilton, William R., *On a New Species of Imaginary Quantities Connected with a Theory of Quaternions*, Proceedings of the Royal Irish Academy, vol 2 (1844), pp 424-434.
4. Stanley, W.S., *Quaternion from Rotation Matrix*, Journal of Guidance and Control, Vol I, No. 3, pp 223-224, American Institute of Aeronautics and Astronautics (AIAA).

E.1 Mathematical Identities

This appendix lists the mathematical identities used in the derivations. The intention is to provide the reader with a convenient reference while following the derivations.

E.2 Trigonometric Identities

This section presents trigonometric identities used in various derivation in the text.

E.2.1 Cosine-Sine Orthogonality

$$cos^2\theta + sin^2\theta = 1$$

E.2.2 Sum and Differences

$$\cos(\alpha + \beta) = \cos\alpha\cos\beta - \sin\alpha\sin\beta$$
$$\cos(\alpha - \beta) = \cos\alpha\cos\beta + \sin\alpha\sin\beta$$

$$\sin(\alpha + \beta) = \sin\alpha\cos\beta + \sin\beta\cos\alpha$$
$$\sin(\alpha - \beta) = \sin\alpha\cos\beta - \sin\beta\cos\alpha$$

E.2.3 Law of Cosines

The law of cosines provides a general solution applicable to any triangle. When used with a right triangle, the law of cosines degenerates to the Pythagorean theorem.

$$c^2 = a^2 + b^2 - 2ab\cos C$$

E.2.4 Law of Sines

The law of sines provides a general solution applicable to any triangle as the equality of the ratio of each side to the sine of its opposing angle.

$$\frac{a}{\sin A} = \frac{b}{\sin B}$$

E.2.5 Half Angle Relationships

$$\cos\frac{\alpha}{2} = \pm\sqrt{\frac{1 + \cos\alpha}{2}}$$

$$\sin\frac{\alpha}{2} = \pm\sqrt{\frac{1-\cos\alpha}{2}}$$

$$\tan\frac{\alpha}{2} = \frac{\sin\alpha}{1+\cos\alpha} = \frac{1-\cos\alpha}{\sin\alpha} = \pm\sqrt{\frac{1-\cos\alpha}{1+\cos\alpha}}$$

E.2.6 Double Angle Relationships

$$\cos 2\theta = 2\cos^2\theta - 1 = \frac{1-\tan^2\theta}{1+\tan^2\theta}$$

$$\sin 2\theta = 2\sin\theta\cos\theta = \frac{2\tan\theta}{1+\tan^2\theta}$$

$$\tan 2\theta = \frac{\sin 2\theta}{1+\cos 2\theta}$$

E.2.7   Hyperbolic Trigonometric Identities

$$\sinh x = \frac{e^x - e^{-x}}{2}$$

$$\cosh x = \frac{e^x + e^{-x}}{2}$$

$$\cosh^2 x - \sinh^2 x = 1$$

$$\cosh^{-1} x = \ln\left(x + \sqrt{x^2 - 1}\right)$$

$$\sinh^{-1} x = \ln\left(x + \sqrt{x^2 + 1}\right)$$

$$\tanh\frac{x}{2} = \frac{\sinh F}{1+\cosh F}$$

References

1. Beyer, William H. (Editor), *CRC Standard Mathematical Tables*, 27th Edition, © 1984 Chemical Rubber Company (CRC) Press, ISBN 0-8493-0627-2.

2. Spiegel, Murray R., *Schaum's Outline Series Theory and Problems of Mathematical Handbook of Formulas and Tables*, © 1968 McGraw-Hill, Inc., ISBN 07-060224-7.
3. Kaplan, Wilfred, and Lewis, D.J., *Calculus and Linear Algebra*, Combined Edition, © 1971 John Wiley & Sons, Inc., ISBN 0-471-45687-X.

F.1 Numerical Integration

Numerical integration is a tool that provides flexible solutions to a variety of space flight dynamics computations. Its use accepts the reality that few practical problems have exact solutions and many analytic solutions in existence have a narrow scope to their applicability. The most dynamic aspect of numerical integration solutions is the ability to tailor a computer implementation at run time to accommodate a wide variety of situations.

The most accurate solutions to trajectory and attitude dynamics predictions come from numerical solutions of second order ordinary differential equations, with time as the independent variable. While the discussion herein focuses on trajectory predictions, the methods are readily adaptable to attitude dynamics.

F.1.1 Newton's Second Law as a Differential Equation

Motion in a trajectory is governed by Newton's Second Law in the form of a differential equation.

$$\ddot{\bar{r}} = \frac{\bar{F}}{m}$$

This equation appears deceptively simple until there is the realization that the force ($\bar{F}$) is not constant. (It becomes further complicated with trajectories that have thrusting maneuvers, since the mass also decreases during the thrusting activity.) For practical trajectory predictions, the net force is a complicated vector function of time, position, and velocity.

$$\bar{F} = \bar{F}(t, \bar{r}, \dot{\bar{r}})$$

The force is definable in inertial Cartesian coordinates. The position, velocity, and acceleration, and force are defined by:

$$\bar{r} = \begin{bmatrix} r_x \\ r_y \\ r_z \end{bmatrix} \qquad \dot{\bar{r}} = \begin{bmatrix} \dot{r}_x \\ \dot{r}_y \\ \dot{r}_z \end{bmatrix} \qquad \ddot{\bar{r}} = \begin{bmatrix} \ddot{r}_x \\ \ddot{r}_y \\ \ddot{r}_z \end{bmatrix} \qquad \bar{F} = \begin{bmatrix} F_x \\ F_y \\ F_z \end{bmatrix}$$

These parameters may be represented in Newton's Second Law with a vector second order differential equation:

$$\ddot{\bar{r}} = \frac{\bar{F}(t, \bar{r}, \dot{\bar{r}})}{m}$$

The solution leverages the form of the time derivatives to facilitate a single integration cycle. A single state vector ($\bar{y}$) is introduced that encapsulates both position and velocity. Its time derivative ($\dot{\bar{y}}$) encapsulates velocity and acceleration.

$$\bar{y} = \begin{bmatrix} \bar{r} \\ \dot{\bar{r}} \end{bmatrix} = \begin{bmatrix} r_x \\ r_y \\ r_z \\ \dot{r}_x \\ \dot{r}_y \\ \dot{r}_z \end{bmatrix} \qquad \dot{\bar{y}} = \begin{bmatrix} \dot{\bar{r}} \\ \ddot{\bar{r}} \end{bmatrix} = \begin{bmatrix} \dot{r}_x \\ \dot{r}_y \\ \dot{r}_z \\ \ddot{r}_x \\ \ddot{r}_y \\ \ddot{r}_z \end{bmatrix}$$

The differential equation can be abstracted as:

$$\dot{\bar{y}} = \bar{f}(t, \bar{y})$$

The solution mimics a Taylor Series in which a function $f(t)$ can approximate its value at offset $t + h$ using an infinite series:

$$f(t + h) = f(t) + c_1 h + c_2 h^2 + c_3 h^3 + \cdots$$

The coefficients c_n are successively higher order derivatives of $f(t)$ as follows:

$$c_n = \frac{1}{n!} \frac{d^n f(t)}{dt^n}$$

A series approximated after a finite number of coefficients has a truncation error consisting of the sum of the higher order terms following the truncation. Higher numbers of terms will provide more accurate approximations. Similarly, a smaller increment interval h will also improve the approximation. Reducing the increment to $h/2$ reduces truncation error by $(1/2)^n$ factor.

Thus, instead of increasing the derivatives beyond the available $\ddot{\bar{r}}$, accuracy is improved by subdividing the integration interval h into m discrete subintervals. The numerical integration becomes an initial value problem, where the starting state $\bar{y}$ is sufficient to predict the state at future (or past) times.

F.1.2 Runge-Kutta Integration

Runge-Kutta (RK) processes are commonly-used numerical integration algorithms for propagating trajectory states. They are single step methods used to compute positions and velocities at a desired time, given an initial trajectory position and velocity state and a model of the forces acting on the spacecraft over time, using the general equation form:

$$\bar{y}_{i+1} = \bar{y}_i + h\varphi(t_i, \bar{y}_i, h)$$

The integration step is denoted by h and φ is the increment function, which is a weighted sum of $\dot{\bar{y}}$ evaluations performed within the interval from t_i to $t_i + h$. The quantities averaged are predictions of $\bar{f}(t, \bar{y})$ evaluated at various stages (or sub-intervals).

A RK method has an order that reflect how accurately the increment function (φ) approximates a Taylor series. An RK method of order "n" (or RKn method) indicates an accuracy in computing $\dot{\bar{y}}_i$ is equivalent to that of an nth-order Taylor series.

F.1.2.1 Runge-Kutta Features

Runge-Kutta methods only require the first derivative of $\bar{y}$ to achieve the equivalent accuracy for which a Taylor series requires n derivatives. Thus, higher order methods have a more accurate approximation over any h interval.

The number of stages internal to interval h is the number of intermediate times in the interval that $\bar{f}(t, \bar{y})$ is evaluated. If the RK order is four or less, the number of stages equals the order.

F.1.2.2 Butcher Matrices

Butcher matrices define the internal workings, including coefficients for a Runge-Kutta interval update. There are three basic Butcher matrices:
1. The "a" matrix is a column vector that defines the fractions of the h interval, defining the steps in which to evaluate $\bar{f}(t, \bar{y})$.
2. The "**b**" matrix is a lower triangular matrix that defines the proportions of update estimates to be applied to the $\dot{\bar{y}}$ results produced from the initial conditions and those of all preceding stages.
3. The "c" matrix is a vector of coefficients to produce the weighted sum of the stage estimates to determine the interval update.

$$[a] = \begin{bmatrix} a_1 \\ a_2 \\ \vdots \\ a_k \end{bmatrix} \quad [b] = \begin{bmatrix} b_{11} & \square & \square & \square \\ b_{21} & b_{22} & \square & \square \\ \vdots & \vdots & \ddots & \square \\ b_{k1} & b_{k2} & \cdots & b_{k,k-1} \end{bmatrix} \quad [c] = \begin{bmatrix} c_1 \\ c_2 \\ \vdots \\ c_k \end{bmatrix}$$

Two characteristics of the Butcher matrices include:
1. The value of any element in the a matrix is the sum of the corresponding row of the **b** matrix.

2. The sum of the c matrix elements is always equal to one to achieve a weighted average.

F.1.2.3 Runge-Kutta Interval Update

Updating over an interval h is a recursive process. There will be an intermediate $\bar{f}_k(t, \bar{y})$ for each row of the $[b]$ matrix. Any $\bar{f}_k(t, \bar{y})$ determined for a row is computed using a weighted sum of the preceding $\bar{f}_k(t, \bar{y})$ evaluations. The weights are the $[b]$ matrix elements in the current row, such that the force is evaluated at $t_i + a_i h$ as follows:

$$\bar{f}_k(t, \bar{y}) = f \left[(t_i + a_i h),\ \bar{y}_i + \sum_{j=1}^{k-1} h b_{k,j} \bar{f}_j \right]$$

Once the intermediate evaluations are made, the final evaluation is a sum of the intermediate $\bar{f}_k(t, \bar{y})$ results, weighted by the $[c]$ matrix, as follows:

$$\bar{y}_{i+1} = \bar{y}_i + h \sum_{k=1}^{n} c_k\, \bar{f}_k(t, \bar{y})$$

F.1.2.4 Runge-Kutta 4 Method

The Butcher matrices for a common fourth order Runge-Kutta method (RK4) are listed below.

$$[a] = \begin{bmatrix} 0 \\ 1/2 \\ 1/2 \\ 1 \end{bmatrix} \qquad [b] = \begin{bmatrix} 0 & 0 & 0 \\ 1/2 & 0 & 0 \\ 0 & 1/2 & 0 \\ 0 & 0 & 1 \end{bmatrix} \qquad [c] = \begin{bmatrix} 1/6 \\ 1/3 \\ 1/3 \\ 1/6 \end{bmatrix}$$

The implementation starts with evaluating the force effects $\bar{f}_1$ at the initial boundary state. Each subsequent force evaluation builds on the preceding force evaluations.

$$\bar{f}_1 = \bar{f}(t_i, \bar{y}_i)$$
$$\bar{f}_2 = \bar{f}\left(t_i + \frac{h}{2},\ \bar{y}_i + h\frac{\bar{f}_1}{2}\right)$$
$$\bar{f}_3 = \bar{f}\left(t_i + \frac{h}{2},\ \bar{y}_i + h\frac{\bar{f}_2}{2}\right)$$
$$\bar{f}_4 = \bar{f}\left(t_i + h,\ \bar{y}_i + h\bar{f}_3\right)$$

The integration step is completed by weighting the $\bar{f}_k$ results with by the $[c]$ matrix coefficients.

$$\bar{y}_{i+1} = \bar{y}_i + h \left(\frac{1}{6} \bar{f}_1 + \frac{1}{3} \bar{f}_2 + \frac{1}{3} \bar{f}_3 + \frac{1}{6} \bar{f}_4 \right)$$

F.1.2.5 The Force Function

Runge Kutta integrators are effectively implemented as generic functions by computing the force in a user-specified function. This is done by including a function pointer as one of the calling arguments.

The example below uses an RK method to propagate a trajectory forward (or backward) in time. The function pointer in this case would point to a force model function that computes the net acceleration on the current trajectory state. The next acceleration includes the gravitational acceleration of the central attracting body, plus any perturbative accelerations.

The force model is most flexible when it maximizes user configurability. Examples of force model user-configurable choices include:
- Central attractive body gravitational model file (i.e., celestial body spherical harmonics coefficients).
    - Degree and order of spherical harmonics
    - Whether to model solid tides (if applicable)
    - Whether to model ocean tides (if applicable)
- Gravitational attraction from secondary celestial bodies (usually treated as point masses, but potentially have spherical harmonics coefficients)
- Atmospheric drag (if applicable)
    - Atmospheric density model
    - Drag model (simple or 3D spacecraft model)
- Radiation pressure
    - Radiation pressure model (simple or 3D spacecraft model)
    - Shadow model (cylindrical or penumbra/umbra)

F.1.2.6 Step Size Adjustment

The RK4 method is straight forward to implement. However, its weakness is with the difficulty in determining an appropriate h step size. While smaller step sizes result in more accurate integrations, a point is reached where there are diminishing returns as the size of h is further reduced. Steps in which h is below the diminishing returns threshold also have correspondingly increased computational cycles due to the increased number of evaluations needed. Thus, it is important to choose an appropriate integration step size.

A further complexity is that the appropriate integration step for one domain portion of a function may be different at a different domain portion. More specifically, the appropriate step size tends to be larger when the force function has lower variability, but smaller when the force function has higher variability. In such instances a dynamic, variable step size is implemented.

The example integrators that follow have methods to dynamically determine the integration step size. Each uses the next higher RK order to estimate the error in the baseline solution. The process first determines the truncation error vector ($\bar{\varepsilon}$) for the current integration error step as the difference between the baseline result ($\bar{y}_{i+1}$) and the result ($\bar{y}_{i+1}^+$) produced by the next higher order:

$$\bar{\varepsilon} = \bar{y}_{i+1} - \bar{y}_{i+1}^+$$

The scalar truncation error (ε) is the maximum of the absolute values of the truncation error's ($\bar{\varepsilon}$) components. The user specifies the tolerance (τ) that the scalar truncation error may not exceed without a step size reduction.

A classic method of step size control is for h to adaptively shrink or grow to an adjusted value (h'), depending on the ratio of the tolerance to the current error as follows:

$$h' = h \left(\frac{\tau}{\varepsilon}\right)^{\frac{1}{1+r}}$$

In the above expression, the r is baseline (i.e., lower) RK order for the method. It is often desirable to apply a dampening factor ($\eta = 0.9$ for example) to reduce the tendency of h to have large oscillations. In practical implementations it is also prudent to have an absolute range ($h_{min} \leq h' \leq h_{max}$) and to truncate the precision of the adjusted (h') value to well above floating point machine precision.

F.1.2.7 Runge-Kutta Fehlberg 4(5) Method

The Runge-Kutta Fehlberg 4(5) (RKF45) Method performs simultaneous 4th and 5th order solutions. The 4th order solution is the result, and the 5th order solution is used for determining step sizing.

$$[a] = \begin{bmatrix} 0 \\ 1/4 \\ 3/8 \\ 12/13 \\ 1 \\ 1/2 \end{bmatrix}$$

The 4th order solution uses the first five subinterval force evaluations. The 5th order solution uses all the 4th order subintervals and adds an additional subinterval force evaluation.

$$
[b] = \begin{bmatrix}
0 & 0 & 0 & 0 & 0 \\
1/4 & 0 & 0 & 0 & 0 \\
3/32 & 9/32 & 0 & 0 & 0 \\
1932/2197 & -7200/2197 & 7296/2197 & 0 & 0 \\
439/216 & -8 & 3680/513 & -845/4104 & 0 \\
-8/27 & 2 & -3544/2565 & 1859/4104 & -11/40
\end{bmatrix}
$$

The weighted sum for the two solutions each has different coefficients. Both solutions use the f_2 intermediate evaluation only to compute the f_3 and subsequent intermediate forces.

$$
[c_4] = \begin{bmatrix}
25/216 \\
0 \\
1408/2565 \\
2197/4104 \\
-1/5 \\
0
\end{bmatrix}
\qquad
[c_5] = \begin{bmatrix}
16/135 \\
0 \\
6656/12825 \\
28561/56430 \\
-9/30 \\
2/55
\end{bmatrix}
$$

F.1.2.8 Runge-Kutta Fehlberg 8(9) Method

The Runge-Kutta Fehlberg 8(9) (RKF89) method performs simultaneous 8th and 9th order solutions. The 8th order solution is the result, and the 9th order solution is used for determining step sizing.

The Runge-Kutta Fehlberg 8(9) method has a 17 element $[a]$ matrix and a 17x16 element $[b]$ matrix. The $[c_8]$ and $[c_9]$ matrices each have 17 elements. Due to the matrix sizes, it would be unwieldy to present their values in matrix form. Instead, their non-zero values are identified below. Note that most floating-point representations such as IEEE-754 64-bit double precision will not accommodate the full precision of most coefficients provided, but the maximum number of digits should be used in the floating-point variables.

Non-zero [a] Matrix Coefficients	
Index	**Coefficient Value**
2	0.44368940376498183109599404281370
3	0.66553410564747274664399106422055
4	0.99830115847120911996598659633083
5	0.3155
6	0.50544100948169068626516126737384
7	0.17142857142857142857142857142857

393

8	0.8285714285714285714285714285714285143
9	0.6654396612101156253495376925586
10	0.2487831796806265206972274560771
11	0.109
12	0.891
13	0.3995
14	0.6005
15	1.0
17	1.0

The 8th order solution uses eight of intermediate force evaluations (i.e., f_1 and f_9 through f_{15}). The 9th order solution uses nine intermediate force evaluations (f_1 and f_9 through f_{14}, plus f_{16} and f_{17}). The f_2 through f_8 intermediate force evaluations are only used for computing the subsequent intermediate force evaluations.

Non-zero [b] Matrix Coefficients	
Index	**Coefficient Value**
(2,1)	0.4436894037649818310959940428137
(3,1)	0.16638352641186818666099776605514
(3,2)	0.49915057923560455998299329816541
(4,1)	0.24957528961780227999149664908271
(4,3)	0.74872586885340683997448994724812
(5,1)	0.20661891163400602426556710393185
(5,3)	0.17707880377986347040380997288319
(5,4)	-0.68197715413869494669377076815048e-1
(6,1)	0.10927823152666408227903890926157
(6,4)	0.40215962642367995421990563690087e-2
(6,5)	0.39214118169078980444392330174325
(7,1)	0.98899281409164665304844765434355e-1
(7,4)	0.35138370227963966951204487356703e-2
(7,5)	0.12476099983160016621520625872489
(7,6)	-0.55745546834989799643742901466348e-1
(8,1)	-0.36806865286242203724153101080691
(8,5)	-0.22273897469476007645024020944166e+1
(8,6)	0.13742908256702910729565691245744e+1
(8,7)	0.20497390027111603002159354092206e+1
(9,1)	0.45467962641347150077351950603349e-1
(9,6)	0.32542131701589147114677469648853
(9,7)	0.28476660138527908888182420573687
(9,8)	0.97837801675979152435868397271099e-2
(10,1)	0.60842071062622057051094145205182e-1
(10,6)	-0.21184565744037007526325275251206e-1
(10,7)	0.1959655726617083195746449066 2983
(10,8)	-0.42742640364817603675144835342899e-2

394

Non-zero [b] Matrix Coefficients	
Index	Coefficient Value
(10,9)	0.174343657368149119965323452558189e-1
(11,1)	0.540597832969319173657785724111182e-1
(11,7)	0.110298255978289265302831276482228
(11,8)	-0.125650085200725564141477637822250e-2
(11,9)	0.367900434775814601363840435663339e-2
(11,10)	-0.577805427709720730408406285718662e-1
(12,1)	0.127324770686671146466451817991602
(12,8)	0.114488050063961053236588757218170
(12,9)	0.287730207096797992776202201849198
(12,10)	0.509453794596113631537358850794656
(12,11)	-0.147996822443725759002421444496400
(13,1)	-0.365267938766167405358485443943338e-2
(13,6)	0.816298960123189197778194212470305e-1
(13,7)	-0.386077356356935064905176943432155
(13,8)	0.308622429246051064504741660252063e-1
(13,9)	-0.580772545283206028158293747335185e-1
(13,10)	0.335986593288849714931434513623223
(13,11)	0.410668804019499586135496227864173
(13,12)	-0.118402459723559855206331561545361e-1
(14,1)	-0.123753579212451432549790961356691e+1
(14,6)	-0.244307685513547853587348613667631e+2
(14,7)	0.547795689327786560504365289911733
(14,8)	-0.444138635334132463749598965693461e+1
(14,9)	0.100131048137132660947926178510221e+2
(14,10)	-0.149957731020517584471709850731421e+2
(14,11)	0.589469485232170136208245396514271e+1
(14,12)	0.173803775034289848776168574405421e+1
(14,13)	0.275123306931667302637586228602762e+2
(15,1)	-0.352608593883345227005029588755881
(15,6)	-0.183961031448482703750441989882311
(15,7)	-0.655701894497416451380068799852511
(15,8)	-0.390861448804398634350255202413101
(15,9)	0.267946467128500229365844232712091
(15,10)	-0.103830229913824908657698585074271e+1
(15,11)	0.166723273242586716647273461685011e+1
(15,12)	0.495519258553159770677329670714411
(15,13)	0.113940011323970632285867381417841e+1
(15,14)	0.513366964246586136881990971915341e-1
(17,1)	-0.135365507861740670804421688899661e+1
(17,6)	-0.183961031448482703750441989882311
(17,7)	-0.655701894497416451380068799852511
(17,8)	-0.390861448804398634350255202413101

| Non-zero [b] Matrix Coefficients ||
Index	Coefficient Value
(17,9)	0.27466285581299925758962207732989
(17,10)	-0.10464851753571915887035188572676e+1
(17,11)	0.16714967667123155012004488306588e+1
(17,12)	0.49523916825841808131186990740287
(17,13)	0.11481836466273301905225795954930e+1
(17,14)	0.41082191313833055603981327527525e-1
(17,16)	1.0

The 8th order solution's weighted sum coefficients are listed below.

| Non-Zero [c₈] Matrix Coefficients ||
Index	Coefficient Value
1	0.32256083500216249913612900960247e-1
9	0.25983725283715403018887023171963
10	0.92847805996577027788063714302190e-1
11	0.16452339514764342891647731842800
12	0.17665951637860074367084298397547
13	0.23920102320352759374108933320941
14	0.39484274604202853746752118829325e-2
15	0.30726495475860640406368305522124e-1

The 9th order model's weighted sum coefficients that differ from those of the 8th order solution are listed below. Note that the 1st coefficient is the difference between the 8th order solution's 1st and 15th coefficient. The 16th and 17th coefficients are equal to the 8th order solution's 15th coefficient.

| [c₉] Matrix Coefficients Differing From [c₈] ||
Index	Coefficient Value
1	$c_8(1) - c_8(15)$
15	0.0
16	$c_8(15)$
17	$c_8(15)$

References

1. Curtis, Howard D., *Orbital Mechanics for Engineering Students*, Fourth Edition, © 2020 Elsevier Ltd., ISBN 978-0-08-102133-0.
2. Fehlberg, Erwin, *Classical Fifth-, Sixth-, Seventh-, and Eighth-Order Runge-Kutta Formulas with Stepsize Control*, NASA TR R-287, 1968.
3. Press, William H., et al., *Numerical Recipes in C, The Art of Scientific Computing*, © 1988 Cambridge University Press, ISBN 0-521-35465-X.

4. McCalla, Thomas R., *Introduction to Numerical Methods and FORTRAN Programming*, © 1967 John Wiley &; Sons, ISBN 0-471-58125-9.

5. Battin, R.H., *An Introduction to the Mathematics and Methods of Astrodynamics*, AIAA Education Series, © 1987 by author, American Institute of Aeronautics and Astronautics (AIAA), ISBN 0-930403-25-8.

6. Gurfil, Pini and Seidelmann, P. Kenneth, *Celestial Mechanics and Astrodynamics: Theory and Practice*, © 2016 Springer-Verlag, ISBN 978-3-662-50368-3.

7. Shampine, L.F. and Gordon, M.K., *Computer Solution of Ordinary Differential Equations, The Initial Value Problem*, © 1975 W. H. Freeman and Company, ISBN 0-7167-0461-7.

Index

www.ingramcontent.com/pod-product-compliance
Lightning Source LLC
Chambersburg PA
CBHW070244230326

41458CB00100B/6087